DRONES

Journey from First to Fifth Generation and Beyond

RIVER PUBLISHERS SERIES IN COMPUTING AND INFORMATION SCIENCE AND TECHNOLOGY

The "River Publishers Series in Computing and Information Science and Technology" covers research which ushers the 21st Century into an Internet and multimedia era. Networking suggests transportation of such multimedia contents among nodes in communication and/or computer networks, to facilitate the ultimate Internet.

Theory, technologies, protocols and standards, applications/services, practice and implementation of wired/wireless networking are all within the scope of this series. Based on network and communication science, we further extend the scope for 21st Century life through the knowledge in machine learning, embedded systems, cognitive science, pattern recognition, quantum/biological/molecular computation and information processing, user behaviors and interface, and applications across healthcare and society.

Books published in the series include research monographs, edited volumes, handbooks and textbooks. The books provide professionals, researchers, educators, and advanced students in the field with an invaluable insight into the latest research and developments.

Topics included in the series are as follows:-

- Artificial intelligence
- Cognitive Science and Brian Science
- Communication/Computer Networking Technologies and Applications
- Computation and Information Processing
- Computer Architectures
- Computer networks
- Computer Science
- Embedded Systems
- Evolutionary computation
- Information Modelling
- Information Theory
- Machine Intelligence
- Neural computing and machine learning
- Parallel and Distributed Systems
- Programming Languages
- Reconfigurable Computing
- Research Informatics
- Soft computing techniques
- Software Development
- Software Engineering
- Software Maintenance

For a list of other books in this series, visit www.riverpublishers.com

DRONES

Journey from First to Fifth Generation and Beyond

Kaushik Kumar
Birla Institute of Technology Mesra, India

Arzu Eren Şenaras
Bursa Uludağ University, Turkey

Chikesh Ranjan
National Institute of Technology Rourkela, India

Onur Mesut Şenaras
Bursa Uludağ University, Turkey

River Publishers

Routledge
Taylor & Francis Group
NEW YORK AND LONDON

Published 2026 by River Publishers
River Publishers
Broagervej 10, 9260 Gistrup, Denmark
www.riverpublishers.com

Distributed exclusively by Routledge
605 Third Avenue, New York, NY 10017, USA
4 Park Square, Milton Park, Abingdon, Oxon OX14 4RN

DRONES: Journey from First to Fifth Generation and Beyond / Kaushik Kumar, Arzu Eren Şenaras, Chikesh Ranjan and Onur Mesut Şenaras

Routledge is an imprint of the Taylor & Francis Group, an informa business

ISBN 978-87-438-1239-5 (hardback)
ISBN 978-87-438-1245-6 (paperback)
ISBN 978-87-438-1242-5 (ebook)
ISBN 978-87-438-1244-9 (e-pub)
ISBN 978-87-438-1241-8 (ebook master)

DOI: 10.1201/9788743812418

Contents

Preface

The authors are pleased to present the book *DRONES – Journey from First to Fifth Generation and Beyond*. The book title has been conceived as a comprehensive and structured resource that brings together the evolution, engineering, applications, and future of drone technology and is designed to serve as a comprehensive academic reference.

Unmanned aerial vehicles (UAVs), popularly known as drones, have redefined the landscape of modern aerospace technology. From their earliest conception as simple radio-controlled aircraft to the complex, intelligent, and autonomous systems of today, drones embody a century of relentless innovation.

The book aims to bridge the gap between fundamental principles and cutting-edge developments, making it valuable to engineering students, researchers, and professionals in aerospace, defense, robotics, and industrial automation. Each chapter builds logically on the previous, tracing the transformation of UAVs through technological generations – each driven by innovation in materials, electronics, communication, artificial intelligence, and autonomy.

This work not only examines drones as flying machines but as integrated systems that combine aerodynamics, control theory, mechatronics, data analytics, and sustainability. The book highlights how UAVs have moved beyond defense applications to revolutionize agriculture, logistics, environmental monitoring, and smart city infrastructure. It also acknowledges the challenges ahead – airspace integration, data privacy, ethical concerns, and the need for robust policy frameworks to ensure safe and sustainable deployment.

The entire book spreads over *nine chapters*, describing the various generations of the prevailing drone technology and what is expected in the future. Each chapter ends with a summarization of the content and some review questions so that the reader can assess their understanding.

Chapter 1, i.e., Overview of Drone Technology, is the introductory chapter presenting a panoramic view of UAV evolution, highlighting their origins in early 20th-century military research and tracing their expansion into civilian, industrial, and research domains. It explains the core principles of drone operation – covering aerodynamics, propulsion, communication, and navigation – along with emerging technologies such as AI, GPS precision, and digital twin integration. The chapter classifies drones into generational stages, setting the conceptual foundation for the chapters that follow.

Chapter 2 introduces the first-generation drones, revisiting the formative era of drone development (1910s–1930s), characterized by manual control, simple fixed-wing structures, and rudimentary radio systems. It introduces pioneering innovations such as the *Aerial Target* (1916), *Kettering Bug* (1918), and *Queen Bee* (1935), emphasizing their contributions to early defense training and reconnaissance. The discussion also covers the foundational technologies – radio communication, mechanical servos, and piston propulsion – that shaped the first UAV architectures.

Chapter 3 moves into second-generation drones. The focus here shifts to the integration of remote control and GPS technologies during the 1980s–1990s. This generation introduced semi-automated navigation, improved endurance, and digital telemetry. The chapter explores how enhanced flight controllers, onboard sensors, and composite materials expanded UAV applications beyond defense – into surveying, mapping, and environmental research. It also highlights early efforts toward autonomous flight through pre-programmed missions.

Chapter 4 marks the maturity of UAV technology, with drones gaining advanced surveillance, reconnaissance, and combat capabilities, grouping them into "Third-generation Drones." This chapter, emerging from the late 1990s to the early 2010s, describes drones such as the Predator and Global Hawk integrated with high-resolution sensors, secure data links, and long-endurance propulsion systems. The chapter discusses the evolution of unmanned combat aerial vehicles (UCAVs), emphasizing the convergence of defense strategy, automation, and ethical considerations in warfare.

Chapter 5 elaborates on fourth-generation drones. The fourth generation represents the shift from purely military use to widespread commercial adoption. With advancements in onboard processing, LiDAR, AI-assisted control, and real-time communication, drones began to perform complex civilian tasks such as precision agriculture, construction monitoring, disaster management, and logistics. This chapter explores the emergence of semi-autonomous flight, multi-sensor fusion, and improved energy efficiency. It also introduces global regulatory frameworks that enable the integration of drones into controlled airspace.

Chapter 6 explores the era of full autonomy, artificial intelligence, and swarm technology, coined as *"Fifth-generation Drones."* Fifth-generation drones are intelligent systems capable of independent decision-making, collaborative missions, and adaptive behavior. Key enabling technologies such as computer vision, edge computing, 5G communication, and blockchain-based security are discussed. The chapter also addresses futuristic applications in smart cities, autonomous logistics, defense swarms, and environmental conservation, positioning drones as integral components of the Industry 5.0 ecosystem.

The last three chapters, i.e., *Chapters 7, 8, and 9*, take the reader to the future of drone technology.

Chapter 7 provides speculations on applications and future prospects of drone technology. It synthesizes the knowledge from previous sections, focusing on global UAV applications, market growth, and future innovations. It discusses how drones are reshaping industries such as agriculture, renewable energy inspection, transportation, and disaster management. It also outlines upcoming trends – such as hybrid aerial mobility, quantum navigation, and sustainable UAV design – emphasizing how drones are evolving into intelligent systems that harmonize with human expertise and digital infrastructure.

Chapter 8 highlights the evolutionary impacts of UAVs on modern systems. This chapter provides an in-depth exploration of how unmanned aerial vehicles (UAVs) have evolved to become pivotal components of modern industrial, social, and technological systems. It emphasizes that drones are not merely aviation tools but catalysts driving systemic transformation across multiple domains. The discussion begins with the historical transition of UAVs from military prototypes to multipurpose platforms serving defense, industry, and society. The chapter elaborates on sectoral expansions – including defense, agriculture, logistics, construction, healthcare, and environmental monitoring – highlighting how drones enhance operational precision, efficiency, and data-driven decision-making. In defense, UAVs have revolutionized surveillance and combat intelligence; in agriculture, they have enabled precision farming and resource optimization; in logistics, they have improved last-mile delivery and global supply chain visibility. The chapter also covers their integration into smart cities, enabling aerial mapping, infrastructure inspection, and real-time traffic management. Further sections discuss societal, economic, and ethical implications, addressing issues of privacy, airspace congestion, and sustainability. The rapid proliferation of UAVs has spurred new industries, reshaped media and filmmaking through aerial cinematography, and contributed to environmental conservation and disaster resilience. Global case studies – spanning from Dubai's air taxis and Zipline's medical delivery drones in Africa to cinematic innovations in Bollywood and Hollywood – illustrate their versatile impact. In conclusion, the chapter portrays UAVs as transformative enablers of the 21st century's digital and industrial revolution. Their influence transcends technology, extending to environmental stewardship, economic empowerment, and human safety. UAVs have evolved into intelligent aerial systems that symbolize humanity's progress toward a more connected, autonomous, and sustainable world.

The concluding chapter of the book, *Chapter 9*, envisions the next era of drone evolution, focusing on emerging trends, futuristic technologies, and global implications of advanced UAV integration. It begins by identifying drones as one of the most transformative technologies of the 21st century – expanding from defense reconnaissance to everyday applications in agriculture, logistics, healthcare, urban management, and planetary exploration. The future of drones, as presented here, is characterized by AI-driven autonomy, swarm intelligence, quantum computing, and 5G/6G-enabled communication. Drones are projected to evolve from independent units to interconnected aerial ecosystems capable of collaborating with ground-based digital infrastructures and smart city networks. The chapter highlights urban air mobility (UAM) as a defining concept of this evolution, envisioning drone-based passenger transport, autonomous cargo corridors, and emergency air logistics operating safely within regulated urban airspaces. Key technological innovations include bio-inspired flight mechanics, hybrid-electric propulsion, and quantum-enhanced navigation systems, all contributing to energy efficiency and climate responsiveness. The chapter also discusses the integration of blockchain for secure communication, AI for predictive maintenance, and IoT for real-time coordination. Moreover, it explores the ethical, regulatory, and environmental challenges that accompany such advancements – emphasizing the importance of governance, cybersecurity, and public trust in widespread adoption. The discussion extends beyond Earth, spotlighting NASA's Ingenuity Mars helicopter and the Dragonfly mission to Titan as milestones in extraterrestrial drone exploration. These missions exemplify how drones are expanding human frontiers in planetary research and astrobiology. Chapter 9 encapsulates the vision of drones as intelligent, networked, and sustainable aerial systems deeply integrated into both terrestrial and extraterrestrial ecosystems. The future trajectory points toward a harmonious coexistence between humans and autonomous aerial agents – where drones function not only as machines of utility but as collaborators in advancing science, mobility, and societal well-being.

This book has been developed with a dual objective: to serve as a comprehensive academic text and as a practical reference for emerging professionals in unmanned systems. Each chapter combines historical insight with technical analysis, enabling readers to appreciate both the legacy and the innovation driving UAV evolution. The inclusion of tables, figures, and review questions further strengthens learning outcomes, making the book suitable for classroom teaching, project reference, and advanced research.

As UAVs continue to shape global economies, defense strategies, and sustainable development initiatives, understanding their technological trajectory becomes indispensable. This book thus aspires to cultivate a generation of engineers and researchers equipped not only with technical knowledge but also with an ethical and societal awareness of drone innovation.

It is our sincere hope that this work inspires curiosity, innovation, and a commitment to responsible technological advancement. The evolution of drone technology is a testament to human creativity and perseverance – qualities that will continue to propel the next frontier of intelligent flight.

Kaushik Kumar
Birla Institute of Technology Mesra, India

Arzu Eren Şenaras
Bursa Uludağ University, Turkey

Chikesh Ranjan
National Institute of Technology Rourkela, India

Onur Mesut Şenaras
Bursa Uludağ University, Turkey

Acknowledgment

The creation of *DRONES – Journey from First to Fifth Generation and Beyond* has been a profound and rewarding journey. This book is the culmination of insights, experiences, and the generous support of numerous individuals and institutions who have contributed in various ways to its development and we would not have been able to complete it without the guidance, cooperation and encouragement of many individuals.

First and foremost, we thank the Almighty for granting us this opportunity and for inspiring us to believe in passion, hard work, and the pursuit of our dreams. In the process of putting this book together it was realized how true this gift of writing is for anyone. This could never have been done without power provided by you and of course, the faith in You, the Almighty.

Our deepest gratitude goes to our families for their unending encouragement and understanding. Your patience during the long hours of writing and your unwavering belief in us provided the emotional support necessary to see this project through to completion. Your words of encouragement and your ability to remind me of the bigger picture have been invaluable, which also helped us to take yet another challenge, which decreases the amount of time we could spend with them. They were our inspiration and motivation and their constant encouragement and late-night brainstorming sessions were invaluable. The unwavering support and insightful feedback helped to shape the direction of this book and to say least, their belief in our vision made all the difference. We would like to thank our parents and grandparents for allowing us to follow our ambitions.

We are deeply indebted to our professional and institutional superiors; their guidance and encouragement have been a constant source of inspiration. Their expertise in communication skills and ability to demystify complex ideas have profoundly influenced the presentational approach taken in this book. Their constructive feedback and unwavering support have been invaluable in shaping the direction and content of this work.

We wish to extend our heartfelt thanks to all of our colleagues, friends in different part of the world for sharing ideas in shaping our thoughts. Your insights and expertise enriched this book in countless ways. It provided thoughtful critique, helpful suggestions, and the benefit of your experiences in the emerging technology. Your willingness to share your knowledge and offer constructive feedback has ensured that this book addresses the practical challenges faced by writers in the scientific community.

We are especially grateful to our students, past and present, whose enthusiasm for learning and eagerness to improve their know-how have been a driving force behind this project. Your questions, challenges, and successes provided the motivation to create a comprehensive resource that serves the needs of aspiring and experienced communicators alike. The workshops, seminars, and one-on-one sessions with you all have been a source of inspiration and learning for us as well.

A special note of appreciation goes to the Reviewers, the Editorial Board Members, Project Development Editor and the complete team of Rivers Publications, whose belief in the importance of this project made its publication possible. We would also like to thank them for their constant and consistent guidance, support and cooperation at all stages of this project. Their effort in every phase, from inception to execution cannot be expressed in words. Your keen eye and thoughtful suggestions helped in bringing clarity and polish to this manuscript and your meticulous attention to detail, editorial expertise, and unwavering commitment to excellence have elevated the quality of this work. A special thank you for providing the occasional push when it was most needed. Thank you for your patience, guidance, and professionalism throughout the publication process. We deeply appreciate your hard work and dedication, support and cooperation rendered to the project.

We also want to acknowledge the contributions of the administrative and technical staff of our institutions for their support to this endeavour. From managing logistics to providing access to resources, your behind-the-scenes efforts ensured the smooth progression of this project.

We owe a debt of gratitude to the larger scientific and academic community. Your work and dedication to advancing knowledge have provided the foundation upon which this book is built. This book is a small contribution to the ongoing dialogue about Drones, and it is my hope that it will inspire others to continue refining the ways we share and disseminate knowledge.

Though this book has been created by individuals located in different corners of the world, it is unified by one shared aspiration: to contribute meaningfully to the future of autonomous aerial systems and to the collective body of knowledge that transcends borders. So we would fail if we do not pat our own backs.

Finally, we are grateful to our readers, whose curiosity and passion are the real motivation behind this work. It is our hope that this book inspires, educates, or simply entertains in the way we intended. Our efforts will come to a level of satisfaction if the students, researchers and professionals concerned with all the fields gets benefitted.

Thank you to everyone who has played a part in this journey. Your contributions, big and small, are deeply appreciated and forever cherished.

With sincere gratitude,

Kaushik Kumar
Birla Institute of Technology Mesra, India

Arzu Eren Şenaras
Bursa Uludağ University, Turkey

Chikesh Ranjan
National Institute of Technology Rourkela, India

Onur Mesut Şenaras
Bursa Uludağ University, Turkey

List of Figures

List of Tables

List of Abbreviations

ADE	Aeronautical Development Establishment
AESA	Active electronically scanned array
AI	Artificial intelligence
AMCA	Advanced Medium Combat Aircraft
AR	Augmented reality
ATOL	Autonomous take-off and landing
AWS	Autonomous weapon systems
BIM	Building information modeling
BLOS	Beyond-line-of-sight
BVLOS	Beyond-visual-line-of-sight
BWB	Blended wing-body
CAS	Close air support
CFRP	Carbon fiber-reinforced polymer
CIA	Central Intelligence Agency
CNN	Convolutional neural network
DaaS	Drone-as-a-service
DEW	Directed-energy weapon
DGCA	Directorate General of Civil Aviation
DoS	Denial-of-service
DRDO	Defence Research and Development Organization
EA	Electronic attack
EASA	European Union Aviation Safety Agency
EoL	End-of-life
ES	Electronic support
eVTOL	Electric vertical take-off and landing
EW	Electronic warfare
FAA	Federal Aviation Administration
FBL	Fly-by-light
FCAS	Future combat air system
FCU	Flight control unit
FDIR	Fault detection, isolation, and recovery
FM	Frequency-modulated

FMV	Full-motion video
GCS	Ground control station
GIS	Geographic information system
GMTI	Ground moving target indication
GPS	Global positioning system
HALE	High-altitude long-endurance
HDI	Human–drone interaction
HMI	Human–machine interface
HSM	Hardware security module
IAI	Israel Aerospace Industries
ICAO	International Civil Aviation Organization
ICRC	International Committee of the Red Cross
IDF	Israeli Defense Forces
IED	Improvised explosive device
IHL	International humanitarian law
IMU	Inertial measurement unit
INS	Inertial navigation systems
IR	Infrared
IRST	Infrared search and track
ISR	Intelligence, surveillance, and reconnaissance
JARUS	Joint Authorities for Rulemaking on Unmanned Systems
KAP	Kite aerial photography
LAWS	Lethal autonomous weapon system
LCAAT	Low-cost attritable aircraft technology
LEO	Low-earth-orbit
LO	Low-observable
LOS	Line-of-sight
MALE	Medium-altitude, long-endurance
MDO	Multi-domain operation
MHC	Meaningful human control
ML	Machine learning
MoD	Ministry of Defense
MUM-T	Manned–unmanned teaming
NCW	Network-centric warfare
NGAD	Next Generation Air Dominance
NGF	Next Generation Fighter
NVDI	Normalized difference vegetation index
NYSDOT	New York State Department of Transportation
PG&E	Pacific Gas & Electric
PGM	Precision-guided munition

PSO	Particle swarm optimization
QKD	Quantum key distribution
RAAF	Royal Australian Air Force
RAF	Royal Air Force
RAM	Radar-absorbent materials
RC	Radio control
RCS	Radar cross-section
RF	Radio frequency
RNN	Recurrent neural network
RTA	Roads and Transport Authority
RTH	Return-to-home
SARA	Synthetic Aperture Radar Antenna
SARM	Search and Rescue Mission
SEAD	Suppression of enemy air defense
SLAM	Simultaneous localization and mapping
SoS	System of systems
SOTIF	Safety of the intended functionality
SUAV	Strategic unmanned air vehicle
TAI	Turkish Aerospace Industries
ToF	Time-of-flight
TPM	Trusted platform module
UAC	United Aircraft Corporation
UAM	Urban air mobility
UAS	Unmanned aerial systems
UAV	Unmanned aerial vehicle
UCAS-D	Unmanned combat air system demonstration
UCAV	Unmanned combat aerial vehicle
USAF	United States Air Force
USGS	United States Geological Survey
UTM	Unmanned traffic management
VO	Visual odometry
VTOL	Vertical take-off and landing
XAI	Explainable artificial intelligence

1

Overview of Drone Technology

1.1 Introduction

Unmanned aerial vehicles (UAVs), commonly known as drones, have emerged as a pivotal innovation in modern aerospace technology, transforming the way aerial operations are conducted across military, commercial, and recreational domains. Initially developed for military reconnaissance and surveillance during the early 20th century, UAVs have since evolved into highly sophisticated systems equipped with advanced sensors, navigation units, and autonomous control capabilities. Their rapid technological growth has enabled applications far beyond defense, reaching industries such as agriculture, logistics, construction, environmental monitoring, and disaster management.

In the commercial sector, drones are revolutionizing data acquisition, aerial mapping, and delivery systems through cost-effective and efficient aerial solutions. The recreational use of drones has also surged, offering enthusiasts access to aerial photography, racing, and hobbyist exploration, supported by improved flight stability and user-friendly interfaces.

The core of drone technology lies in the seamless integration of aerodynamics, propulsion, communication, navigation, and artificial intelligence. Continuous advancements in battery technology, miniaturized sensors, GPS precision, and autonomous flight algorithms are driving the next generation of UAVs toward higher endurance, reliability, and intelligence (Choi, et al. 2023).

This chapter provides a comprehensive overview of drone technology, tracing its historical evolution, classification, design elements, and operational principles. It also highlights key enabling technologies and emerging trends that continue to shape the future of UAVs as indispensable tools in defense, industry, and daily life.

DOI: 10.1201/9788743812418-1

1.2 The Impact of Drone Technology

Drone technology has emerged as a transformative force with far-reaching implications across military, commercial, and civilian domains. By combining advancements in aerodynamics, electronics, communication systems, and artificial intelligence, drones have evolved from simple remote-controlled aircraft into intelligent, autonomous platforms capable of performing complex tasks with precision, speed, and reliability (Singhal, 2021). Their growing influence signifies not just technological progress but also a structural shift in how industries operate, governments manage security, and societies engage with the aerial dimension of modern innovation.

1.2.1 Drones in military applications

Drones have profoundly transformed military strategy and defense operations by introducing a new dimension of intelligence, surveillance, and reconnaissance (ISR) capabilities. Modern militaries rely heavily on UAVs for missions that require persistence, precision, and safety in high-risk environments. Drones such as the *Predator*, *Global Hawk*, and *Heron* are equipped with high-resolution cameras, thermal imaging sensors, radar systems, and real-time communication links that provide commanders with continuous situational awareness.

Unmanned combat aerial vehicles (UCAVs) have further extended drone functionality to include precision-guided strikes, minimizing collateral damage and reducing the exposure of human pilots to hostile zones. Drones also play an essential role in logistics, transporting supplies, conducting damage assessments, and supporting humanitarian operations in conflict zones. Additionally, swarm drone systems and AI-enabled autonomous coordination are emerging as critical assets in electronic warfare, decoy operations, and tactical defense scenarios.

Figure 1.1 illustrates the diverse and strategic roles of drones in military applications, emphasizing their transformation into critical assets for modern defense operations. The figure highlights how unmanned aerial vehicles (UAVs) are deployed in surveillance, reconnaissance, combat, logistics, and intelligence missions, enhancing battlefield awareness and operational efficiency. Equipped with high-resolution cameras, thermal sensors, and communication modules, drones conduct real-time surveillance and gather crucial intelligence, supporting rapid tactical decisions. Armed variants, known as unmanned combat aerial vehicles (UCAVs), perform precision strikes on high-value targets while minimizing pilot risk. Additionally, drones

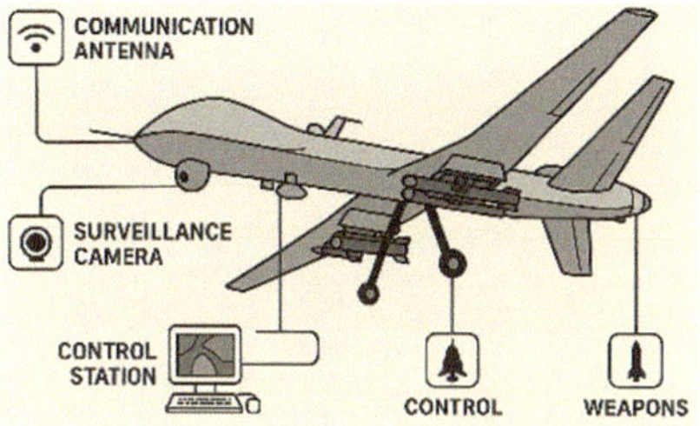

Figure 1.1 Strategic utilization of drones in modern warfare.

are utilized for logistics support, delivering essential supplies to inaccessible or dangerous zones, and for search and rescue (SAR) missions, aiding in locating personnel and assessing damage in post-conflict environments.

1.2.2 Commercial and civilian use

In the civilian and industrial sectors, drone technology has become synonymous with efficiency, safety, and innovation. Industries leverage UAVs for real-time data collection, precision mapping, and infrastructure monitoring. In agriculture, drones equipped with multispectral and hyperspectral sensors assist in yield estimation, irrigation planning, and pest management, leading to improved productivity and resource optimization. In construction and infrastructure, UAVs are revolutionizing surveying, inspection of bridges and pipelines, and project monitoring, reducing both cost and human risk.

Environmental and emergency management applications also benefit immensely from drones, assisting in wildfire detection, flood monitoring, and disaster relief by providing rapid situational data and enabling the delivery of essential supplies to inaccessible areas. In logistics, companies such as Amazon and Zipline are pioneering drone-based delivery systems that promise faster, contactless transportation of goods, medical supplies, and vaccines. In media and entertainment, drones have redefined cinematography, allowing for aerial perspectives previously achievable only through costly manned aircraft. Furthermore, the increasing adoption of UAVs in urban planning, traffic analysis, and law enforcement demonstrates their growing integration into smart city frameworks.

Figure 1.2 illustrates the expanding role of drones across commercial and civilian sectors, emphasizing how unmanned aerial vehicles (UAVs) have become essential tools in industries that demand precision, efficiency, and accessibility. At the center of the figure is a multi-rotor drone, representing the technological versatility that supports diverse applications, including

Figure 1.2 Commercial and civilian applications of drone technology.

aerial photography, infrastructure inspection, agriculture, and environmental monitoring.

In aerial photography, drones equipped with high-resolution cameras enable filmmakers, journalists, and surveyors to capture dynamic perspectives and generate detailed visual data. In infrastructure inspection, UAVs are employed to assess bridges, power lines, pipelines, and industrial plants, providing real-time imagery while minimizing human risk and maintenance costs. Agricultural applications leverage drones for precision farming, monitoring crop health, soil moisture, and pesticide distribution, enhancing yield and sustainability. Lastly, environmental monitoring utilizes drones for mapping forest cover, analyzing pollution patterns, and tracking wildlife movements, supporting ecological conservation and disaster response.

1.2.3 Global influence and market growth

The global impact of drone technology extends beyond technical innovation; it is driving new economic models and reshaping industries. The UAV market, once limited to military procurement, now represents one of the fastest-growing sectors in aerospace and robotics. According to international market forecasts, the global drone market is expected to exceed USD 150 billion by 2035, with significant growth in Asia-Pacific, North America, and Europe. This expansion is fueled by advancements in autonomy, battery life, artificial intelligence, and regulatory modernization that facilitate safe operation within shared airspace.

Governments and private industries are investing heavily in drone research and ecosystem development, including training, manufacturing, and maintenance infrastructure. Startups and academic institutions are collaborating on innovations in unmanned traffic management (UTM), beyond-visual-line-of-sight (BVLOS) operations, and AI-driven swarm coordination. These developments are positioning drones as central components of the Industry 5.0 paradigm, where human collaboration, smart automation, and sustainability intersect.

Globally, drones are influencing sectors such as environmental monitoring, renewable energy inspection, border management, and urban logistics, reflecting a shift toward data-driven decision-making and eco-efficient operations. As nations incorporate UAVs into policies and digital strategies, the technology is not merely enhancing operational capacity; it is redefining global connectivity, economic competitiveness, and the future of autonomous mobility.

1.3 Technological Foundations

1.3.1 Early developments in aerial systems

The evolution of drone technology is rooted in early experimentation with unmanned flight that began in the early 20th century. Initial developments were driven by military interests seeking to perform reconnaissance and training missions without risking human pilots. Early systems such as the "Aerial Target" (1916) and the "Queen Bee" (1935) laid the groundwork for remotely piloted aircraft through basic radio control mechanisms and gyroscopic stabilization. The Cold War era saw significant advances in unmanned systems, particularly in surveillance and target drones, as nations recognized the strategic importance of persistent aerial observation. These formative years established the structural, aerodynamic, and control principles that continue to underpin modern UAV designs (Blom, 2010). Table 1.1 presents the chronological evolution of early aerial systems, tracing the progression from simple observation balloons and kites to the first radio-controlled aircraft. This table highlights the pioneering innovations that established the technological foundations of unmanned aerial vehicles (UAVs), emphasizing the gradual development of propulsion, remote control, and automation that shaped the future of drone technology.

1.3.2 Breakthroughs in electronics and navigation

The rapid progress of drone technology in recent decades has been largely enabled by breakthroughs in miniaturized electronics, embedded systems, and navigation technologies. The integration of lightweight sensors,

Table 1.1 Early developments in aerial systems.

Period/year	Development/ innovation	Developer/country	Key features/technological advancement	Significance/contribution
Before 1900	Tethered Balloons and Kites for Aerial Observation	France, the United Kingdom, and China	Used for early reconnaissance and weather observation; manually operated.	Marked the earliest attempts at unmanned aerial observation for military and scientific purposes.
1916	*Aerial Target*	A. M. Low (United Kingdom)	Radio-controlled powered aircraft; experimental air defense drone.	First practical demonstration of remote-controlled flight.
1917	*Ruston Proctor Aerial Target*	Ruston Proctor Aircraft Co. (United Kingdom)	Small pilotless monoplane with early radio control.	Among the first successful powered UAV prototypes.
1918	*Kettering Bug*	Charles Kettering (United States)	Unmanned aerial torpedo with pre-set mechanical control.	Early example of automated flight and precursor to cruise missiles.
1935	*Queen Bee*	De Havilland Aircraft Co. (United Kingdom)	Radio-controlled biplane for military gunnery training.	First operational and mass-produced UAV; origin of the term "drone."
1939–1945	*Radioplane OQ-2*	Reginald Denny Industries (United States)	Catapult-launched, radio-controlled target drone.	First UAV to be mass-produced and used for training during WWII.
1950s	*Ryan Firebee Series*	Ryan Aeronautical (United States)	Jet-powered drone with long-range radio guidance.	Set the foundation for Cold War reconnaissance UAVs.
1960s	*Ryan Model 147 "Lightning Bug"*	USAF / Ryan Aeronautical (United States)	Advanced reconnaissance UAV with camera payloads.	Used in the Vietnam War; demonstrated operational UAV capability.
1970s	*IAI Scout* (Prototype Phase)	Israel Aerospace Industries (Israel)	Real-time video transmission and remote control.	Introduced UAVs as integral tools for modern surveillance and combat support.

microcontrollers, and efficient propulsion systems has drastically reduced size and cost while enhancing reliability and performance. The introduction of the global positioning system (GPS) revolutionized drone navigation, allowing for precise positioning, autonomous waypoint tracking, and stable flight operations. Inertial measurement units (IMUs), gyroscopes, accelerometers, and magnetometers further improved flight stability and control accuracy. Advances in wireless communication, particularly in real-time data transmission and remote telemetry, have facilitated long-range operations and seamless integration with ground control systems. Together, these innovations form the technological backbone of modern UAV platforms, enabling diverse civilian and defense applications.

1.3.3 Role of artificial intelligence and autonomy

Artificial intelligence (AI) has emerged as a defining force in the latest generation of drone technology, shifting systems from remote-controlled platforms to fully autonomous aerial entities. AI algorithms enable drones to process sensor data, recognize patterns, and make real-time decisions without human intervention. Machine learning enhances flight optimization, object detection, obstacle avoidance, and mission planning, allowing UAVs to adapt dynamically to complex environments. Computer vision technologies support precision navigation and autonomous landing, while swarm intelligence principles enable coordinated operations among multiple drones for surveillance, mapping, or logistics. Furthermore, AI integration with cloud computing and edge processing expands analytical capabilities, allowing for real-time data interpretation and predictive maintenance.

1.4 Generation breakdown

The evolution of drone technology can be categorized into five distinct generations, each marking a significant advancement in design, control, and operational capability. These generations reflect the transition from simple remotely piloted aircraft to fully autonomous aerial systems integrated with artificial intelligence and networked operations.

1.4.1 First generation: Manual drones

The first generation of drone technology marks the formative stage in the evolution of unmanned aerial systems (UAS), characterized by rudimentary design, limited functionality, and entirely manual control. These early drones, developed primarily during the early to mid-20th century, served as experimental platforms to test the concept of flight without an onboard pilot. Their primary use was in the military domain, especially for target practice and reconnaissance during wartime, where they acted as substitutes for manned aircraft in hazardous or repetitive missions [Simic Milas, et al. 2018].

Technologically, first-generation drones were manually operated through basic radio frequency control systems that allowed limited maneuverability within a short range. They lacked the sophisticated avionics, sensors, and navigation systems seen in later generations. Flight stability and precision were entirely dependent on the skill of the ground operator, as no automated stabilization or feedback systems existed. Structurally, these drones were simple, lightweight, and often constructed from wood, fabric, or early metallic alloys, resembling small fixed-wing aircraft.

One of the earliest known examples was the "Aerial Target" (1916), developed by A. M. Low in the United Kingdom, which demonstrated the feasibility of radio-controlled flight. This was followed by the "Queen Bee" (1935), a radio-controlled target drone used by the Royal Navy for anti-aircraft gunnery training. The term "drone" itself is believed to have originated from this model, inspired by the sound produced by its engine, akin to that of a male bee.

Although their technological capabilities were limited, first-generation manual drones laid the foundation for modern unmanned flight. They introduced the essential principles of remote piloting, control mechanisms, and system integration that would later evolve into advanced autonomous systems. This generation marked the beginning of a transformative journey – from human-dependent remote aircraft to the intelligent, automated drones that define contemporary aerospace innovation.

Figure 1.3 illustrates the structure and fundamental components of a first-generation manual drone, representing the earliest stage of unmanned aerial vehicle (UAV) development. These drones were primarily used during the early to mid-20th century for military purposes such as target practice, reconnaissance, and pilot training. The figure highlights key components, including the propeller, which provides thrust for forward motion; the fuselage, which houses the engine and basic control systems; the landing gear, which supports takeoff and landing; the radio antenna, responsible for receiving manual control signals from the ground operator; and the tail assembly, which ensures flight stability and directional control. Constructed with simple fixed-wing designs and lightweight materials, these drones operated solely through manual radio control, without the aid of sensors, GPS, or automation. As shown in the figure, the design emphasizes basic aerodynamics

Figure 1.3 Manual drone (first generation UAV)

and mechanical simplicity, reflecting the technological limitations of the time while laying the groundwork for future innovations in autonomous and intelligent UAV systems.

1.4.2 Second generation: Remote control and GPS integration

The second generation of drones marked a transformative period in the evolution of unmanned aerial systems, moving from basic manual operation to semi-automated and more reliable flight control. Emerging prominently during the 1980s and 1990s, this era witnessed significant technological improvements, primarily through the introduction of remote control systems, onboard electronics, and global positioning system (GPS) technology. These advancements collectively enhanced flight precision, stability, and mission endurance, expanding drone applications beyond military training into surveillance, research, and civilian domains.

Technically, second-generation drones incorporated radio-controlled interfaces that allowed operators to manage altitude, direction, and speed over greater distances. The integration of GPS navigation enabled autonomous waypoint flight paths, automatic return-to-home functions, and improved spatial accuracy. Drones in this generation typically featured an onboard flight controller, inertial sensors, and communication modules that facilitated real-time telemetry and data feedback to the operator. The materials used for construction shifted toward lightweight composites and improved aerodynamics, allowing longer flight durations and higher payload capacities.

This generation also introduced early forms of camera payloads and sensor integration, making drones capable of performing tasks such as aerial mapping, reconnaissance, and environmental observation. The use of digital communication links improved command reliability, while advances in power systems, particularly with lithium-based batteries, enhanced operational efficiency.

Figure 1.4 illustrates the architecture and key components of a second-generation drone, highlighting the integration of GPS modules, radio control receivers, and flight controllers. The diagram depicts the interaction between the onboard systems and the ground control station, emphasizing how satellite-based navigation and electronic stabilization enabled more accurate and automated flight paths. This figure reflects the technological evolution that bridged the gap between manual drones and modern autonomous UAV platforms.

1.4.3 Third generation: Surveillance and combat capabilities

The third generation of unmanned aerial systems (UAS) represents a decisive shift from utility and navigation improvements toward operational maturity and mission-centric design, with platforms engineered specifically for sustained intelligence, surveillance, reconnaissance (ISR), and strike roles. Emerging in earnest from the late 1990s through the 2010s, this generation is characterized by long-endurance airframes, higher-altitude operations, robust payload integration, secure datalinks, and hardened avionics that enable persistent, command-driven missions in contested environments.

Technically, third-generation drones combine several key subsystems optimized for operational effectiveness. High-efficiency propulsion and larger fuel/power reserves provide multi-hour to multi-day endurance; stabilized gimbal sensors (EO/IR, synthetic aperture radar) allow precision imaging and moving-target indication; multi-band datalinks and anti-jamming electronics ensure resilient communications; and onboard mission computers support data fusion and basic automation for loitering, target tracking, and pattern recognition. Weaponized variants (UCAVs and armed MALE systems) add weapon management suites, secure fire-control links, and mission planning tools that allow coordinated ISR-to-strike workflows with reduced

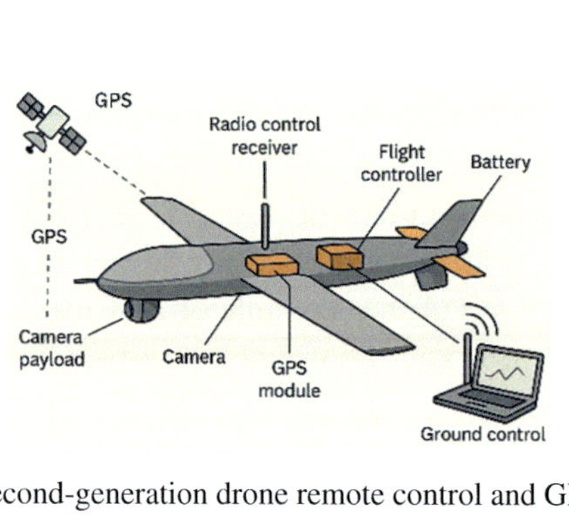

Figure 1.4 Second-generation drone remote control and GPS integration.

sensor-to-shooter latency. Additionally, third-generation systems often include electronic-warfare payloads, signals-intelligence receivers, and modular payload bays to rapidly reconfigure for different roles.

Operationally, this generation widened the use of UAS from tactical reconnaissance to strategic persistent surveillance, precision strike, border and maritime patrol, and coordination with manned assets. Capabilities such as automatic target detection and cueing, beyond-line-of-sight (BLOS) operations, and airborne loitering munitions increased responsiveness and reduced risk to personnel. At the same time, the proliferation of these capabilities introduced technical and non-technical challenges: sophisticated counter-UAS measures (electronic jamming, directed-energy, and kinetic interceptors), complex rules of engagement, legal and ethical concerns over autonomous use of force, and the need for secure supply chains and hardened software to resist cyber intrusion.

Figure 1.5 is a labeled schematic showing a medium-altitude, long-endurance (MALE) platform fitted with a stabilized EO/IR gimbal, SAR antenna, secure datalink, mission computer, fuel tank for extended endurance, and a weapon station. Arrows indicate data flow from sensors to the mission computer and then to the ground control station or weapon release authority. A secondary inset shows a typical mission sequence: persistent surveillance → automatic detection → human authorization → precision engagement.

1.4.4 Fourth generation: Semi-autonomy and commercial applications

The fourth generation of drone technology signifies a critical transition from military-dominated platforms to widespread commercial and civilian adoption, driven by advancements in automation, sensing, and data analytics.

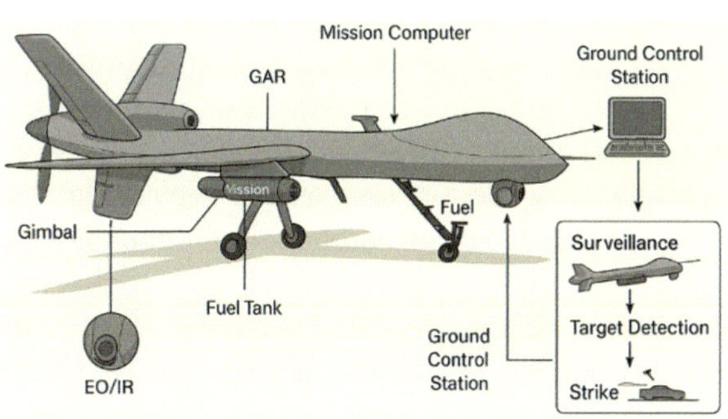

Figure 1.5 Third-generation drone ISR and combat capabilities.

Emerging in the early 2010s, this generation introduced semi-autonomous flight capabilities, where drones could perform pre-programed missions and adaptive maneuvers with minimal human intervention. The integration of on-board processors, real-time data links, and advanced control algorithms allowed drones to execute complex operations with enhanced precision, reliability, and safety.

Technologically, fourth-generation drones are equipped with multi-sensor fusion systems including GPS, LiDAR, ultrasonic sensors, cameras, and inertial measurement units (IMUs) that enable obstacle detection, terrain following, and autonomous navigation in diverse environments. These systems are managed through flight controllers and AI-assisted software that stabilize flight paths and make corrective decisions based on sensor inputs. Improved battery technology, particularly lithium-polymer (Li-Po) and lithium-ion systems, extended flight times and supported heavier payloads such as high-resolution cameras, multispectral sensors, and delivery packages.

The most transformative aspect of this generation is its expansion into commercial, industrial, and public-sector applications. In agriculture, drones became indispensable for precision farming, monitoring crop health, mapping fields, and applying fertilizers or pesticides with pinpoint accuracy. In construction and infrastructure, UAVs enabled aerial surveying, structural inspections, and 3D mapping. Environmental monitoring, disaster management, and public safety operations leveraged drones for real-time situational awareness, search and rescue, and post-disaster damage assessment. The logistics and delivery sector also began exploring UAV integration for last-mile transport, with prototypes successfully demonstrating automated package delivery.

From a regulatory perspective, this generation coincided with the establishment of civil aviation frameworks by organizations such as the FAA (Federal Aviation Administration), EASA (European Union Aviation Safety Agency), and DGCA (Directorate General of Civil Aviation, India). These frameworks introduced airspace classifications, pilot certification, and safety guidelines that allowed drones to be integrated safely into national airspaces.

The fourth-generation drones symbolize the convergence of autonomy, digitalization, and industry-focused innovation. They bridge the gap between fully manual systems and next-generation intelligent UAVs by enabling adaptive control, autonomous missions, and vast commercial scalability. This generation positioned drones as indispensable tools across industries, transforming them from specialized military devices into mainstream technological assets that support economic growth, sustainability, and digital transformation.

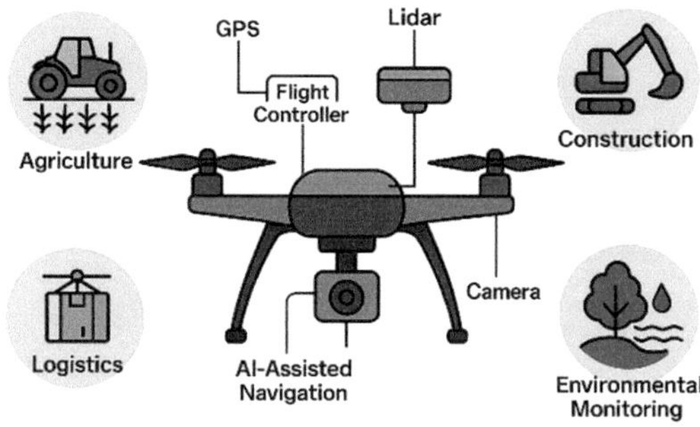

Figure 1.6 Fourth-generation drone semi-autonomy and commercial applications.

Figure 1.6 illustrates a quadcopter-based fourth-generation drone equipped with multi-sensor modules, GPS, LiDAR, and on-board cameras. The figure highlights the integration of a flight controller and an AI-assisted navigation system, enabling semi-autonomous operation. Various application domains, such as agriculture, construction, environmental monitoring, and logistics, are represented through icons surrounding the drone, showcasing its multifunctional commercial role and technological versatility.

1.4.5 Fifth generation: Full autonomy, AI, and swarm technology

The fifth generation of drone technology represents the pinnacle of unmanned aerial system evolution, characterized by full autonomy, artificial intelligence (AI) integration, and swarm coordination. Unlike previous generations that relied on human intervention or semi-autonomous control, fifth-generation drones are capable of independent decision-making, adaptive learning, and collaborative mission execution. This advancement has transformed UAVs from remotely operated tools into intelligent, networked systems capable of operating in complex and dynamic environments with minimal human oversight.

At the core of this generation is AI-driven autonomy, enabled by machine learning algorithms, computer vision, and advanced sensor fusion. These systems process real-time data from multiple sources, including cameras, LiDAR, radar, and infrared sensors, to recognize patterns, avoid obstacles, and optimize mission strategies autonomously. AI enables drones to perform complex tasks such as target recognition, formation flying, environmental mapping, and predictive navigation without relying on pre-programed instructions.

A defining feature of this generation is swarm technology, inspired by collective biological behaviors such as bird flocking and insect colonies. In

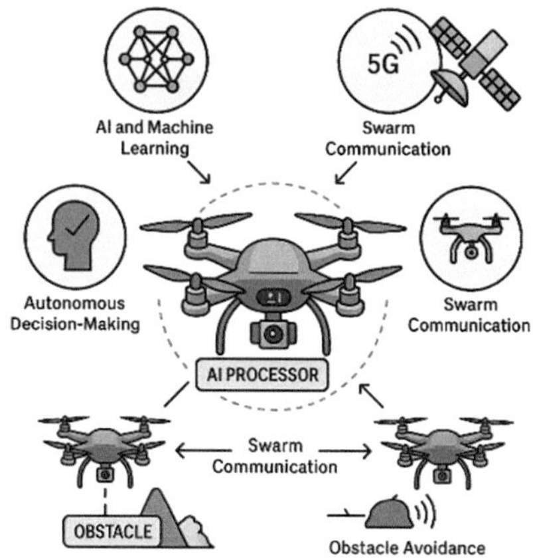

Figure 1.7 Fifth-generation drone full autonomy, AI, and swarm technology.

swarm operations, multiple drones communicate and coordinate through distributed intelligence networks, allowing them to share information, allocate tasks, and adapt collectively to environmental changes or mission demands. This capability is particularly transformative in applications such as large-scale surveillance, disaster response, search and rescue, and coordinated combat missions, where multi-agent collaboration enhances efficiency, resilience, and coverage.

The integration of 5G connectivity, edge computing, and cloud-based AI platforms further expands the operational capabilities of fifth-generation drones. These technologies support high-speed data transmission, low-latency communication, and real-time processing, enabling large-scale swarm deployment and intelligent traffic management in urban airspaces. Additionally, quantum navigation systems and blockchain-based communication security are being explored to enhance reliability and prevent cyber threats in autonomous operations.

Applications of fifth-generation drones extend across defense, logistics, smart cities, precision agriculture, and environmental sustainability. Autonomous delivery systems, infrastructure monitoring swarms, and defense-oriented reconnaissance units exemplify how intelligent UAVs are reshaping global industries. Governments and the private sectors are now focusing on integrating autonomous aerial ecosystems with regulatory frameworks that ensure safety, ethics, and airspace management.

Figure 1.7 illustrates a quadcopter-based fourth-generation drone equipped with multi-sensor modules, GPS, LiDAR, and onboard cameras. The figure highlights the integration of a flight controller and an AI-assisted navigation system, enabling semi-autonomous operation. Various application domains, such as agriculture, construction, environmental monitoring, and logistics, are represented through icons surrounding the drone, showcasing its multifunctional commercial role and technological versatility.

1.5 Summary

The chapter provides a comprehensive understanding of the evolution, structure, and impact of unmanned aerial vehicles (UAVs), tracing their transformation from manually operated systems to highly intelligent, autonomous platforms. It explores the technological foundations that have shaped drone development, covering advancements in aerodynamics, electronics, navigation, and artificial intelligence, while outlining their generational progression from early military prototypes to modern AI-driven swarms. The chapter emphasizes the broad spectrum of drone applications across defense, industry, and society, highlighting their role in surveillance, logistics, agriculture, infrastructure monitoring, and disaster management. Furthermore, it underscores the growing economic and strategic significance of drones in global markets, as well as the integration of autonomy and data-driven decision-making that defines the latest generation of UAVs.

Review Questions

1. What are unmanned aerial vehicles (UAVs), and how do they differ from manned aircraft?
2. Describe the early motivations behind the development of unmanned aerial vehicles.
3. How did military needs drive the initial innovation of UAV systems?
4. In what ways have drones evolved from military to commercial and recreational applications?
5. What technological factors contributed to the rapid growth of drone usage in recent decades?
6. How has the accessibility of drone technology impacted civilian industries?
7. Discuss the societal implications of drone proliferation in public and private sectors.

8. What are the key advantages of UAVs compared to traditional aerial platforms?
9. What are the primary functions of drones in military reconnaissance and surveillance?
10. How have drones changed modern warfare and defense strategy?
11. What are unmanned combat aerial vehicles (UCAVs), and what roles do they serve?
12. Describe how drones enhance situational awareness and battlefield intelligence.
13. In what ways are drones transforming commercial sectors such as agriculture and logistics?
14. How are UAVs applied in environmental monitoring and disaster management?
15. Explain the role of drones in infrastructure inspection and urban planning.
16. What are the challenges of integrating drones into civilian airspace?
17. Discuss the global economic impact of drone technology and its market growth.
18. How do international regulations influence the development of the UAV industry?
19. What are the current trends shaping the global drone market?
20. How has the widespread adoption of drones influenced employment and innovation ecosystems?
21. What were the earliest aerial systems that inspired the development of drones?
22. How did radio control contribute to the early progress of UAVs?
23. What breakthroughs in electronics enabled the miniaturization of drone systems?
24. Describe the importance of GPS in revolutionizing UAV navigation and control.
25. How do sensors and inertial measurement units (IMUs) enhance flight stability?

26. In what ways has artificial intelligence transformed drone autonomy and performance?
27. Explain how computer vision and machine learning are used in modern UAV systems.
28. What is the role of cloud computing and edge processing in real-time drone operations?
29. What defines a first-generation drone, and what were its primary limitations?
30. How did second-generation drones benefit from GPS and remote-control integration?
31. What advancements distinguished third-generation drones from their predecessors?
32. How did surveillance and combat capabilities evolve during the third generation?
33. Describe the emergence of semi-autonomous features in fourth-generation drones.
34. What commercial industries have benefited most from fourth-generation UAV applications?
35. How did advancements in sensors and AI contribute to semi-autonomy in drones?
36. What distinguishes fifth-generation drones in terms of autonomy and intelligence?
37. Explain the concept of swarm technology and its operational advantages.
38. How do fifth-generation drones utilize AI for real-time decision-making and coordination?
39. Discuss how 5G and advanced communication networks enable swarm operations.
40. Summarize the overall evolution of drone generations from manual control to full autonomy.

2

First-generation Drones

2.1 Introduction

The first generation of drones represents the foundational stage in the evolution of unmanned aerial systems (UAS), marking the earliest efforts to develop aircraft that could be controlled remotely without an onboard pilot. Emerging during the early to mid-20th century, these pioneering systems were primarily developed for military training, target simulation, and reconnaissance, long before drones became synonymous with modern intelligence and commercial applications. Their development was motivated by the need to reduce pilot risk in dangerous missions and to explore the potential of remotely piloted flight for defense and research purposes.

Technologically, first-generation drones were manually operated through rudimentary radio control systems, which transmitted limited signals from a ground station to the aircraft. These systems relied on basic mechanical servomechanisms and analog controls, which provided minimal precision and a short operational range. Lacking electronic stabilization, navigation sensors, or onboard processors, these early UAVs required continuous human control. The aircraft's performance and stability were heavily dependent on the operator's skill and line-of-sight visibility, as there were no telemetry systems to provide feedback during flight.

Structurally, these early drones were designed with simple fixed-wing configurations, typically made from lightweight materials such as wood, fabric, and early aluminum alloys. Propulsion was achieved using small piston engines or early internal combustion systems. Aerodynamic efficiency was minimal but sufficient for short-range, low-altitude missions. The designs were often modeled after small conventional aircraft but stripped of cockpit space to accommodate control systems and fuel.

One of the earliest prototypes, the Aerial Target (1916), developed by A. M. Low in the United Kingdom, demonstrated the first attempt at creating a pilotless aircraft controlled via radio waves. It was followed by the Queen Bee (1935), a radio-controlled target drone developed by the Royal Navy

DOI: 10.1201/9788743812418-2

and Royal Air Force for anti-aircraft gunnery practice. The term "drone" itself is believed to have originated from this aircraft, named after the male bee known for its buzzing sound, similar to that produced by the aircraft's engine. These pioneering platforms provided valuable insight into remote flight mechanics and control stability, setting the stage for future unmanned systems (Pliatsios, et al. 2021).

The operational use of first-generation drones was largely confined to military applications, including target towing, artillery training, and limited reconnaissance missions. Although their navigation capabilities were extremely primitive and their endurance limited, they provided defense forces with an invaluable tool for simulation and experimentation. These early UAVs also sparked growing interest among engineers and researchers, demonstrating that remote flight control could evolve into a viable and strategic technology.

Despite their technological simplicity, the first generation of drones played a critical role in establishing the conceptual and technical framework for modern UAV design. They introduced the principles of remote communication, control architecture, and stability management concepts that would later evolve into the sophisticated autonomous systems used today. In essence, these early systems transformed the theoretical idea of pilotless aircraft into a tangible reality, providing the first stepping stone toward the modern era of intelligent, data-driven aerial robotics.

Thus, the first generation of drones symbolizes not only the birth of unmanned flight but also the beginning of a technological revolution that continues to shape global defense, commercial, and research landscapes. These early experiments laid the groundwork for subsequent generations of drones that would incorporate automation, artificial intelligence, and advanced navigation, transforming simple manually controlled aircraft into the highly capable autonomous systems that define the field today.

2.1.1 Definition

First-generation drones refer to the earliest form of unmanned aerial vehicles (UAVs) developed primarily during the early to mid-20th century, characterized by manual radio-controlled operation, limited range, and rudimentary flight stability. These drones were designed mainly for military target practice, surveillance, and experimental purposes, serving as prototypes for future unmanned flight systems. They lacked onboard intelligence, sensors, and automated navigation, operating entirely under human control from the ground via analog radio transmitters.

2.1.2 Basic characteristics of first-generation drones

The first generation of drones represents the earliest stage in the evolution of unmanned aerial systems (UAS), characterized by simple mechanical structures and entirely manual control mechanisms. These drones were primarily used during the early to mid-20th century for military training, target simulation, and limited reconnaissance missions. Although their technological sophistication was minimal, they introduced the fundamental principles of remote-controlled flight, forming the foundation for subsequent UAV advancements.

- *Manual control*: First-generation drones were fully operated through basic radio control systems, requiring constant human input. The operator, using a handheld transmitter, controlled the aircraft's altitude, direction, and speed without any assistance from stabilization or feedback systems. This mode of operation demanded significant skill and concentration, as the aircraft's flight stability depended entirely on human precision. For example, the *Aerial Target (1916)*, an early British UAV prototype, was manually guided through simple radio commands, demonstrating the earliest form of remote-controlled flight.
- *Limited range and endurance*: These drones had a very limited operational range and flight endurance due to weak radio transmission power and small onboard fuel reserves. They could only be operated within a direct line of sight, typically over short distances, as any signal loss would immediately result in loss of control. Flight durations rarely exceeded a few minutes, which restricted their utility to experimental or short training missions.
- *Fixed-wing design*: Structurally, most first-generation drones featured a fixed-wing airframe, closely resembling miniature aircraft. Their aerodynamic design was functional rather than optimized for performance, focusing primarily on maintaining lift and basic maneuverability. The wings were non-adjustable, and aerodynamic control was achieved through rudimentary mechanical linkages connected to the control surfaces.
- *Lightweight construction*: Early drones were constructed from wood, fabric, and early aluminum alloys, materials chosen for their light weight and ease of fabrication. This lightweight design made them easier to launch and maneuver, but also limited their structural integrity and payload capacity. The *Queen Bee (1935)*, for instance, utilized a

wooden fuselage covered with fabric to reduce cost and weight, setting a standard for early UAV design simplicity.

- *Mechanical simplicity*: The mechanical systems of these drones were extremely basic, consisting of servo-driven linkages and analog control mechanisms. They lacked avionics, processors, or any automated components. As a result, pilots had to compensate manually for turbulence, wind, and other aerodynamic disturbances, making the operation highly challenging.
- *Absence of sensors or autonomy*: First-generation drones were devoid of any sensing or autonomous capabilities. They operated without gyroscopes, altimeters, or navigational aids such as GPS. Navigation depended entirely on the operator's visual observation and manual corrections during flight, often making long or complex missions impractical.

Figure 2.1 illustrates the fundamental design and operational characteristics of first-generation drones, showcasing their simplicity and manual mode of operation. These early UAVs were primarily fixed-wing aircraft constructed from lightweight materials such as wood, canvas, or early aluminum alloys, emphasizing structural efficiency over endurance. The figure highlights key features, including manual control, where operators relied entirely on basic radio transmitters to steer and stabilize the drone in flight. The limited range and endurance of these systems were constrained by weak radio signals and small fuel capacities, restricting operations to short, line-of-sight distances.

The illustration also emphasizes the mechanical simplicity of first-generation drones – lacking sensors, automation, or feedback systems. Control surfaces such as rudders and elevators were managed via direct servo linkages,

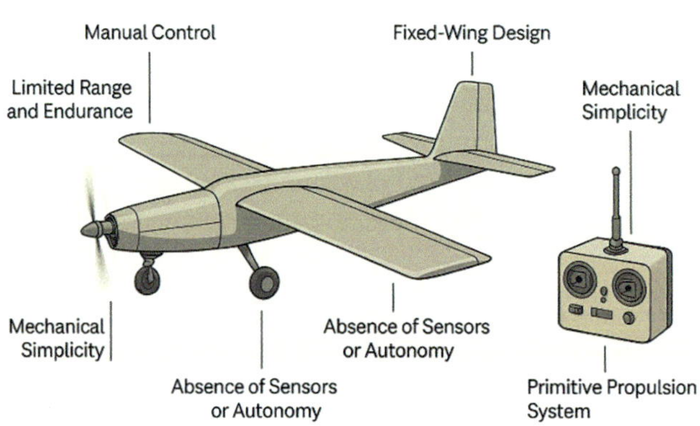

Figure 2.1 Basic characteristics of first-generation drones.

while propulsion depended on primitive piston or combustion engines. The absence of sensors or autonomous systems made flight stability and navigation fully dependent on the operator's visual tracking and skill.

2.2 Early Concepts of Unmanned Aerial Vehicles (UAVs)

The early concepts of unmanned aerial vehicles (UAVs) emerged from humanity's enduring desire to achieve controlled flight without risking human lives. Long before drones became technologically feasible, the idea of remotely guided or pilotless aircraft had already taken shape through both military vision and scientific experimentation. These early concepts laid the intellectual and engineering groundwork for the modern UAVs we know today, blending innovation in aerodynamics, communication, and control mechanisms (Newcome, 2004).

The first practical concepts appeared in the early 20th century, largely driven by military research. During World War I, several nations recognized the potential of unmanned aircraft for reconnaissance and target training. One of the earliest and most influential examples was the "Aerial Target" (1916) developed by A. M. Low in the United Kingdom. This radio-controlled aircraft was among the first attempts to create a pilotless vehicle guided remotely through wireless signals. Although its operational performance was limited, it demonstrated the feasibility of remote-controlled flight.

In the United States, similar developments were taking place. The Kettering Bug (1918), designed by Charles Kettering and funded by the US Army, was an experimental aerial torpedo capable of carrying explosives over pre-set distances using a simple mechanical timer. While it lacked remote control or guidance, it is often considered a precursor to modern cruise missiles and UAVs, representing a key step in the conceptual transition from manned to unmanned aerial systems.

By the 1930s, the evolution of radio technology enabled more sophisticated experiments. The British "Queen Bee" (1935), a radio-controlled target aircraft derived from the de Havilland Tiger Moth, was the first successful operational UAV used for military training. Its success was so influential that the term "*drone*" was coined from its name, inspired by the buzzing sound of its engine, resembling that of a male bee. The Queen Bee demonstrated how remote-controlled aircraft could be used repeatedly and safely for defense training and research, solidifying the concept of UAVs in military practice.

These early innovations shared several defining characteristics: manual radio control, limited range and stability, and experimental propulsion systems. They lacked the automation, sensors, and navigation capabilities of modern UAVs but introduced essential design principles such as control through wireless communication, aerodynamic optimization for stability, and the separation of the pilot from the aircraft.

In essence, the early concepts of UAVs were born out of experimentation, necessity, and ingenuity. They represented a period of technological exploration where inventors and military engineers sought to extend the boundaries of flight and safety. Although constrained by the technology of their time, pioneers such as A. M. Low and Charles Kettering established the conceptual and mechanical foundations of unmanned aviation. Their visionary work paved the way for future generations of drones, transforming what began as experimental curiosities into one of the most revolutionary technologies in aerospace history.

Table 2.1 summarizes the early conceptual evolution of unmanned aerial vehicles (UAVs), tracing their development from simple tethered observation systems to the first functional radio-controlled aircraft. The table highlights how each stage of progress from manual aerial observation using balloons and kites to powered and guided unmanned flight contributed to shaping the foundation of modern UAV technology.

The period between 1898 and 1945 marked the establishment of key technological principles such as wireless communication, mechanical automation, and radio-guided flight. Early innovators, including Nikola Tesla, A. M. Low, and Charles Kettering, introduced the fundamental concepts of remote control and autonomous operation. The Queen Bee (1935) and Radioplane OQ-2 (1939) represent milestones where UAVs transitioned from experimental devices to practical military tools.

2.3 Historical Development

The historical development of first-generation drones reflects the pioneering efforts that transformed the concept of unmanned flight from theoretical experimentation into practical aerial systems. This period, spanning from the early 20th century to the late 1930s, established the foundation for modern unmanned aerial vehicles (UAVs) through progressive advancements in radio control, mechanical design, and propulsion technologies.

The origins of UAV development can be traced back to early 20th-century wartime experimentation, particularly during World War I, when engineers and scientists sought to create pilotless aircraft to reduce risks to human pilots during reconnaissance and combat operations. The British engineer A. M. Low played a pivotal role with his creation of the Aerial Target (1916), one

Table 2.1 Early concepts of unmanned aerial vehicles (UAVs).

Period/year	Concept/ development	Inventor/ developer	Technological principle	Significance/ contribution
Pre–1900s	Tethered balloons and kites	Various (France, China, United Kingdom)	Manual aerial observation platforms using tethered lift.	Represented the earliest unmanned systems used for reconnaissance and meteorological observation.
1898	Radio-controlled boat prototype	Nikola Tesla (United States)	Wireless control through electromagnetic transmission.	Introduced the principle of remote control, laying the groundwork for future UAV communication systems.
1916	*Aerial Target*	A. M. Low (United Kingdom)	Radio-guided powered aircraft.	First demonstration of controlled unmanned flight; prototype for future UAVs.
1917	*Ruston Proctor Aerial Target*	A. M. Low/Ruston Proctor Aircraft Co. (United Kingdom)	Early radio-controlled powered monoplane.	Among the first successful applications of remote-controlled powered flight.
1918	*Kettering Bug*	Charles Kettering (United States)	Mechanically pre-programed flight using engine revolution counters.	Pioneered automated flight distance control; precursor to modern cruise missiles.
1935	*Queen Bee*	De Havilland Aircraft Co. (United Kingdom)	Radio-controlled training aircraft.	First operational UAV used for gunnery practice; origin of the term "drone."
1939–1945	*Radioplane OQ-2*	Reginald Denny Industries (United States)	Radio-controlled, catapult-launched aerial target.	First mass-produced UAV, marking the transition from experimental to practical drone systems.

of the first radio-controlled aircraft designed to intercept enemy zeppelins. Although the technology of the time limited its success, it demonstrated that remote-controlled powered flight was achievable (Cuerno-Rejado, et al. 2016).

Following this, the Ruston Proctor Aerial Target (1917), also developed under Low's supervision, became one of the earliest functioning pilotless aircraft capable of maintaining controlled flight. Around the same time in the United States, Charles Kettering introduced the Kettering Bug (1918), an unmanned aerial torpedo powered by a small engine and guided by a mechanical pre-set control system. The Kettering Bug is historically recognized as a precursor to modern cruise missiles, introducing the concept of automated flight distance control and pre-programed flight paths (Valavanis, et al. 2007).

The interwar period (1920s–1930s) saw further refinement of UAV technology as military organizations continued to explore unmanned aircraft for training and target simulation. This led to the development of the De Havilland DH.82B "Queen Bee" (1935) in the United Kingdom, a radio-controlled biplane derived from the Tiger Moth trainer aircraft. The Queen Bee became the first operational and mass-produced drone, used extensively by the Royal Navy and Royal Air Force for anti-aircraft gunnery practice. Its success not only validated the reliability of radio-controlled flight but also introduced the term "*drone*," derived from its buzzing engine sound.

By the late 1930s, UAV technology had progressed from experimental prototypes to functional training and reconnaissance systems, as demonstrated by the Radioplane OQ-2 (1939) in the United States. Developed by Reginald Denny Industries, it became the first UAV to be produced on an industrial scale, marking a turning point where drones transitioned from one-off experimental machines to standardized, operational tools (Keane, et al. 2013).

Thus, the first generation of drones (1910s–1930s) represents the formative stage of unmanned aviation, a period defined by innovation, experimentation, and gradual technical maturity. Each developmental milestone, from the Aerial Target to the Radioplane OQ-2, contributed a key technological element radio communication, automation, or mass production – that laid the groundwork for the advanced and autonomous UAV systems that would emerge in later decades. Table 2.2 outlines the historical development of first-generation drones, showcasing the technological and functional evolution of early unmanned aerial systems from 1916 to the late 1930s. Each entry highlights how innovators such as A. M. Low, Charles Kettering, and Reginald Denny contributed to the advancement of remote-controlled flight through pioneering experiments in radio communication, mechanical automation, and aerial propulsion.

Table 2.2 Historical development of first-generation drones.

Year/period	Drone/ system name	Developer/ country	Technological features	Purpose/ application	Significance/ contribution
1916	*Aerial Target*	A. M. Low (United Kingdom)	Early radio-controlled powered aircraft; fixed-wing design.	Experimental air defense drone intended to intercept enemy zeppelins.	First practical attempt at remotely controlled flight; established radio-guided flight concept.
1917	*Ruston Proctor Aerial Target*	Ruston Proctor Aircraft Co. under A. M. Low (United Kingdom)	Monoplane with radio control system and piston engine propulsion.	Pilotless aircraft for military experimentation.	One of the earliest successful demonstrations of powered, unmanned flight.
1918	*Kettering Bug*	Charles F. Kettering (United States)	Mechanical pre-programed control using engine revolution counters.	Unmanned aerial torpedo designed to deliver explosives.	Introduced automated flight control; precursor to cruise missile technology.
1935	*Queen Bee*	De Havilland Aircraft Co. (United Kingdom)	Radio-controlled biplane based on the Tiger Moth trainer aircraft.	Anti-aircraft gunnery training and target simulation.	First operational and mass-produced UAV; origin of the term "drone."
1939–1945	*Radioplane OQ-2*	Reginald Denny Industries (United States)	Catapult-launched, radio-controlled drone powered by a small piston engine.	Target drone for US Army Air Corps gunnery practice.	First UAV to be mass-produced; established drones as practical training tools.

2.3.1 Early experiments and innovations

The roots of unmanned aerial vehicle (UAV) development can be traced back to the pre-World War I era, when scientists and engineers first began experimenting with the idea of pilotless flight. During the late 19th and early 20th centuries, inventors explored ways to use emerging technologies, particularly radio communication and lightweight engines, to achieve remotely controlled flight. These early efforts were largely experimental and lacked the precision or reliability of later systems, but they established the fundamental concepts that would define UAV design for decades.

The concept of unmanned flight began to take form with the development of automatic and remotely guided aircraft. Innovators such as Nikola Tesla were among the first to envision remotely controlled vehicles. In 1898, Tesla demonstrated a radio-controlled boat, proving that wireless control of machines was possible, a breakthrough that later inspired aerial applications.

By the early 20th century, as aviation rapidly advanced following the Wright brothers' first flight (1903), engineers began to explore whether aircraft could be flown without onboard pilots. These early experiments culminated during World War I, when several nations invested in the creation of pilotless aircraft for military purposes. The "Aerial Target" (1916) developed by A. M. Low in Britain and the Kettering Bug (1918) in the United States were landmark projects in this period. The Aerial Target was one of the first radio-controlled aircraft, while the Kettering Bug used a simple gyroscopic guidance system and mechanical timer to deliver explosives over short distances. Although their operational success was limited, these prototypes proved the feasibility of unmanned flight and set the stage for future UAV development.

These early 20th-century experiments reflected the intersection of mechanical innovation, emerging radio technology, and military necessity. While constrained by the lack of reliable power sources and communication systems, they introduced critical ideas such as remote piloting, automated control, and aerial targeting that became the foundation for modern drones.

2.3.2 The role of balloons and kites in aerial surveillance

Long before the advent of powered flight, balloons and kites played an instrumental role in the early history of aerial observation and unmanned flight concepts. As early as the 18th and 19th centuries, military strategists recognized the value of elevated platforms for reconnaissance and communication. These early aerial systems, though not truly "unmanned aircraft" in the modern sense, served as precursors to UAVs by demonstrating how aerial vantage points could be used for surveillance and intelligence gathering.

Observation balloons were first employed during the French Revolutionary Wars (1794) when the French army used the "l'Entreprenant" hydrogen balloon to observe enemy troop movements at the Battle of Fleurus. Throughout the American Civil War (1861–1865) and later conflicts, tethered balloons became standard tools for battlefield reconnaissance, allowing armies to map terrain, direct artillery fire, and monitor enemy positions from above. These balloons were often manned but laid the conceptual groundwork for remote aerial observation, a defining function of later drones.

Kites also played a significant role in the early development of aerial imaging and meteorological research. In the late 19th century, Arthur Batut (1888) and Lawrence Hargrave (1890s) experimented with kite aerial photography (KAP) attaching cameras to kites to capture images from the sky. These unmanned photographic experiments represented some of the earliest attempts to gather data from the air without direct human presence onboard. By the early 1900s, meteorologists used kite systems to collect atmospheric data, temperature, and pressure at various altitudes establishing the principle of remote sensing, which would later become integral to UAV-based environmental monitoring.

2.4 Technological Foundations

The technological foundations of first-generation drones were established during the early 20th century when engineers and military scientists began experimenting with remotely piloted aircraft. Although primitive compared to today's UAVs, these early systems introduced the core engineering principles that shaped the evolution of unmanned flight namely, remote communication, aerodynamic design, propulsion systems, and mechanical control mechanisms.

At the heart of these early drones was the radio control (RC) system, which allowed human operators to guide an aircraft from the ground using wireless signals. This technology, still in its infancy at the time, relied on simple transmitters and receivers that could modulate a few basic flight commands such as turning, climbing, or descending. The lack of frequency stability and the limited range of early radio transmitters restricted operational distance, but this innovation represented the first step toward remote-controlled aviation.

In terms of aerodynamic design, first-generation drones primarily adopted fixed-wing configurations, similar to conventional airplanes. Their structure consisted of a main fuselage, wings, a tail assembly, and control surfaces, including rudders and elevators. The design focused on achieving basic lift and stability rather than agility or efficiency. The airframes were typically constructed from wood, canvas, and early lightweight metals, chosen for ease of fabrication and cost-effectiveness.

Propulsion technology for these drones was equally rudimentary. Most relied on small internal combustion or piston engines, which provided limited thrust suitable for short-range flights. Propeller-driven systems were manually started and offered minimal power regulation during operation. Despite their simplicity, these engines marked the beginning of motorized unmanned flight and introduced the principle of integrating propulsion with wireless control.

Mechanically, these drones operated through servo-based linkages and analog control systems that manipulated flight surfaces. Control was achieved through direct radio input, without any stabilization assistance or onboard computing. As a result, flight stability depended entirely on the operator's skill, weather conditions, and visual tracking capabilities.

While these aircraft lacked any form of automation or sensory feedback, their experiments laid the groundwork for key UAV technologies that would emerge later such as autopilot systems, telemetry, and sensor integration. The Aerial Target (1916) and Queen Bee (1935) exemplified the application of these foundational technologies, combining basic radio control with powered propulsion and fixed-wing designs.

The technological foundations of first-generation drones reflect a fusion of mechanical ingenuity and emerging electronic control, driven by early experimentation in wireless communication and aeronautical engineering. These pioneering systems established the core technological architecture propulsion, control, communication, and structure that remains central to all UAV designs today. For students and researchers, studying these foundational technologies provides valuable insight into how simple mechanical systems evolved into the complex, intelligent, and autonomous drones that define modern aerial engineering.

2.4.1 Principles and components of first-generation drones

The first generation of drones was built upon fundamental principles of remote control, aerodynamics, and mechanical propulsion. Although technologically primitive, these early UAVs demonstrated that controlled, pilotless flight was achievable through coordinated systems of communication, propulsion, and structural design. Their operation depended entirely on human-controlled inputs, transmitted via basic radio signals, and executed through mechanical linkages connected to flight surfaces.

2.4.1.1 Fundamental principles of operation

- *Remote control principle*: The first-generation drones operated through basic radio frequency (RF) control, where an operator transmitted commands using a ground-based controller. The aircraft's onboard receiver interpreted these signals and activated servomechanisms to adjust control surfaces such as the rudder, elevator, and ailerons. This established the foundation for modern UAV communication systems.
- *Aerodynamic lift and stability*: Like conventional aircraft, these drones relied on fixed-wing aerodynamics. Lift was generated as air moved across the wings, while the tail and control surfaces provided balance and directional stability. Pilots manually adjusted the aircraft's pitch, roll, and yaw to maintain flight.
- *Propulsion mechanism*: Small piston or internal combustion engines powered a propeller, producing thrust to overcome drag and maintain forward motion. These engines were manually started and mechanically simple, representing the earliest examples of motorized UAV propulsion.
- *Line-of-sight operation*: Drones were flown strictly within visual range, as there were no onboard sensors, telemetry, or navigation aids. The operator had to maintain direct visual contact throughout the mission for manual control and recovery.

2.4.1.2 Key components of first-generation drones

1. *Propeller*: Provided thrust for forward motion using a small piston engine.
2. *Engine (piston engine)*: The main power source, typically a single-cylinder internal combustion engine.
3. *Fuselage*: The central body of the drone housing the engine, fuel system, and control electronics.
4. *Fixed wings*: Generated lift through aerodynamic design, enabling sustained flight.
5. *Tail assembly (vertical and horizontal stabilizers)*: Provided stability and control in pitch and yaw.

6. *Radio receiver and antenna*: Received control signals from the ground operator, converting them into mechanical actions.
7. *Mechanical control linkages (servomechanisms)*: Translated radio commands into physical movements of control surfaces.
8. *Landing gear*: Allowed take-off and landing on simple runways or open terrain.
9. *Fuel tank*: Stored a small quantity of fuel to power short-duration missions.

2.4.1.3 Operational concept

The operator transmitted signals using a handheld radio controller, guiding the drone in real-time. The onboard receiver converted these signals into servo movements, which controlled flight direction. The drone's propulsion system maintained constant thrust, while the operator adjusted flight attitude through rudder and elevator inputs. Stability was achieved through constant manual corrections, as no autopilot or gyroscopic systems existed at the time.

Figure 2.2 illustrates the fundamental design and operational principles of a first-generation fixed-wing drone, highlighting its structural elements and control mechanisms. These early unmanned aerial vehicles were manually operated and relied on basic radio communication systems to transmit control signals from a ground operator. The ground control transmitter sent commands via radio waves, received by the onboard radio receiver and antenna, which then activated the mechanical control linkages connected to the rudder, elevator, and ailerons.

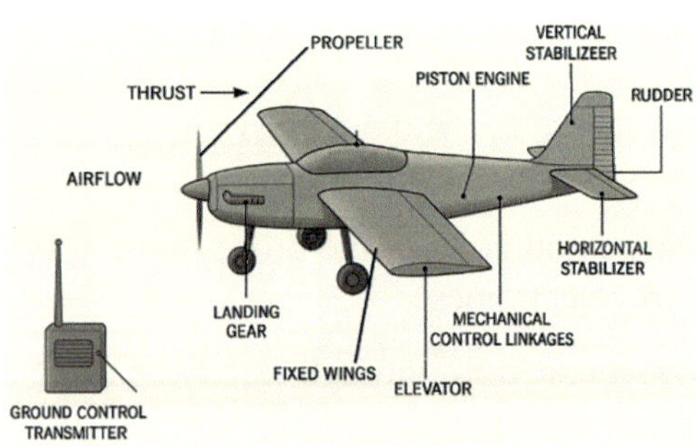

Figure 2.2 Principles and components of a first-generation drone.

The fuselage housed the fuel tank, radio receiver, and engine, which powered the propeller to generate forward thrust. Lift was produced as air flowed over the fixed wings, while the horizontal and vertical stabilizers at the tail provided aerodynamic balance and directional control. The landing gear enabled take-off and landing on simple runways or open terrain.

Arrows in the figure indicate the directions of thrust and airflow, emphasizing the aerodynamic forces acting on the aircraft during flight. The diagram clearly demonstrates how early drones functioned through a manual, line-of-sight operation, where stability and navigation depended entirely on the operator's input and visual observation.

2.5 Military Applications of First-generation Drones

The first generation of drones was developed primarily for military purposes, marking the earliest use of unmanned systems in defense operations. These drones emerged during the early to mid-20th century, a period when nations sought safer and more efficient methods for training, reconnaissance, and target simulation. Although limited in technological sophistication, first-generation drones provided an invaluable foundation for future military UAV programs by demonstrating the strategic potential of pilotless flight in combat and support roles. Table 2.3 summarizes the major military applications of first-generation drones, illustrating their early use in training, reconnaissance, combat experimentation, and technological research. These early UAVs were primarily developed to improve defense readiness and gunnery precision, reduce pilot risk, and explore the possibilities of pilotless aerial operations (Mayer, 2015).

2.5.1 Target practice and gunnery training

The most prominent military use of first-generation drones was for anti-aircraft gunnery and pilot training. These drones served as moving aerial targets, allowing soldiers to practice shooting at dynamic airborne objects without risking human pilots. This function was critical for improving the accuracy of anti-aircraft artillery during World War I and the interwar period. A notable example is the de Havilland DH.82B "Queen Bee" (1935), developed by the British Royal Navy and Royal Air Force. It was a radio-controlled aircraft used to train gunners in tracking and targeting moving objects. The "Queen Bee" became so well known that the term "*drone*" was later derived from its name, symbolizing pilotless aerial vehicles.

Table 2.3 Military applications of first-generation drones.

Application area	Drone model/ example	Country of use	Purpose/ mission role	Key features/ technology used	Significance/ contribution
Target practice and gunnery training	*Queen Bee (1935)*	United Kingdom	Served as a moving aerial target for anti-aircraft gunnery training.	Radio-controlled biplane; reusable; operated by the Royal Navy and RAF.	First operational and mass-produced drone; improved training safety and accuracy.
Reconnaissance and surveillance	*Aerial Target (1916)*, *Ruston Proctor Aerial Target (1917)*	United Kingdom	Designed for experimental remote reconnaissance and interception missions.	Early radio control system; basic aerial stability; line-of-sight operation.	Introduced the concept of unmanned surveillance and data gathering.
Aerial bombing and torpedo delivery	*Kettering Bug (1918)*	United States	Functioned as an unmanned aerial torpedo intended to deliver explosives over pre-set distances.	Mechanical timer for automatic engine cutoff; pre-programed flight path.	First automated flight and guided munitions concept; precursor to modern cruise missiles.
Experimental combat simulation	*Ruston Proctor Aerial Target (1917)*	United Kingdom	Tested remote operation and control accuracy under combat-like conditions.	Early radio-guided control surfaces; piston engine propulsion.	Demonstrated feasibility of remotely operated combat aircraft.
Military research and technological development	Various prototypes (1916–1939)	United Kingdom and United States	Conducted as research tools for studying aerodynamics, propulsion, and control mechanisms.	Integration of servo linkages, control actuators, and analog radio systems.	Provided experimental data that advanced aviation engineering and control theory.

2.5.2 Reconnaissance and surveillance missions

Although early drones lacked advanced cameras or sensors, military researchers began experimenting with remote reconnaissance using lightweight cameras or observation payloads. These efforts aimed to gather intelligence without endangering pilots. Early prototypes like the Aerial Target (1916) in the United Kingdom and the Kettering Bug (1918) in the United States were designed for short-range missions to observe enemy movements or deliver basic aerial imagery. While their accuracy and endurance were limited, these projects introduced the concept of unmanned intelligence gathering, which would later evolve into modern aerial surveillance.

2.5.3 Aerial bombing experiments

Another early military application of drones was in aerial bombing and guided munitions. The Kettering Bug, developed by Charles Kettering during World War I, was one of the earliest examples of an unmanned aerial torpedo. It was pre-programed using a mechanical timer to release its payload after flying a predetermined distance. Although it lacked real-time control and accuracy, the project demonstrated the potential of unmanned aircraft as a precision-strike weapon, paving the way for later developments in cruise missiles and combat UAVs.

2.5.4 Experimental combat simulations

First-generation drones also played a role in experimental combat simulations, where militaries tested radio-controlled systems for tactical operations. These early trials explored the feasibility of using drones to simulate enemy aircraft or act as decoys to divert enemy fire. Such experiments highlighted the advantages of deploying unmanned systems in risky scenarios, reducing pilot casualties and improving defense strategies.

2.5.5 Propaganda and research tools

In addition to direct combat training, early drones were used as research and demonstration tools to showcase national technological progress. Governments and military laboratories utilized these aircraft to experiment with remote control technologies, aerodynamics, and radio communications, which later influenced the design of both military UAVs and civilian aircraft.

2.6 Developments and Pioneers of First-generation Drones

The first generation of drones emerged from a period of rapid technological experimentation during the early 20th century, driven by the vision of scientists, engineers, and military strategists who sought to achieve flight without human pilots. This period, extending from World War I through the 1930s, witnessed remarkable progress in radio control, aerodynamics, and propulsion systems. The early innovators and prototype models of this era laid the essential groundwork for all subsequent UAV advancements, transforming the idea of pilotless aircraft from theoretical speculation into practical reality.

2.6.1 Contributions of early innovators in UAV technology

The pioneering phase of unmanned aviation was shaped by a group of inventors and engineers whose contributions defined the fundamental principles of drone operation. Their work focused on radio control, mechanical guidance, and automated flight stability, concepts that continue to influence UAV design today.

- *A. M. Low (United Kingdom)*
  - Often regarded as one of the founding figures of UAV technology, British engineer Archibald Montgomery Low developed the "Aerial Target" (1916) one of the earliest radio-controlled aircraft. Designed for military use during World War I, the Aerial Target aimed to intercept enemy zeppelins. Although it never saw combat, it successfully demonstrated the feasibility of remote-controlled flight. Low's pioneering work in radio communication and guidance systems established the basic operational framework for later drones.
- *Charles F. Kettering (United States)*
  - American inventor Charles Kettering, working with the US Army Signal Corps during World War I, developed the "Kettering Bug" (1918), a pilotless aerial torpedo powered by a small engine. It was pre-programed to fly a set distance using a mechanical control system and then release an explosive payload. Though its guidance accuracy was limited, the Kettering Bug is widely considered a precursor to both cruise missiles and autonomous UAVs, representing a major step toward automation in aerial warfare.

- *Nikola Tesla (Serbia/United States)*
  - Although not directly involved in drone construction, Nikola Tesla's early experiments in wireless control provided the conceptual foundation for UAV communication systems. His 1898 demonstration of a radio-controlled boat proved that machines could be operated remotely using electromagnetic signals a breakthrough that inspired later innovators to apply similar principles to aircraft.
- *Reginald Denny (United States)*
  - Actor and aviator Reginald Denny played a critical role in advancing radio-controlled aircraft in the 1930s. He established the Reginald Denny Industries company, which produced the Radioplane OQ-2, the first mass-produced radio-controlled target drone used by the US Army during World War II. Denny's efforts marked the transition from experimental prototypes to industrial-scale UAV production.
- *De Havilland and the Royal Navy (United Kingdom)*
  - The de Havilland Aircraft Company, in collaboration with the Royal Navy, developed the DH.82B "Queen Bee" (1935) a modified, radio-controlled version of the Tiger Moth biplane. It became the first truly operational UAV used for anti-aircraft gunnery training, and its success popularized the term "*drone.*" The Queen Bee demonstrated the military potential of reusable unmanned systems, solidifying UAV technology as a key research field in aviation.

2.6.2 Key first-generation drone models

2.6.2.1 Aerial Target (1916)

The Aerial Target (1916) represents one of the earliest and most significant milestones in the history of unmanned aerial vehicle (UAV) development. It was developed by A. M. Low in the United Kingdom during World War I under the supervision of the Royal Flying Corps and the Royal Navy. Designed as an experimental radio-controlled aircraft, the Aerial Target aimed to explore the feasibility of using unmanned aircraft for air defense operations, particularly as a means to intercept enemy zeppelins and aircraft without endangering pilots.

Technically, the Aerial Target was a small, powered fixed-wing aircraft equipped with a radio control system that received commands from a ground-based transmitter. The radio receiver on board was connected to servo-operated control surfaces – such as the rudder and elevators – that allowed the operator to guide the aircraft remotely. This innovation made the Aerial Target the first successful attempt at a powered, remotely controlled flight, establishing the foundational principle of radio-guided control in aviation.

Although it never entered active military service due to limited range, signal instability, and early technological constraints, the project was highly influential. It demonstrated that remote piloting was technically achievable, proving that aircraft could be controlled from the ground without a human pilot onboard (Campbell, 2008).

The Aerial Target (1916) thus holds historical importance as a pioneering step in UAV technology, setting the stage for subsequent developments such as the *Kettering Bug (1918)* and *Queen Bee (1935)*. It symbolized the beginning of the transition from human-operated flight to the concept of unmanned and autonomous aerial systems a revolutionary idea for its time.

Figure 2.3 illustrates one of the earliest experimental unmanned aerial vehicles (UAVs), developed under A. M. Low in the United Kingdom during World War I. Designed as a radio-controlled, pilotless aircraft, the Aerial Target featured a lightweight wooden airframe, a front-mounted propeller driven by a small piston engine, and basic control surfaces operated through radio-transmitted signals. It was intended for defensive missions, particularly to intercept enemy zeppelins, and served as a proof of concept for remotely controlled powered flight. Although its limited range and stability prevented operational deployment, the Aerial Target marked a historic breakthrough in aviation technology, establishing the core principles of radio-guided control that would later evolve into modern UAV systems.

Figure 2.3 The Aerial Target (1916) – the world's first radio-controlled unmanned Aircraft.

2.6.2.2 Kettering Bug (1918)

The Kettering Bug (1918), developed by Charles F. Kettering in the United States, was one of the most remarkable and ambitious projects in early unmanned aviation. Conceived during World War I, it was designed as an unmanned aerial torpedo capable of carrying an explosive payload to a predetermined target. The aircraft was built by the US Army Signal Corps in collaboration with the Dayton-Wright Airplane Company, marking one of the earliest large-scale military efforts to create an automated flying weapon.

The Kettering Bug operated on a mechanically pre-set guidance system. Instead of real-time radio control, it used a simple yet ingenious mechanical timer that counted engine revolutions to estimate distance traveled. Once the pre-set number of revolutions was reached, the engine would shut off, and the wings would detach, allowing the drone to glide toward the target with its 200-pound explosive payload. This innovation represented the first known attempt to automate flight and payload delivery, introducing the fundamental concept of pre-programed flight control that would later evolve into modern autopilot and missile guidance systems.

Constructed primarily from wood, cardboard, and fabric, the Kettering Bug had a wingspan of about 12 feet (3.7 meters) and was powered by a 40-horsepower, four-cylinder Ford engine, giving it a range of approximately 75 miles (120 kilometers). Although it was never used in active combat before the end of World War I, its successful test flights demonstrated that unmanned aircraft could accurately follow a pre-determined course without human intervention.

The significance of the Kettering Bug lies in its role as a technological bridge between early experimental UAVs and modern cruise missiles. It introduced the concept of autonomous flight based on mechanical programing, influencing later developments in aerial weapons, navigation systems, and automatic control theory. In essence, the Kettering Bug was a pioneering precursor to today's precision-guided munitions, illustrating how early innovation in mechanical automation laid the groundwork for the sophisticated unmanned systems that define modern warfare.

Figure 2.4 illustrates the *Kettering Bug (1918)*, one of the earliest examples of an autonomous aerial torpedo, developed by Charles F. Kettering in the United States during World War I. The aircraft featured a lightweight wooden and fabric fuselage, twin wings, and a small piston engine that powered a wooden propeller. Its flight path was controlled by a mechanically pre-set guidance system, which used an internal gyroscope and a revolution counter to determine the distance flown before cutting the engine, allowing the drone to dive onto its target. Designed to deliver explosives without risking

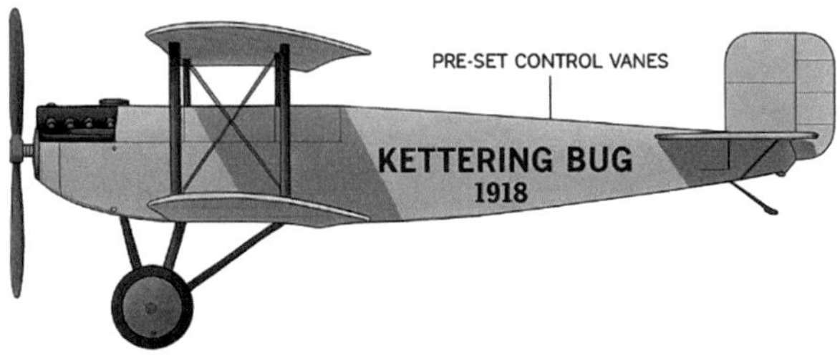

Figure 2.4 The Kettering Bug (1918) – a mechanically programed unmanned aerial torpedo.

human pilots, the Kettering Bug marked a groundbreaking innovation in early unmanned flight. Though it never saw combat use, this design introduced the principles of automated navigation and pre-programed control, making it a direct precursor to modern cruise missile technology (Stoer, 2001).

2.6.2.3 Queen Bee (1935)

The Queen Bee (1935), developed by the De Havilland Aircraft Company in the United Kingdom, stands as a landmark in the history of unmanned aviation. Designed as a radio-controlled target drone for gunnery training, it was the first fully operational and mass-produced unmanned aerial vehicle (UAV) used by both the Royal Navy and the Royal Air Force (RAF). Its introduction marked the transition from experimental UAV concepts to practical, large-scale military application, demonstrating the effectiveness of drones as reusable aerial targets.

Technically, the Queen Bee was based on the De Havilland DH.82 Tiger Moth, a two-seat biplane widely used as a pilot trainer. The aircraft was modified by fitting a radio control system that allowed ground operators to manage the aircraft's engine power, flight direction, and altitude remotely. A backup cockpit was retained, enabling a pilot to manually take control when needed, making it versatile for both manned and unmanned operation. The drone could take off, perform aerial maneuvers, and land under remote control a remarkable achievement for its time (Dancey, 2014).

The Queen Bee was primarily used for anti-aircraft gunnery practice, allowing military personnel to train with realistic moving targets without endangering pilots. Hundreds of these drones were produced between 1935 and 1943, operating from coastal bases and aircraft carriers across Britain. Their reliability, cost-effectiveness, and operational success made them a standard tool in military training programs throughout the pre-war and World War II periods.

The significance of the Queen Bee extends beyond its military value it is historically recognized as the origin of the term "drone." The aircraft's buzzing engine sound, reminiscent of a male bee, led Royal Navy personnel to nickname it "Queen Bee." This term was later adopted generically to describe all pilotless aerial vehicles.

The Queen Bee (1935) was not only the first operational and mass-produced UAV but also a defining milestone in drone technology. It proved the practicality of radio-controlled flight for real-world military training, established the feasibility of reusable unmanned systems, and contributed a lasting term "*drone*" to the global aerospace vocabulary. Its success paved the way for the rapid evolution of UAVs during and after World War II, bridging the gap between early experimental prototypes and the advanced autonomous drones of later generations.

Figure 2.5 depicts the *Queen Bee (1935)*, the world's first operational and mass-produced unmanned aerial vehicle (UAV) developed by the De Havilland Aircraft Company in the United Kingdom. Based on the DH.82 Tiger Moth trainer aircraft, the Queen Bee was adapted for radio-controlled flight and used extensively by the Royal Navy and Royal Air Force for anti-aircraft gunnery training. The figure highlights its wooden biplane structure, open cockpit (retained for manual operation during testing), and onboard radio receiver system connected to servomechanisms for controlling the rudder and elevator. The Queen Bee could be launched from both land and ship-based platforms, providing reusable, cost-effective aerial targets. Its success demonstrated the practicality of remotely operated aircraft and introduced the term "drone," derived from its buzzing engine sound. This innovation established the foundation of UAV operational training, marking a milestone in the history of unmanned flight.

Figure 2.5 The Queen Bee (1935) – the first operational radio-controlled drone.

2.6.2.4 Radioplane OQ-2 (1939)

The Radioplane OQ-2 (1939), developed by Reginald Denny Industries in the United States, represents a major technological and industrial milestone in the evolution of unmanned aerial vehicles (UAVs). Created as a radio-controlled target drone for the US Army Air Corps, it was specifically designed to provide realistic moving targets for anti-aircraft gunnery training. The OQ-2 became the first mass-produced UAV in history, symbolizing the beginning of the industrialization and operational deployment of drones during World War II.

Reginald Denny, a British actor, aviator, and model aircraft enthusiast, played a pioneering role in transforming radio-controlled flight from a hobbyist pursuit into a military technology. His company, Reginald Denny Industries, collaborated with the US Army and engineer Walter Righter, who designed a reliable two-cylinder, two-stroke piston engine to power the drone. The result was a compact, lightweight, and cost-effective aircraft that could be launched by catapult and recovered via parachute, enabling repeated use (Jeler, 2020).

Technically, the OQ-2 was a small fixed-wing aircraft with a wingspan of approximately 12 feet (3.6 meters) and powered by a 6-horsepower engine capable of speeds up to 85 mph (137 km/h). It was controlled by a radio transmitter, allowing the ground operator to guide it through flight maneuvers such as climbs, dives, and turns closely simulating enemy aircraft movements for live-fire training exercises. The system's simplicity, durability, and ease of operation made it ideal for large-scale military use.

Between 1941 and 1945, over 15,000 Radio plane OQ-2 drones were produced, marking the first instance of UAV mass production in aviation history. These drones were used extensively across US training bases, significantly improving the accuracy and efficiency of anti-aircraft crews. The manufacturing success of the OQ-2 also established Radio plane Company (later acquired by Northrop) as a key contributor to the emerging drone industry.

The significance of the Radioplane OQ-2 lies not only in its widespread operational use but also in its role in standardizing UAV production and design principles. It proved that unmanned aircraft could be manufactured affordably, operated reliably, and deployed on a national scale for defense applications. This success encouraged continued investment in drone research, directly influencing post-war UAV development for reconnaissance and surveillance.

Figure 2.6 illustrates the *Radioplane OQ-2 (1939)*, recognized as the first mass-produced unmanned aerial vehicle (UAV) developed by Reginald

Figure 2.6 The Radioplane OQ-2 (1939) – the first mass-produced target drone.

Denny Industries in the United States. Designed for the US Army Air Corps, this compact, fixed-wing drone served primarily as a radio-controlled target aircraft for anti-aircraft gunnery training during World War II. The image highlights its streamlined fuselage, tricycle landing gear, and small piston engine with a front-mounted propeller, reflecting its lightweight yet durable construction. Operated via radio signals from the ground, the OQ-2 could be launched by catapult and recovered by parachute, allowing repeated use. Its production of over 15,000 units marked the beginning of UAV industrialization, demonstrating that unmanned systems could be manufactured at scale and operated reliably. The *Radioplane OQ-2* thus stands as a landmark in drone history, bridging early experimental UAVs and the operational aerial platforms that followed.

2.6.2.5 Ruston Proctor Aerial Target (1917)

The Ruston Proctor Aerial Target (1917) stands as one of the earliest and most influential achievements in the history of unmanned aerial vehicle (UAV) development. Built by the Ruston Proctor Aircraft Company in the United Kingdom, under the technical supervision of A. M. Low, this aircraft was designed as a pilotless, radio-controlled platform for defense experimentation during World War I. The project was part of Britain's pioneering efforts to explore the potential of remotely controlled aircraft for military applications, particularly in intercepting and neutralizing enemy zeppelins and aircraft without risking human pilots.

Technically, the Ruston Proctor Aerial Target was a small, powered monoplane equipped with an early radio control system, allowing it to be guided from the ground through transmitted commands. It was powered by an ABC Gnat engine, which provided sufficient thrust for short-duration flights. The design incorporated basic flight control surfaces rudders and elevators connected to servo mechanisms that responded to wireless control signals. The aircraft's airframe was constructed primarily from lightweight wood and fabric, maintaining aerodynamic simplicity and structural integrity suitable for experimental purposes.

The operational principle of the Aerial Target was straightforward yet groundbreaking: a ground-based operator used a radio transmitter to send coded pulses to the aircraft's receiver, which in turn activated the servomechanisms controlling its flight direction and altitude. Although the aircraft's range and stability were limited by the era's primitive radio technology, several successful test flights were conducted in 1917, confirming the feasibility of powered, unmanned flight under radio control (Wolf, 2021).

The significance of the Ruston Proctor Aerial Target lies in its status as one of the earliest successful demonstrations of a powered, pilotless aircraft. It not only validated A. M. Low's theoretical concepts of radio-guided flight but also established the foundation for subsequent UAV projects, including the later "Aerial Target" prototypes and other first-generation drones. Furthermore, it represented a key technological bridge between balloon-based aerial systems and the more advanced autonomous aircraft that would emerge in later decades.

Figure 2.7 shows the *Ruston Proctor Aerial Target (1917)*, an early radio-controlled pilotless aircraft developed in the United Kingdom under the supervision of A. M. Low and constructed by the Ruston Proctor Aircraft Company. Designed during World War I, this drone was intended for defensive experimentation as part of Britain's early efforts to create unmanned interceptors. The aircraft featured a lightweight wooden biplane structure with fabric-covered wings, a small piston engine, and a propeller-driven design. Its flight control was achieved through radio-transmitted signals

Figure 2.7 The Ruston Proctor Aerial Target (1917) – early pilotless flight experiment.

that activated servomechanisms linked to the rudder and elevators. Though limited by the primitive radio and power technology of its time, the Ruston Proctor Aerial Target became one of the first successful demonstrations of powered unmanned flight, validating the feasibility of radio-guided control and laying the groundwork for future developments in UAV systems.

2.7 Limitations of First-generation Drones

Despite their historical importance and pioneering role in the evolution of unmanned aerial vehicles (UAVs), first-generation drones were constrained by numerous technological, operational, and practical limitations. These early systems represented the infancy of drone innovation relying on rudimentary engineering, unstable control mechanisms, and unreliable communication systems. While they proved that pilotless flight was possible, their limited performance prevented them from achieving large-scale operational success.

- *Limited range and endurance*

  First-generation drones suffered from short flight durations and limited operational range, largely due to weak radio transmitters and small onboard fuel capacities. Early radio control systems could only transmit signals over a few kilometers, forcing drones to remain within the operator's line of sight. Additionally, inefficient engines and lightweight airframes restricted their endurance to just a few minutes of flight, making long-distance missions impossible.

- *Lack of stability and control precision*

  The absence of onboard stabilization systems, sensors, or autopilots made these drones difficult to maneuver. Control inputs were transmitted manually by the operator, often with noticeable delays and inaccuracies. Even slight wind disturbances could cause severe instability or crashes. Without gyroscopic or electronic feedback systems, maintaining smooth and controlled flight was a major challenge.

- *Primitive communication systems*

  Early drones relied on basic radio frequency (RF) control, which was prone to signal interference, low power output, and limited bandwidth. The lack of secure or multi-channel communication often led to loss of control mid-flight. Additionally, the analog transmission systems of the time could not support complex commands or telemetry data, restricting drones to simple directional control.

- *Minimal payload capacity*

  Constructed from lightweight materials such as wood, canvas, and early metal alloys, first-generation drones had very limited payload-carrying capability. This restricted their use to basic tasks like target practice, as they could not accommodate cameras, sensors, or heavy explosive payloads effectively. As a result, their functionality remained confined to experimental and training purposes.

- *No autonomy or navigation systems*

  First-generation UAVs operated solely under manual control, with no onboard systems for autonomous flight, navigation, or decision-making. They lacked essential instruments such as altimeters, gyroscopes, and compasses, making them heavily dependent on the operator's visual tracking. This absence of automation limited their operational reliability and mission complexity.

- *High crash and failure rates*

  Due to their mechanical simplicity and technological limitations, early drones had high rates of failure and crashes. Many prototypes were lost during testing because of poor stability, signal loss, or engine malfunction. Each crash represented a significant financial and technical setback, as parts had to be manually rebuilt and recalibrated.

- *Environmental and weather sensitivity*

  First-generation drones were extremely vulnerable to weather conditions, including strong winds, humidity, and temperature fluctuations. Their light structures and low engine power made them unsuitable for operation under adverse environmental conditions, further limiting their practical usability.

- *Lack of real-time feedback and data transmission*

  These early drones had no telemetry or real-time feedback systems, preventing operators from monitoring performance parameters like altitude, speed, or orientation. Without data transmission, adjustments were made purely by observation, resulting in delayed reactions and inaccurate flight control.

- *Limited military and commercial utility*

  Although they demonstrated the feasibility of unmanned flight, first-generation drones lacked the technological maturity for large-scale military or commercial application. Their use was confined mainly to anti-aircraft target practice and basic experimental purposes, offering limited tactical or operational value.

- *Technological immaturity and high development costs*

  The technology of the early 20th century was not advanced enough to support reliable UAV systems. Components such as radio transmitters, lightweight engines, and control servos were expensive and difficult to produce in quantity. As a result, most projects remained experimental prototypes rather than deployable systems.

2.8 Case Study: Early Drone Use in Military Exercises

The use of first-generation drones in early military exercises marked a pivotal moment in defense innovation, showcasing the potential of unmanned aerial vehicles (UAVs) as practical tools for training, reconnaissance, and experimentation. Although limited in sophistication, these early systems demonstrated that remotely operated aircraft could enhance military preparedness, improve targeting accuracy, and reduce risks to human pilots.

One of the most notable examples comes from the United Kingdom in the mid-1930s, with the deployment of the De Havilland DH.82B *Queen Bee* during large-scale anti-aircraft training operations. Developed under the direction of the British Air Ministry, the Queen Bee was a radio-controlled version of the Tiger Moth biplane, designed to simulate enemy aircraft for anti-aircraft gunnery practice. These training exercises provided a revolutionary alternative to using manned aircraft for live-fire drills, as they allowed gunners to practice engaging moving aerial targets safely and repeatedly.

During these exercises, the Queen Bee was launched from coastal stations and guided remotely from the ground using a radio transmitter operated by a trained controller. The aircraft could climb, turn, dive, and perform evasive maneuvers, offering realistic flight behavior that challenged gun crews to adjust aim and timing. Operators closely coordinated with gunners to measure accuracy and effectiveness. In cases where the drone was hit, recovery operations were simple and cost-effective due to its lightweight wooden construction and low production cost.

The success of the Queen Bee exercises validated the concept of using drones as reusable training platforms and underscored the safety and efficiency benefits of pilotless systems. By 1939, more than 400 Queen Bee units had been produced and operated by both the Royal Navy and Royal Air Force (RAF), forming an essential part of Britain's pre-war defense training infrastructure. These exercises not only improved anti-aircraft proficiency but also demonstrated the strategic advantages of unmanned flight, leading to continued investment in UAV technology.

Across the Atlantic, similar developments took place in the United States with the Radioplane OQ-2 program during World War II. The US Army Air Corps conducted extensive training exercises using these small, mass-produced target drones to enhance gunnery precision. Operated via radio control and powered by compact piston engines, the OQ-2 offered consistent performance and was easily recoverable by parachute. These drones became central to large-scale military training efforts, with tens of thousands produced for use across multiple bases.

The operational use of drones in these early exercises provided critical empirical data on flight control reliability, radio communication stability, and human-machine coordination. It also allowed defense organizations to refine their approaches to range control, safety protocols, and weapon calibration.

2.9 Summary

The chapter represents the formative stage in the evolution of unmanned aerial vehicles (UAVs), spanning from the early 1910s to the late 1930s. This era marked the transition from conceptual experimentation to practical application, driven by wartime innovation and early advances in radio control, aerodynamics, and propulsion. Early UAVs were manually operated, fixed-wing aircraft with simple mechanical linkages, lightweight wooden or canvas airframes, and basic radio systems, lacking sensors or autonomy. The foundations of unmanned flight emerged through pioneering concepts such as tethered balloons, kites, and wireless control experiments by innovators like A. M. Low, Charles Kettering, and Reginald Denny. Key developments included the *Aerial Target (1916)* and *Ruston Proctor Aerial Target (1917)* in the United Kingdom, the *Kettering Bug (1918)* in the United States, and later operational drones such as the *Queen Bee (1935)* and *Radioplane OQ-2 (1939)*. These systems were primarily used for gunnery training, reconnaissance, and experimental bombing, showcasing the potential of unmanned flight in defense operations. Despite limitations in range, stability, and control precision, first-generation drones laid the technological and conceptual groundwork for modern UAV design, establishing key principles in radio communication, automation, and aerial control that would guide future developments in military and civilian drone technology.

Review Questions

1. What is meant by "First-Generation Drones," and during which period where they developed?
2. Describe the fundamental purpose behind the development of early unmanned aerial vehicles (UAVs).
3. Define the basic characteristics of first-generation drones in terms of design and control.
4. How did early drone technology differ from modern UAV systems?
5. What role did early aviation materials play in the construction of first-generation UAVs?
6. Explain why first-generation drones are considered the foundation of modern UAV technology.
7. What were the primary limitations of early drone flight control systems?
8. Why was manual radio control essential to first-generation drone operations?
9. Discuss how tethered balloons and kites contributed to the early concepts of unmanned aerial observation.
10. How did Nikola Tesla's experiments in wireless control influence UAV technology?
11. What was the significance of early aerial observation methods in shaping UAV design philosophy?
12. Identify the major differences between passive aerial systems (like kites) and powered early UAVs.

13. What technological developments made the transition from manned to unmanned aircraft possible?
14. Summarize the key historical stages in the development of first-generation drones.
15. What factors during World War I led to the experimentation with unmanned aircraft?
16. How did the interwar period contribute to advancements in UAV engineering and testing?
17. Compare the development of UAV technology in the United Kingdom and the United States during the early 20th century.
18. Why was the "Aerial Target (1916)" considering a milestone in remote-controlled flight history?
19. How did the *Ruston Proctor Aerial Target (1917)* improve upon earlier UAV concepts?
20. Describe the main principles underlying the operation of first-generation drones.
21. What were the key components of early UAVs and their respective functions?
22. How did propulsion and control systems evolve in first-generation UAV designs?
23. Explain how servo mechanisms and analog radio systems contributed to drone functionality.
24. What were the typical materials used in constructing first-generation UAV airframes, and why?
25. List and describe the major military applications of first-generation drones.
26. How did drones enhance gunnery and anti-aircraft training during the 1930s?
27. What was the purpose of using UAVs for reconnaissance and surveillance missions?
28. How did the *Kettering Bug (1918)* serve as a precursor to modern cruise missiles?

29. Explain how early UAVs were used in experimental combat simulations.
30. What was the significance of UAVs in improving pilot safety during training exercises?
31. Who was A. M. Low, and what was his contribution to early drone technology?
32. Describe the key innovations introduced by the *Aerial Target (1916)* and *Ruston Proctor Aerial Target (1917)*.
33. What were the defining features and uses of the *Kettering Bug (1918)*?
34. Why is the *Queen Bee (1935)* considering the first operational and mass-produced UAV?
35. How did the *Radioplane OQ-2 (1939)* mark the beginning of UAV industrialization?
36. Discuss the contributions of Reginald Denny to the advancement of drone technology.
37. What lessons were learned from the technological limitations of early UAVs?
38. Identify and explain the major technical and operational limitations of first-generation drones.
39. How did environmental factors such as wind and visibility affect early UAV performance?
40. In what ways did the experiences from early drone experiments influence the development of second-generation UAVs?

3

Second-generation Drones

3.1 Introduction

The second generation of drones represents a pivotal era in the evolution of unmanned aerial vehicles (UAVs), marking the transition from basic, manually operated machines of the early 20th century to more advanced, radio-controlled and semi-autonomous systems. Spanning from the 1950s through the early 2000s, this generation was characterized by rapid advancements in electronics, navigation, propulsion, and communication technologies, transforming drones from simple military training tools into versatile aerial platforms with applications in defense, science, and industry (Kardasz, et al. 2016).

During this period, drones evolved from analog radio-operated prototypes to aircraft equipped with digital control systems, real-time telemetry, and global positioning system (GPS) technology, enabling precise navigation and mission repeatability. The introduction of flight stabilization systems, onboard sensors, and data transmission links allowed drones to operate more efficiently and with greater accuracy, significantly expanding their operational capabilities. These innovations made UAVs invaluable not only in military reconnaissance and surveillance but also in civilian domains such as environmental monitoring, aerial mapping, and scientific research.

The historical context of this evolution is closely tied to the geopolitical tensions of the Cold War and post-Vietnam War era, where drones became critical instruments for intelligence gathering, target observation, and battlefield reconnaissance. Nations such as the United States, Israel, and the Soviet Union invested heavily in UAV research, resulting in breakthrough platforms like the *Ryan Model 147 "Lightning Bug," IAI Scout*, *RQ-2 Pioneer*, and *CL-289*. These systems demonstrated that unmanned aircraft could perform complex missions reliably, safely, and repeatedly, reshaping both military strategy and aerospace engineering.

The second generation also marked the emergence of civilian and commercial drone applications, as researchers began utilizing UAVs for

DOI: 10.1201/9788743812418-3

agriculture, disaster assessment, meteorological studies, and aerial photography. The integration of GPS, compact cameras, and miniaturized electronics made drones more accessible and multifunctional.

3.1.1 Definition

The second generation of drones refers to the era of unmanned aerial vehicle (UAV) development spanning from the 1950s to the early 2000s, characterized by the introduction of advanced radio control, onboard electronic systems, and early integration of satellite-based navigation such as GPS. This generation marked the transition from manually operated, first-generation drones to semi-autonomous aerial platforms capable of executing pre-programed missions with improved stability, range, and reliability.

Unlike their predecessors, second-generation drones incorporated digital communication links, flight control computers, and sensor-based stabilization, allowing for enhanced precision and reduced operator dependency. These UAVs were no longer confined to experimental or training roles but were deployed for real-time surveillance, reconnaissance, target tracking, and data collection across military and civilian sectors. The period also saw the gradual expansion of drone technology into scientific research, environmental monitoring, and commercial applications, laying the groundwork for the intelligent, AI-driven UAVs that followed.

In essence, the second generation of drones represents the technological and operational evolution from purely manual control systems to remotely guided, sensor-assisted, and GPS-enabled platforms, establishing drones as practical tools for both defense and non-military missions.

3.1.2 Basic characteristics of second generation of drones

The basic characteristics of second-generation drones reflect a major leap in technology, design, and operational flexibility. The following features distinguish them from the simpler, mechanically operated first-generation UAVs:

- *Advanced radio control systems*: Equipped with multi-channel, frequency-modulated (FM) transmitters and receivers, enabling precise and interference-resistant control over long distances.
- *GPS-based navigation*: Early integration of the global positioning system (GPS) allowed drones to follow pre-defined routes, conduct waypoint navigation, and return to launch points autonomously.

- *Digital communication and telemetry*: Real-time transmission of flight data, imagery, and sensor feedback to ground control stations enhanced situational awareness and mission efficiency.
- *Flight stabilization and onboard electronics*: Inclusion of inertial measurement units (IMUs), gyroscopes, and flight controllers improved aerial stability and allowed semi-autonomous flight under variable weather conditions.
- *Improved airframe and propulsion systems*: Use of lightweight composite materials and efficient electric or turbine engines increased endurance, payload capacity, and maneuverability.
- *Enhanced payload capabilities*: Integration of high-resolution cameras, thermal sensors, and reconnaissance modules for surveillance, mapping, and target acquisition.
- *Semi-autonomous operation*: Ability to execute partially automated flight missions with reduced operator input, marking the first step toward autonomous UAV systems.
- *Multi-domain applications*: Expanded use in military reconnaissance, environmental studies, meteorology, cartography, and agriculture, demonstrating drones' versatility beyond defense operations.

Figure 3.1 illustrates the basic characteristics of second-generation drones, highlighting their technological evolution from manually operated systems to semi-autonomous aerial platforms. The figure shows key components such as the GPS module, flight controller, radio receiver, and stabilization

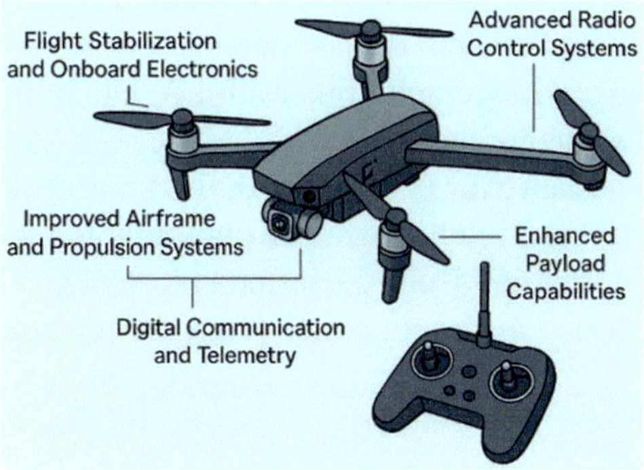

Figure 3.1 Basic characteristics of second-generation drones.

sensors (gyroscopes and IMUs) that enhance flight precision and stability. It also depicts the integration of digital telemetry and communication links between the drone and the ground control station (GCS), enabling real-time data exchange and mission monitoring. Advanced propulsion systems and lightweight composite airframes improved endurance and maneuverability, while the addition of high-resolution cameras and sensor payloads expanded functionality in reconnaissance, mapping, and environmental monitoring. Overall, the figure represents the defining features of GPS navigation, onboard electronics, and remote communication that transformed drones into reliable, multi-role aerial systems during the second-generation era.

3.2 Historical Development of Second-generation Drones

The historical development of second-generation drones represents a transformative era in unmanned aerial vehicle (UAV) evolution, spanning from the 1950s to the early 2000s, when drone technology advanced from basic, manually operated systems to semi-autonomous platforms equipped with digital communication, electronic stabilization, and satellite-based navigation. This period was marked by a convergence of military necessity, rapid progress in electronics, and the introduction of computing and GPS technologies, which collectively revolutionized drone design and application (Singh, et al. 2025).

The origins of second-generation drones can be traced back to the post-World War II era, when military research institutions in the United States, the Soviet Union, and the United Kingdom began repurposing first-generation target drones for more sophisticated reconnaissance and surveillance missions. The Cold War (1947–1991) intensified these developments as both superpowers sought new methods of intelligence gathering without endangering pilots. Early jet-powered drones like the Ryan Firebee series (1950s) demonstrated extended range and higher altitude capabilities, becoming the foundation for subsequent reconnaissance UAVs.

During the Vietnam War (1955–1975), the United States pioneered the use of UAVs for real-time intelligence and photographic reconnaissance. The Ryan Model 147 "Lightning Bug" performed more than 3000 missions, capturing aerial images and assessing bomb damage deep within hostile territories. These missions proved the tactical and operational value of unmanned aircraft in modern warfare. Similarly, in Israel, the development of the IAI Scout (1970s–1980s) marked a breakthrough in UAV technology, introducing real-time video transmission, data telemetry, and electronic warfare capabilities. The Scout's battlefield success established Israel as a global leader in drone innovation.

By the 1980s and 1990s, UAVs evolved into more compact, reliable, and sensor-rich platforms. The introduction of microprocessors and solid-state electronics allowed for the miniaturization of flight control systems, while inertial navigation systems (INS) and global positioning system (GPS) modules provided precision navigation and automated flight control. The RQ-2 Pioneer (1986), developed jointly by the US Navy and Marine Corps, exemplified these advancements, featuring live video transmission, GPS-based route tracking, and reliable remote operation. Similarly, the CL-289 (1990), a collaborative project between Germany and Canada, showcased jet-powered propulsion and terrain-following radar, further refining autonomous navigation capabilities.

The post-Cold War period (1990s–early 2000s) witnessed the expansion of drone applications beyond military use. The enhanced reliability and reduced cost of electronic components enabled UAVs to enter civilian sectors such as agriculture, environmental monitoring, meteorology, cartography, and aerial photography. Lightweight designs, electric propulsion, and GPS integration allowed for precise, long-duration missions suitable for scientific and commercial operations.

Table 3.1 presents the chronological development of second-generation drones, tracing their evolution from early Cold War reconnaissance systems to advanced, GPS-guided aerial platforms of the late 20th century. Each listed UAV reflects a milestone in unmanned technology, demonstrating progress in navigation, propulsion, data transmission, and automation. The table highlights how these innovations transformed drones from experimental military tools into reliable, multi-role aerial systems, laying the foundation for the autonomous and AI-integrated UAVs of the modern era.

3.3 Principles and Components of Second-generation Drones

The second generation of drones marked a defining shift in UAV design philosophy and functionality, introducing new aerodynamic, electronic, and communication principles that laid the foundation for modern unmanned systems. Unlike the mechanically simple first-generation UAVs, these drones incorporated microelectronics, digital telemetry, and GPS-based navigation, allowing for improved control precision, stability, and operational range. The combination of remote piloting, semi-autonomous flight, and real-time data

Table 3.1 Historical development of second-generation drones.

Period/Year	Drone model/ system	Developer/ country	Technological features	Primary applications	Significance/ contribution
1950s	*Ryan Firebee Series*	Ryan Aeronautical (United States)	Jet-powered target and reconnaissance drone; improved radio control and endurance.	Reconnaissance, training, and target simulation.	Foundation of Cold War UAV programs; established reliable jet-powered UAVs.
1960s–1970s	*Ryan Model 147 "Lightning Bug"*	US Air Force/ Ryan Aeronautical (United States)	High-altitude reconnaissance; onboard cameras; long-range communication link.	Real-time aerial surveillance during the Vietnam War.	First large-scale operational use of UAVs for combat intelligence missions.
1970s–1980s	*IAI Scout*	Israel Aerospace Industries (Israel)	Equipped with live video transmission, telemetry, and electronic warfare systems.	Battlefield reconnaissance and intelligence.	Introduced real-time data transmission; proved UAV combat effectiveness.
1986	*RQ-2 Pioneer*	US Navy and Marine Corps (United States)	GPS-assisted navigation, real-time video feed, and compact airframe.	Artillery spotting, reconnaissance, and coastal surveillance.	Enhanced operational accuracy; first major US Navy UAV success.
1990	*CL-289*	Dornier (Germany) and Canadair (Canada)	Jet-powered, GPS navigation, and terrain-following radar.	Tactical battlefield surveillance and mapping.	Demonstrated autonomous flight with GPS and advanced sensors.
1990s–2000s	*RQ-11 Raven / Civilian Mini-UAVs*	AeroVironment (United States)/ Multiple Developers	Electric propulsion, digital control systems, compact sensors, GPS-based autopilot.	Tactical reconnaissance, environmental studies, and mapping.	Represented the miniaturization and diversification of UAV technology for dual-use (military and civilian).

feedback made this generation a major step toward an intelligent, autonomous aerial platform.

3.3.1 Principles of operation

- *Remote control and data link communication*: The core operating principle of second-generation drones relied on radio frequency (RF) data links for communication between the drone and the ground control station (GCS). Operators transmitted flight commands using multi-channel transmitters, while telemetry data such as altitude, position, and system status was sent back in real time. This two-way communication established the foundation for networked UAV operations (Kagawa, et al 2017).
- *GPS-based autonomous navigation*: The introduction of the global positioning system (GPS) allowed drones to follow predefined flight paths and return automatically to the launch point. GPS integration enabled waypoint navigation, hover stability, and precise position tracking, which reduced pilot workload and improved mission repeatability.
- *Electronic flight stabilization*: Equipped with inertial measurement units (IMUs) and gyroscopic sensors, these drones could automatically correct roll, pitch, and yaw deviations. This stabilization system ensured smooth flight even in turbulent conditions and was one of the first steps toward autopilot technology.
- *Digital telemetry and real-time feedback*: The digital revolution introduced telemetry systems capable of transmitting video, sensor data, and system diagnostics to the ground station. This advancement allowed for real-time situational awareness, critical for both reconnaissance and scientific missions.
- *Payload integration and mission flexibility*: Second-generation drones were designed to accommodate a range of modular payloads, including cameras, infrared sensors, and scientific instruments. This flexibility made UAVs suitable for diverse missions across military, environmental, and commercial domains.

3.3.2 Major components of second-generation drones

The second generation of drones introduced a significant leap in unmanned aerial vehicle (UAV) engineering through the integration of advanced electronic, aerodynamic, and navigational systems. Unlike the simple, manually controlled drones of the first generation, these UAVs were equipped with sophisticated control, communication, propulsion, and payload technologies that enabled enhanced stability, longer endurance, and semi-autonomous flight capabilities. Each component within these systems played a vital role in ensuring operational precision, mission versatility, and improved performance under varied environmental and tactical conditions.

- *Airframe*: The airframe of second-generation drones was primarily constructed using lightweight composite materials such as carbon fiber, fiberglass, and aluminum alloys to maximize structural strength while minimizing weight. These aerodynamic designs were optimized for endurance, agility, and reduced drag, ensuring smoother flight dynamics and improved fuel efficiency. The shift from wooden or fabric structures of first-generation UAVs to composites marked a turning point in drone durability and performance.
- *Propulsion system*: Propulsion systems included piston engines, turbojets, or electric motors, depending on mission requirements. These systems provided higher power-to-weight ratios and allowed for extended flight times. Improved engine efficiency and reliability supported longer missions with greater payload capacity, enabling drones to operate at higher altitudes and speeds for reconnaissance, surveillance, and mapping operations.
- *Flight controller (FCU)*: The flight control unit (FCU) functioned as the UAV's central processing system, or "brain." It interpreted commands from the operator or autopilot system and coordinated control surface movements to maintain stability and direction. By integrating data from various sensors such as gyroscopes, accelerometers, and barometers, the FCU enabled semi-autonomous flight, route correction, and altitude control, significantly reducing pilot workload.
- *Global positioning system (GPS) module*: The inclusion of GPS technology revolutionized drone navigation. The GPS module allowed for autonomous waypoint navigation, precise geolocation, and automatic return-to-home (RTH) features. With satellite-based guidance, drones could perform pre-programed missions, follow exact flight paths, and execute precise landings without continuous manual control, enhancing mission reliability and efficiency.
- *Inertial measurement unit (IMU)*: The IMU integrates gyroscopes and accelerometers to monitor the drone's motion, orientation, and angular

velocity. This system played a critical role in flight stabilization, compensating for turbulence or sudden changes in wind conditions. By continuously feeding data to the flight controller, the IMU maintained smooth and steady flight even under challenging environmental conditions.

- *Communication and telemetry system*: Second-generation drones featured robust communication and telemetry links for real-time data transmission between the UAV and the ground control station (GCS). These systems used radio frequencies or satellite communication to transmit flight data, video feeds, and sensor readings. The two-way communication link enabled operators to monitor mission progress, adjust flight parameters, and receive live imagery or telemetry feedback from the drone.
- *Power source*: The power system supplied energy to propulsion and electronic subsystems. Depending on the UAV design, drones used fuel tanks, batteries, or hybrid systems. Fuel-based propulsion offered higher endurance for military and long-range missions, while battery-powered systems became popular for shorter, quieter, and environmentally friendly operations. The efficiency of the power source directly influenced range, payload, and flight duration.
- *Payload module*: The payload module was a mission-specific component designed to carry sensors, cameras, or scientific instruments. These payloads enabled applications in surveillance, reconnaissance, cartography, meteorology, and environmental monitoring. Modular payload systems allowed quick reconfiguration, transforming the same UAV for different missions, thus enhancing operational flexibility and cost-effectiveness.
- *Ground control station (GCS)*: The ground control station served as the command center for UAV operation. It provided an interface for flight control, mission planning, and data visualization. Equipped with telemetry displays, communication consoles, and live video feeds, the GCS allowed operators to monitor flight performance, adjust navigation parameters, and coordinate multi-UAV operations in real time.

Table 3.2 summarizes the essential components and functional architecture of second-generation drones, highlighting the integration of structural, electronic, and communication systems. Each component contributed to improved stability, precision, and mission versatility, marking the transition

Table 3.2 Components and functions of second-generation drones.

Component	Description/function	Significance
Airframe	Constructed using lightweight composite materials such as carbon fiber or fiberglass; optimized for aerodynamic efficiency.	Provided durability, increased endurance, and reduced aerodynamic drag, enhancing flight stability and performance.
Propulsion system	Included piston engines, small turbines, or electric motors, depending on mission type and payload requirements.	Enhanced power-to-weight ratio, improved speed, and extended operational endurance for diverse missions.
Flight controller (FCU)	The onboard microprocessor that processes sensor input and translates pilot or automated commands into control actions.	Served as the UAV's "brain," enabling semi-autonomous and stable flight operations through real-time data processing.
Global positioning system (GPS) module	Integrated satellite navigation receiver enabling waypoint navigation, autonomous flight paths, and return-to-home functions.	Allowed precise, repeatable, and autonomous flight, significantly improving navigational accuracy and mission reliability.
Inertial measurement unit (IMU)	Contained gyroscopes and accelerometers to measure motion, orientation, and angular velocity.	Enhanced flight stability, reduced operator input, and enabled smooth flight under variable environmental conditions.
Communication and telemetry system	Established a two-way data link between the UAV and ground control station (GCS) using radio or satellite communication.	Enabled real-time transmission of telemetry, video, and mission data for enhanced situational awareness.
Power source (battery or fuel tank)	Provided electrical or fuel energy required for propulsion, avionics, and onboard systems.	Determined overall flight endurance, payload capacity, and mission duration.
Payload module	Carried modular equipment such as cameras, infrared sensors, or scientific instruments for mission-specific applications.	Enabled flexible, multifunctional use in surveillance, reconnaissance, mapping, and environmental research.
Ground control station (GCS)	Operator interface that displayed real-time flight data, mission parameters, and video feeds.	Allowed mission planning, monitoring, route adjustments, and overall UAV coordination.

from manually operated UAVs to semi-autonomous, GPS-guided aerial platforms. The synergy between the flight controller, navigation modules, and telemetry systems established the technological framework for modern unmanned aerial systems (UAS).

3.3.3 System integration and functional operation

In operation, the ground control station (GCS) transmits control signals to the UAV via radio or satellite link, commanding throttle, altitude, and navigation. The flight controller unit (FCU) interprets these inputs and adjusts the propulsion and control surfaces based on feedback from the IMU and GPS systems. Data from onboard cameras and sensors is transmitted back to the GCS for real-time analysis. This closed-loop system enabled semi-autonomous mission execution with minimal manual intervention, greatly improving operational safety and efficiency.

The integration of electronic control systems, navigation aids, and modular payloads marked a major technological leap over earlier UAV generations. These systems not only improved flight performance but also introduced mission adaptability, allowing drones to switch between reconnaissance, surveillance, and research roles with minimal modification.

Figure 3.2 illustrates the principles and components of second-generation drones, highlighting the integration of advanced electronic systems, GPS navigation, and real-time communication links that enabled semi-autonomous flight. The figure shows the central role of the flight control unit (FCU), which processes input from the inertial measurement unit (IMU) and GPS module to maintain stability, altitude, and directional control. It also depicts the communication and telemetry system, which provides a two-way data link between the drone and the ground control station (GCS) for live video,

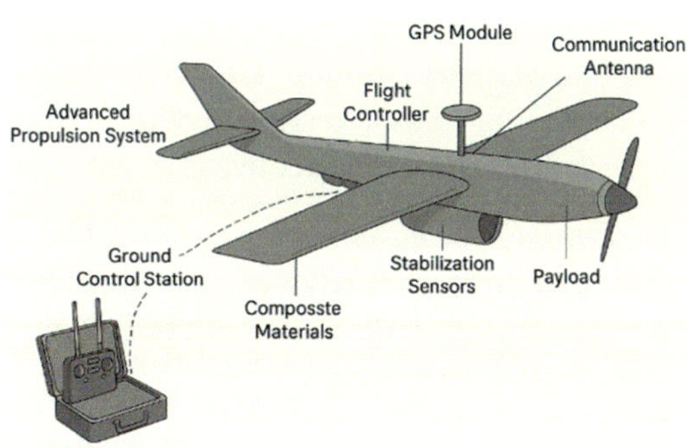

Figure 3.2 Integrated architecture of second-generation UAV systems.

telemetry, and command exchange. The airframe and propulsion system, built from lightweight composite materials, ensure endurance and maneuverability, while the payload module houses sensors or cameras for reconnaissance and monitoring missions. Overall, the figure demonstrates how the coordinated operation of navigation, control, propulsion, and communication systems transformed second-generation drones into reliable, multi-role aerial platforms capable of precise and efficient mission execution.

3.4 Applications of Second-generation UAV Systems

The second generation of unmanned aerial vehicle (UAV) systems marked a transformative period in the evolution of drone technology, expanding their use from specialized military operations to a wide range of civilian, scientific, and industrial applications. Between the 1950s and early 2000s, the combination of improved propulsion systems, digital communication, GPS navigation, and real-time telemetry allowed drones to become more reliable, versatile, and mission-capable. These advancements enabled UAVs to play an increasingly vital role in reconnaissance, surveillance, data acquisition, and environmental monitoring, establishing their importance in both strategic defense and non-military sectors.

3.4.1 Military and defense applications

- *Reconnaissance and surveillance*: Second-generation drones were extensively deployed for battlefield reconnaissance and intelligence gathering, especially during the Cold War and Vietnam War. UAVs such as the *Ryan Model 147* "*Lightning Bug*" and *IAI Scout* provided real-time imagery and electronic intelligence (ELINT), allowing armed forces to monitor enemy movements and assess damage after strikes without risking pilots.
- *Target practice and gunnery training*: UAVs like the *Ryan Firebee* were used as aerial targets for missile testing and anti-aircraft gunnery exercises, offering reusable and cost-effective platforms for training.
- *Electronic warfare (EW) and communication relay*: Drones equipped with electronic countermeasure payloads and signal relay systems supported communication networks and disrupted enemy radar or missile systems during combat operations.
- *Border patrol and coastal surveillance*: UAVs such as the *RQ-2 Pioneer* were employed by the US Navy and Marine Corps for maritime

reconnaissance and coastal defense, using real-time video feeds to detect and track vessels.

3.4.2 Scientific and environmental applications

- *Meteorological and atmospheric studies*: Equipped with sensors for temperature, pressure, and humidity measurement, UAVs were used in weather monitoring, storm tracking, and atmospheric sampling, providing valuable data for climate research.
- *Environmental and wildlife monitoring*: Drones facilitated aerial observation of wildlife habitats, deforestation, and environmental degradation, supporting conservation and ecological research efforts.
- *Disaster management and emergency response*: UAVs were employed in post-disaster assessments for mapping affected areas, locating survivors, and assessing infrastructure damage, especially in regions inaccessible to manned aircraft.

3.4.3 Civilian and commercial applications

- *Aerial photography and mapping*: The inclusion of high-resolution cameras and GPS systems allowed UAVs to be used for topographical mapping, urban planning, and infrastructure documentation.
- *Agriculture and precision farming*: UAVs began to be adopted in agricultural monitoring, analyzing crop health, irrigation patterns, and pest infestations using multispectral imaging and environmental sensors.
- *Surveying and land inspection*: In civil engineering and land development, drones were utilized to inspect bridges, pipelines, and construction sites, reducing manual labor and safety risks.
- *Scientific research and exploration*: Drones provided cost-effective tools for geological surveys, oceanographic research, and remote sensing, supporting data collection in hazardous or hard-to-reach environments.

3.4.4 Communication and technological applications

- *Data transmission and relay operations*: Second-generation drones incorporated digital telemetry systems capable of transmitting real-time data between remote stations, making them effective in field research and mobile network expansion.

- *Technology demonstration platforms*: UAVs served as testbeds for developing autopilot systems, sensor integration, and flight control algorithms, which directly contributed to the design of third-generation and autonomous drones.

Figure 3.3 illustrates the diverse range of applications of second-generation unmanned aerial vehicle (UAV) systems, highlighting their evolution from military reconnaissance tools to versatile platforms for scientific, environmental, and civilian operations. The figure is divided into four main sectors: military and defense, scientific and environmental, civilian and commercial, and communication and technology, each representing the expanding functionality of drones during this era. In the military domain, UAVs were employed for reconnaissance, surveillance, and electronic warfare, providing real-time intelligence and situational awareness during conflicts such as the Vietnam and Gulf Wars. The scientific and environmental applications segment showcases UAV use in atmospheric research, weather observation, wildlife monitoring, and disaster assessment, offering valuable data for research and humanitarian efforts. The civilian and commercial section highlights drones in aerial photography, mapping, agriculture, and infrastructure inspection, where GPS integration and imaging systems revolutionized industrial efficiency. Finally, the communication and technology segment demonstrates the role of UAVs as platforms for data relay, flight automation testing, and system integration research.

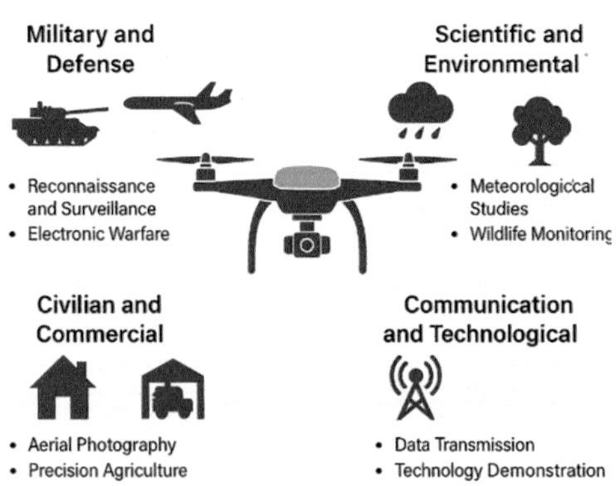

Figure 3.3 Applications of second-generation UAV systems.

3.5 Developments and Pioneers of Second-generation Drone Models

The development of second-generation drone models marked a period of extraordinary technological progress, driven by advancements in aeronautics, electronics, and computing, as well as the growing strategic need for reliable, pilotless aerial systems. Spanning from the 1950s through the early 2000s, this generation witnessed the transformation of unmanned aerial vehicles (UAVs) from basic radio-controlled aircraft to sophisticated reconnaissance and surveillance platforms capable of long-range, semi-autonomous operations. This evolution was shaped by visionary engineers, defense organizations, and aerospace industries that laid the groundwork for the modern UAV systems used today.

3.5.1 Evolution of second-generation UAV development

The post-World War II era catalyzed significant research into UAVs, as global powers sought safer, more efficient ways to collect intelligence without risking pilots. Early developments focused on enhancing flight control systems, endurance, and payload capabilities, resulting in the first operationally effective UAVs. The Cold War became a defining period for UAV innovation, as both the United States and the Soviet Union invested heavily in aerial reconnaissance technologies.

By the 1960s and 1970s, the integration of jet propulsion, radio telemetry, and high-resolution imaging systems transformed UAVs into indispensable tools for intelligence gathering. The Vietnam War accelerated this trend, as UAVs demonstrated their ability to conduct real-time reconnaissance, damage assessment, and electronic warfare missions. Simultaneously, nations such as Israel began pioneering smaller, more efficient UAVs designed for real-time video transmission and battlefield coordination.

The 1980s and 1990s marked the refinement and operational maturity of UAVs through the incorporation of GPS navigation, digital flight controllers, and telemetry systems. This period also saw a diversification of drone applications, including civilian uses in mapping, meteorology, and environmental monitoring, establishing UAVs as multi-domain platforms.

3.5.2 Key pioneers and notable drone models

3.5.2.1 The Ryan Firebee (1951)

The Ryan Firebee (1951), developed by Ryan Aeronautical Company (United States), stands as one of the most significant milestones in the history of second-generation unmanned aerial vehicles (UAVs). Originally designed as a jet-powered target drone for the US Air Force and Navy, the Firebee quickly evolved into a versatile aerial platform capable of fulfilling reconnaissance, surveillance, and research missions. Powered by a small turbojet engine and built with an aerodynamically efficient airframe, it offered high speed, altitude capability, and endurance far superior to earlier piston-driven drones (Wyatt, 2013).

The Firebee's technological innovation lay in its radio-controlled flight system, which allowed precise, remote piloting and recovery through parachute deployment. Later versions incorporated onboard cameras, telemetry systems, and radar reflectors, making it an adaptable tool for military training, weapon testing, and intelligence gathering during the Cold War. Its jet propulsion system marked a shift from experimental, short-range UAVs to long-endurance, reusable aircraft, capable of realistic flight simulations for combat training (Hall, 2014).

Over time, the Firebee family expanded into several variants, including the BQM-34 Firebee, which saw extensive use in reconnaissance operations during the Vietnam War. These upgrades integrated improved navigation controls, automatic flight stabilization, and the ability to carry payloads for electronic warfare (EW) missions (Keane, et al. 2013).

Figure 3.4 shows the Ryan Firebee (1951), a pioneering jet-powered unmanned aerial vehicle (UAV) developed by Ryan Aeronautical Company for the US Air Force. The figure highlights the aircraft's sleek, aerodynamic fuselage, mid-mounted wings, and bright red-orange finish – features that enhance visibility and stability during flight operations. Designed initially as a target drone, the Firebee's turbojet propulsion system and radio-controlled guidance allowed for high-speed, high-altitude performance that was unprecedented for its time. Later variants were equipped with onboard cameras and telemetry systems, expanding their role into reconnaissance and electronic

Figure 3.4 Ryan Firebee (1951) – the first jet-powered target and reconnaissance drone.

warfare. The illustration captures its compact yet powerful design, symbolizing the shift from early piston-powered drones to modern, reusable jet-powered UAVs capable of realistic flight simulation and intelligence missions. The Ryan Firebee thus represents a key technological milestone in the evolution of second-generation drones, combining performance, reusability, and mission adaptability.

3.5.2.2 Ryan Model 147 "Lightning Bug" (1964)

The Ryan Model 147 "Lightning Bug" (1964), developed by Ryan Aeronautical Company (United States), represents one of the most significant advancements in the history of unmanned aerial reconnaissance. Evolving from the successful Ryan Firebee platform, the Lightning Bug was a modified, jet-powered UAV specifically designed for strategic intelligence gathering during the Vietnam War and the Cold War. It was equipped with high-resolution cameras, sensor payloads, and autonomous navigation systems capable of flying long-range, pre-programed missions deep into hostile territories without risking human pilots (Hall, 2014).

The Lightning Bug introduced several groundbreaking innovations for its time, including altitude control systems, data telemetry, and programmable flight paths, allowing it to conduct precise reconnaissance missions under radio silence. It could operate at altitudes exceeding 60,000 feet and cover thousands of kilometers, capturing photographic and infrared imagery critical for battle damage assessment, enemy troop movement analysis, and air defense mapping. Recovery was achieved using parachutes, with mid-air retrieval often performed by helicopters or aircraft (Gettinger, 2024).

During the Vietnam War, the Lightning Bug flew more than 3000 missions, gathering intelligence over Vietnam, China, and North Korea. Its success demonstrated the strategic value of UAVs in modern warfare, especially in high-risk reconnaissance operations. Later variants, such as the Model 147H and 147S, incorporated advanced sensor packages and improved endurance, making them precursors to the autonomous surveillance UAVs of the 1980s and beyond (Prisacariu, 2022).

Figure 3.5 illustrates the Ryan Model 147 "Lightning Bug," a modified, jet-powered reconnaissance drone developed by Ryan Aeronautical Company for the US Air Force in 1964. Based on the Firebee platform, this UAV was designed for strategic intelligence missions during the Vietnam War, capable of long-range, high-altitude operations exceeding 60,000 feet. The figure highlights the aircraft's aerodynamic body, swept wings, and centrally mounted jet engine, optimized for stability and endurance. Equipped with high-resolution cameras and autonomous navigation systems, the Lightning

Figure 3.5 Ryan Model 147 "Lightning Bug" (1964) – the first combat-proven reconnaissance drone.

Bug could follow pre-programed routes deep into enemy territory, collecting vital reconnaissance imagery before being recovered by parachute. Its bright red-orange color improved visibility during retrieval missions. The Ryan Model 147 demonstrated the first large-scale operational use of UAVs for combat intelligence, establishing drones as indispensable tools in military reconnaissance and paving the way for future autonomous surveillance systems.

3.5.2.3 IAI Scout (1979)

The IAI Scout (1979), developed by Israel Aerospace Industries (IAI), stands as one of the most influential UAVs of the second generation, marking a major breakthrough in real-time aerial reconnaissance and surveillance. Designed during a period of growing military demand for flexible, unmanned systems, the Scout was among the first UAVs to successfully integrate real-time video transmission, data telemetry, and electronic warfare (EW) capabilities into a single operational platform. Compact, reliable, and highly maneuverable, it set the standard for modern tactical UAVs and firmly established Israel as a global leader in unmanned aerial technology.

The Scout featured a lightweight airframe constructed primarily from composite materials, improving endurance and operational efficiency. It was powered by a small piston engine and equipped with a two-blade propeller, allowing flight durations of over 7 hours and a range exceeding 200 kilometers. Its stabilized electro-optical payload enabled high-resolution day and night imaging, while the telemetry and communication systems provided real-time data transmission to ground control stations. The Scout could take off and land from short, unprepared runways, offering remarkable deployment flexibility in field operations.

Operationally, the Scout proved its effectiveness during the 1982 Lebanon War, where it was used extensively for battlefield reconnaissance, artillery spotting, and electronic intelligence (ELINT). It provided live imagery of enemy movements, coordinated air and artillery strikes, and dramatically improved situational awareness for ground forces. This success demonstrated the tactical advantage of UAVs in coordinated military operations, inspiring numerous other countries to adopt similar systems (Zaloga, 2011).

Figure 3.6 depicts the IAI Scout (1979), one of the most influential UAVs of the second generation, developed by Israel Aerospace Industries (IAI). The figure highlights the Scout's twin-boom tail configuration, straight high-mounted wings, and distinctive front-mounted propeller driven by a small piston engine. The compact and lightweight airframe, made of composite materials, enhanced endurance and operational efficiency, while the spherical electro-optical camera payload mounted beneath the nose enabled high-resolution real-time imaging. The illustration also shows the antenna housing atop the fuselage, representing its advanced data telemetry and communication system that allowed real-time video transmission to ground control stations. Operationally, the Scout was used for battlefield reconnaissance, target tracking, and artillery adjustment, with remarkable success during the 1982 Lebanon War, where it provided live imagery that revolutionized tactical decision-making. Overall, the figure demonstrates the Scout's innovative blend of endurance, real-time data capability, and maneuverability, which set a new standard for modern tactical UAVs and influenced future UAV designs worldwide.

3.5.2.4 RQ-2 Pioneer (1986)

The RQ-2 Pioneer (1986), developed jointly by Israel Aerospace Industries (IAI) and AAI Corporation (United States), represents one of the most

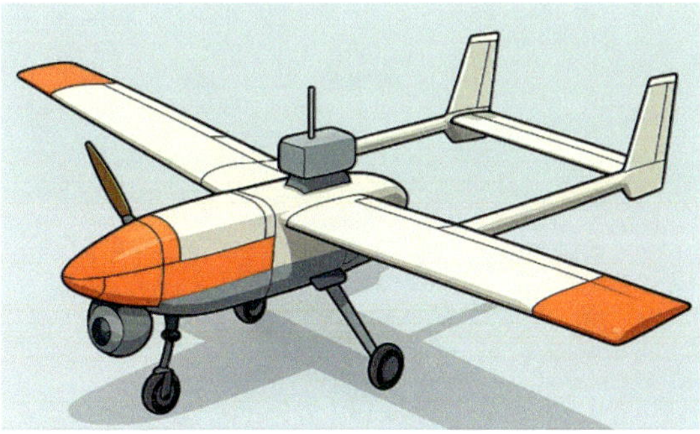

Figure 3.6 IAI Scout (1979) – the first tactical real-time reconnaissance UAV.

successful and operationally significant second-generation unmanned aerial vehicles (UAVs). Designed for the US Navy and Marine Corps, the Pioneer was based on the Israeli IAI Scout platform but incorporated major improvements in flight control, navigation, and payload technology, enabling it to serve as a tactical reconnaissance and surveillance system. Compact, reliable, and versatile, the Pioneer was instrumental in demonstrating the military effectiveness of UAVs during real-world combat operations in the late 20th century.

The Pioneer featured a twin-boom airframe with a pusher propeller configuration and was powered by a two-cylinder piston engine that provided a maximum endurance of 5 to 6 hours and a range of over 185 kilometers. Its lightweight composite body enhanced maneuverability, while the stabilized electro-optical/infrared (EO/IR) camera payload enabled high-quality, day-and-night imaging. Equipped with GPS-assisted navigation and real-time data telemetry, it could transmit live video and mission data directly to ground or ship-based control stations. The ground control station (GCS) was equipped with digital displays and tracking systems, allowing operators to monitor reconnaissance footage, coordinate artillery strikes, and support naval surveillance missions (Major, 2012).

The RQ-2 Pioneer gained historical significance during the 1991 Gulf War, where it provided real-time battlefield intelligence, locating enemy positions, and guiding naval gunfire with unprecedented accuracy. Its performance was so effective that it reportedly prompted Iraqi soldiers to surrender to a UAV during combat, a testament to its psychological and tactical impact. Beyond warfare, the Pioneer was also utilized for coastal patrol, target acquisition, and border surveillance, showcasing its operational versatility.

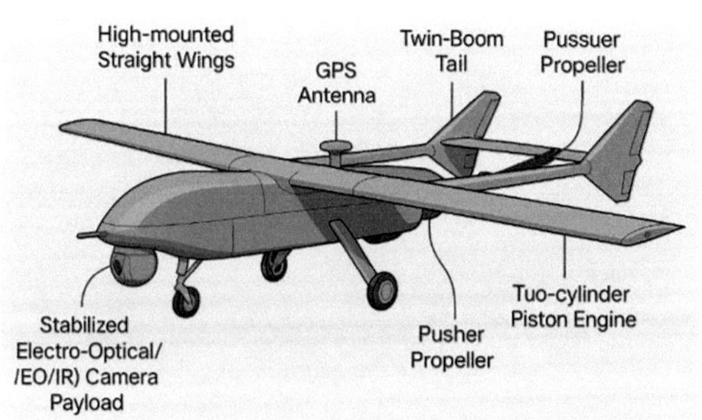

Figure 3.7 RQ-2 Pioneer (1986) – tactical reconnaissance and surveillance UAV.

Figure 3.7 illustrates the RQ-2 Pioneer (1986), a joint development between Israel Aerospace Industries (IAI) and AAI Corporation (United States), designed for the US Navy and Marine Corps. The diagram shows its twin-boom tail configuration, pusher propeller, and high-mounted straight wings, optimized for stable, long-endurance flight. The electro-optical/infrared (EO/IR) payload located under the fuselage provided real-time imaging for reconnaissance and artillery spotting, while the GPS antenna enabled accurate navigation and waypoint control. Powered by a two-cylinder piston engine, the Pioneer achieved endurance of up to six hours and a range of over 185 kilometers. Its ability to transmit live video feeds to ground and ship-based stations revolutionized naval surveillance and tactical intelligence during the Gulf War, where it guided naval gunfire and gathered battlefield imagery. The figure encapsulates the Pioneer's role as a combat-proven tactical UAV, combining GPS navigation, real-time telemetry, and reliable control, solidifying its legacy as a foundational model in modern UAV development.

3.5.2.5 CL-289 (1990)

The CL-289 (1990) represents one of the most advanced UAV systems of the second generation, developed through a collaboration between Dornier GmbH (Germany) and Canadair (Canada) for NATO forces. Designed primarily for tactical reconnaissance and battlefield surveillance, the CL-289 integrated a high degree of automation, GPS navigation, and terrain-following radar technology, making it one of the first UAVs capable of autonomous route tracking and target imaging under all weather and visibility conditions. It was engineered to deliver fast, accurate, and near-real-time battlefield intelligence, demonstrating the growing sophistication and reliability of drone systems at the close of the Cold War era.

The CL-289 featured a jet-powered propulsion system, allowing it to reach speeds of up to 720 km/h and operate at altitudes between 300 and 9000 meters. Its composite airframe design, optimized for high aerodynamic efficiency, could carry a modular sensor payload that included infrared line scanners, optical cameras, and radar imaging systems. These sensors enabled the CL-289 to capture high-resolution imagery over large battlefield areas, supporting target identification and damage assessment. The UAV was launched using a mobile catapult system and recovered via parachute and airbags, allowing for quick redeployment and operational flexibility in the field (Uhl, 1990).

A notable innovation of the CL-289 was its autonomous flight control system, which used pre-programed GPS waypoints and an inertial navigation system (INS) for flight path accuracy without continuous operator input. This

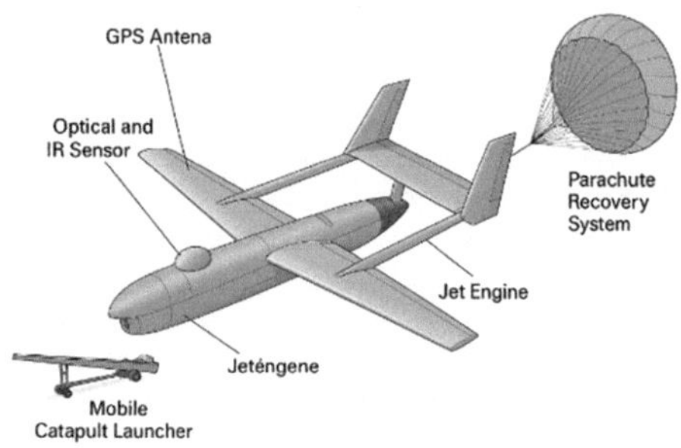

Figure 3.8 CL-289 (1990) – jet-powered tactical reconnaissance UAV.

allowed it to conduct low-altitude, terrain-hugging reconnaissance missions while maintaining stealth and avoiding radar detection. The drone's ground control system could process and analyze captured imagery immediately after recovery, enabling rapid dissemination of intelligence to command units.

Operationally, the CL-289 was employed by several NATO member nations, including Germany, France, and Canada, and saw extensive service in Balkan and Kosovo reconnaissance missions during the 1990s. It proved especially effective in providing detailed battlefield imagery for tactical planning and post-strike evaluation, significantly reducing the risk to manned aircraft in hostile environments.

Figure 3.8 illustrates the CL-289 (1990), a jet-powered tactical reconnaissance UAV developed collaboratively by Dornier GmbH (Germany) and Canadair (Canada) for NATO operations. The diagram highlights key structural and functional components, including the optical and infrared (IR) sensors, GPS antenna, and jet engine integrated within a sleek, composite airframe designed for high-speed, low-altitude reconnaissance. The mobile catapult launcher is shown at the front, used for rapid field deployment, while the parachute recovery system at the rear demonstrates its safe, reusable design. The CL-289's autonomous navigation system enabled pre-programed flight paths using GPS and inertial guidance, allowing precise, terrain-following reconnaissance missions under all weather conditions.

3.5.2.6 RQ-11 Raven (2003)

The RQ-11 Raven (2003), developed by AeroVironment Inc. (United States), marks the culmination of second-generation UAV technology and the transition toward portable, hand-launched, miniaturized drone systems. Designed primarily for tactical reconnaissance, surveillance, and target acquisition, the Raven became one of the most widely used small unmanned aerial vehicles in the world, revolutionizing battlefield intelligence and situational awareness at the unit level. Unlike its predecessors, which required complex launch systems or runways, the RQ-11 could be deployed by a single operator in the field, making it an essential tool for ground troops, Special Forces, and border patrol units.

The RQ-11 Raven features a lightweight airframe constructed from composite materials and a wingspan of just 1.4 meters, allowing it to be carried and launched by hand. It is powered by a quiet electric motor, providing an operational endurance of approximately 60–90 minutes and a range of up to 10 kilometers. The drone is equipped with modular payloads, including electro-optical and infrared (EO/IR) cameras, which transmit live video and telemetry data to a portable ground control station (GCS) in real time. Its GPS-based autopilot system allows for fully autonomous flight, including pre-programed routes, waypoint navigation, and automatic return-to-base functions (Van Bourgondien, 2012).

Operationally, the RQ-11 Raven became a critical asset during the Iraq and Afghanistan conflicts, where it provided real-time reconnaissance and surveillance for platoon-level operations. It was used for monitoring enemy movement, identifying ambush sites, and assessing terrain before troop advancement. The drone's low acoustic signature and lightweight design made it ideal for stealthy reconnaissance missions in complex terrains such as urban environments and mountainous regions.

The Raven's user-friendly interface and modular system design allowed rapid deployment and easy maintenance, contributing to its widespread adoption by over 30 military forces worldwide, as well as by civil defense, border security, and law enforcement agencies. It became a model for future miniature UAV development, influencing systems such as the RQ-20 Puma and Wasp III, which further enhanced endurance and imaging capabilities.

Figure 3.9 depicts the RQ-11 Raven (2003), a hand-launched tactical unmanned aerial vehicle developed by AeroVironment Inc. (United States) for real-time reconnaissance and surveillance. The figure highlights key components such as the forward-mounted EO/IR camera, GPS antenna, and electric motor with a rear pusher propeller, mounted on a lightweight airframe with high-mounted wings for stability and endurance. The Gimbal Control Unit beneath the nose ensures camera stabilization during flight, while the ground control station (GCS), shown to the right, provides portable command and telemetry capabilities for mission monitoring. Designed for ease of use, the Raven can be launched by hand, controlled autonomously via GPS waypoints, and recovered with minimal equipment. The figure effectively demonstrates the Raven's compact, field-ready design and its critical role in

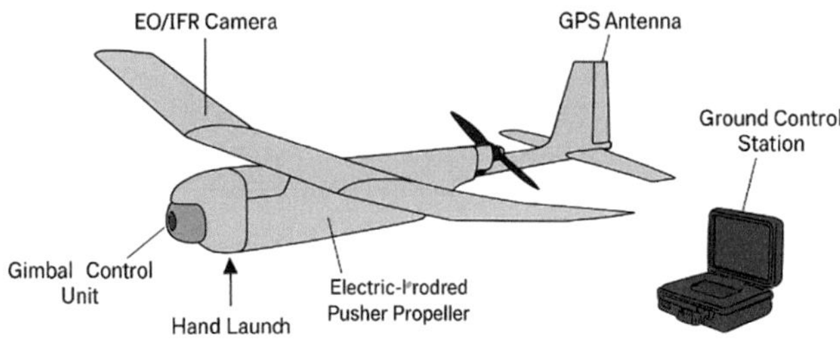

Figure 3.9 RQ-11 Raven (2003) – hand-launched tactical UAV.

providing platoon-level tactical intelligence through live video and telemetry data transmission in modern military operations.

Table 3.3 summarizes the major developments and pioneers of second-generation drone models, highlighting the technological breakthroughs that defined UAV evolution between the 1950s and early 2000s.

3.6 Challenges and Limitations of Second-generation Drones

While the second generation of drones marked a major technological leap in the evolution of unmanned aerial systems (UAS), this era also faced several technical, operational, and strategic challenges that constrained performance and mission scope. Despite advancements in navigation, communication, and payload integration, UAVs of this period were still in a transitional phase, bridging the gap between manually operated first-generation systems and the fully autonomous, AI-driven drones of later decades. These challenges underscored the technological limitations of the time and highlighted the need for further innovation in automation, endurance, and system reliability.

3.6.1 Technical limitations

- *Limited endurance and range*: Second-generation UAVs were often restricted by fuel efficiency and power limitations. Jet- and piston-engine drones, such as the *RQ-2 Pioneer* or *IAI Scout*, offered endurance between 4 to 8 hours, adequate for tactical missions but insufficient for long-duration strategic reconnaissance.
- *Dependence on line-of-sight communication*: Most drones are operated using radio frequency (RF) communication, which requires a direct line-of-sight connection between the UAV and the ground control

Table 3.3 Key developments and pioneers of second-generation drone models.

Model/year	Developer/country	Key technological features	Primary role/ application	Significance
Ryan Firebee (1951)	Ryan Aeronautical/ United States	Jet-powered UAV with improved aerodynamics and endurance; radio-controlled flight system.	Target training and reconnaissance.	One of the first reusable UAVs; the foundation for later surveillance platforms.
Ryan Model 147 "Lightning Bug" (1964)	US Air Force/Ryan Aeronautical	Equipped with cameras and sensors for high-altitude, long-range missions.	Strategic reconnaissance in the Vietnam War.	First combat-proven UAV; established drones as vital intelligence tools.
IAI Scout (1979)	Israel Aerospace Industries/Israel	Real-time video transmission, telemetry, and electronic warfare payloads.	Battlefield reconnaissance and target tracking.	Pioneered small UAV technology; widely exported for military use.
RQ-2 Pioneer (1986)	US Navy and Marine Corps/United States	GPS-assisted navigation, stabilized camera systems, and compact airframe.	Naval reconnaissance and artillery spotting.	Key success in naval operations; first US UAV to transmit live battlefield imagery.
CL-289 (1990)	Dornier (Germany) and Canadair (Canada)	Jet propulsion, GPS navigation, and terrain-following radar.	Tactical battlefield surveillance and mapping.	Enhanced autonomous flight; used by NATO forces for reconnaissance.
RQ-11 Raven (2003)	AeroVironment/ United States	Lightweight electric UAV with digital autopilot and portable design.	Tactical reconnaissance and surveillance.	Demonstrated miniaturization and ease of deployment in modern warfare.

station. Terrain obstructions or electromagnetic interference frequently disrupted signals, leading to data loss or mission termination.

- *Limited autonomy*: Although early GPS integration improved navigation, drones still relied heavily on human operators. Autonomous decision-making, obstacle avoidance, and adaptive flight planning were largely absent, limiting operational flexibility.
- *Sensor and payload constraints*: The sensor resolution, stabilization, and data transmission rates were limited by the analog technology of the era. Weather conditions such as cloud cover or wind turbulence often degrade imaging quality, particularly in optical payloads.
- *Navigation and control accuracy*: Early GPS modules lacked precision and robustness against jamming or signal loss. Inertial systems provided some redundancy, but drift errors accumulated over time, affecting positional accuracy during long missions.

3.6.2 Operational and logistical challenges

- *Complex launch and recovery systems*: Drones like the *CL-289* required catapult-assisted launches and parachute recovery, demanding specialized equipment and trained crews. These methods increased operational costs and deployment times, especially in combat zones.
- *Maintenance and reliability issues*: UAV systems of this generation require frequent calibration, component replacement, and post-flight inspection. Limited durability of engines, control surfaces, and electronics led to high maintenance-to-flight-hour ratios.
- *Weather sensitivity*: Second-generation UAVs lacked advanced flight control algorithms for turbulence correction or wind compensation. Operations were often grounded due to rain, fog, or strong crosswinds, affecting mission continuity.
- *Limited real-time data processing*: Although telemetry enabled real-time video transmission, data processing capabilities were minimal. Most image interpretation and analysis occurred post-flight, delaying decision-making in time-sensitive missions.
- *Training and operational complexity*: The lack of automation meant that skilled operators and technicians were required for every flight operation, making UAV missions resource-intensive and limiting large-scale deployment.

3.6.3 Strategic and regulatory limitations

- *Restricted civilian integration*: The use of drones outside the military was minimal due to regulatory uncertainty and airspace restrictions. Civil aviation authorities lacked frameworks for UAV certification, airworthiness, and safety compliance.
- *High production and operational costs*: The cost of developing and maintaining drones such as the *RQ-2 Pioneer* or *CL-289* was substantial. These systems required sophisticated materials, propulsion units, and support infrastructure, limiting adoption by smaller organizations or research institutions.
- *Limited multi-mission flexibility*: Many UAVs were designed for specific mission profiles, such as reconnaissance or target simulation. Payloads were not easily interchangeable, reducing adaptability for diverse applications.
- *Vulnerability to electronic countermeasures*: With increasing radar and electronic warfare capabilities, second-generation UAVs were susceptible to jamming, signal spoofing, or interception, compromising mission integrity and data security.

3.7 Case Study: Operational Deployment and Impact of Second-generation Drones

The following case studies illustrate the technological evolution, strategic value, and real-world deployment of second-generation drones through their use in key military operations and reconnaissance missions. These case studies highlight how UAVs transitioned from experimental platforms to indispensable tactical *assets* capable of transforming battlefield awareness, mission precision, and data-driven decision-making.

3.7.1 Case study 1: Ryan Model 147 "Lightning Bug" – reconnaissance in the Vietnam War (1964–1975)

The Ryan Model 147 "Lightning Bug," a derivative of the Ryan Firebee, was among the first UAVs to demonstrate large-scale operational success during the Vietnam War. Deployed by the US Air Force under the *Buffalo Hunter Program*, these drones conducted more than 3000 reconnaissance missions, collecting photographic intelligence deep within enemy territories, including Vietnam, Laos, China, and North Korea.

Each Lightning Bug was equipped with high-resolution film cameras, radar sensors, and telemetry systems, capable of autonomous, pre-programed flight at altitudes above 60,000 feet. After mission completion, drones were recovered by parachute and retrieved mid-air by helicopters or aircraft to preserve the valuable reconnaissance film.

The data collected by these UAVs was instrumental in battle damage assessment, strategic target identification, and air defense mapping. The Lightning Bug's success showcased the feasibility of pilotless reconnaissance, significantly reducing human risk. This program validated UAV technology as a cost-effective, survivable alternative to manned spy aircraft, influencing future drone programs such as the RQ-2 Pioneer and Global Hawk.

Impact:

- Introduced the concept of reusable UAVs for long-range intelligence.
- Demonstrated high-altitude, autonomous navigation with precision camera payloads.
- Established drones as critical tools for intelligence, surveillance, and reconnaissance (ISR).

3.7.2 Case study 2: IAI Scout – real-time intelligence in the Lebanon War (1982)

The IAI Scout, developed by Israel Aerospace Industries, was the first UAV to integrate real-time video transmission and electronic warfare (EW) payloads into a combat-proven platform. During the 1982 Lebanon War, the Scout provided live imagery of enemy movements and coordinated strikes with manned aircraft and artillery units, revolutionizing the use of drones in tactical warfare.

Deployed extensively by the Israeli Defense Forces (IDF), the Scout demonstrated unprecedented battlefield efficiency. It transmitted live video feeds to command centers via telemetry links, enabling real-time decision-making and target correction. The UAV's endurance (over 7 hours) and reliability allowed continuous surveillance of critical regions without exposing pilots to hostile fire.

The Scout's performance led to global recognition of UAVs as essential intelligence tools. Its success directly inspired international collaboration with the United States, leading to the development of the RQ-2 Pioneer, which adapted Scout's design for naval and ground operations.

Impact:

- Pioneered real-time battlefield surveillance and target coordination.
- Validated UAV integration into combined arms operations.
- Served as a foundational model for United States and NATO reconnaissance UAVs.

3.7.3 Case study 3: RQ-2 Pioneer – operational success in the Gulf War (1991)

The RQ-2 Pioneer, jointly developed by IAI and AAI Corporation, played a decisive role in the 1991 Gulf War, becoming one of the first UAVs to provide live battlefield video to tactical commanders. Operating from naval vessels and desert bases, the Pioneer conducted artillery spotting, target tracking, and coastal surveillance, transmitting real-time imagery directly to command posts.

The UAV's reliability and operational endurance (up to 6 hours) made it a vital reconnaissance tool. In one remarkable event, Iraqi soldiers reportedly surrendered to an unarmed Pioneer drone, recognizing its role in directing naval bombardments an event often cited as the first instance of enemy surrender to a UAV.

The Pioneer's successful performance under combat conditions validated the operational concept of unmanned tactical reconnaissance, influencing future designs like the RQ-5 Hunter and RQ-7 Shadow.

Impact:

- First UAV to deliver live video feeds from the battlefield to command units.
- Enhanced accuracy of naval and artillery strikes.
- Demonstrated UAV reliability under active combat conditions.

3.7.4 Case study 4: CL-289 – NATO's autonomous reconnaissance platform (1990s)

The CL-289, developed by Dornier GmbH (Germany) and Canadair (Canada), was one of the most advanced autonomous UAVs of its time. Used by NATO forces during the Balkan conflicts of the 1990s, the CL-289 performed terrain-following reconnaissance missions using GPS and inertial navigation systems (INS) to capture radar and infrared imagery.

Launched from mobile catapult systems and recovered by parachute, the CL-289's jet propulsion and high-speed flight allowed deep-penetration missions in hostile airspace. The system's rapid data recovery and image analysis capability provided commanders with near-real-time intelligence for target assessment and mission planning.

Impact:

- Introduced autonomous flight with terrain-following radar and GPS navigation.
- Enabled high-speed, high-altitude reconnaissance across contested areas.
- Influenced the development of third-generation UAVs emphasizing autonomy and precision.

3.8 Summary

This chapter represents a pivotal era in the evolution of unmanned aerial systems (UAS), defined by the integration of radio control, GPS navigation, onboard electronics, and real-time telemetry that transformed early manual drones into semi-autonomous reconnaissance and surveillance platforms. This generation saw major advancements in airframe design, propulsion efficiency, and sensor technology, enabling extended endurance and improved mission precision. Historically, these UAVs evolved from post-World War II research to Cold War and Vietnam War reconnaissance missions, laying the groundwork for tactical and scientific applications. Models such as the Ryan Firebee (1951), Ryan Model 147 "Lightning Bug" (1964), IAI Scout (1979), RQ-2 Pioneer (1986), CL-289 (1990), and RQ-11 Raven (2003) exemplified innovations in propulsion, data transmission, and GPS-based control, proving drones' strategic value in both military and civilian domains. Despite challenges such as limited endurance, dependence on line-of-sight communication, and regulatory constraints, second-generation UAVs revolutionized intelligence gathering, environmental monitoring, and real-time battlefield operations. Through successful deployments in conflicts like Vietnam, Lebanon, and the Gulf War, these drones demonstrated the tactical and operational viability of unmanned flight establishing the technological and functional foundation for the autonomous, AI-driven UAVs that define the modern era.

Review Questions

1. Define the second generation of drones and describe its time period of development.
2. What were the key technological innovations that distinguished second-generation drones from first-generation UAVs?
3. Explain the role of radio control and GPS integration in improving drone flight operations.
4. Discuss the historical context that led to the rise of second-generation UAVs during the post-World War II era.
5. How did the Cold War contribute to the advancement of unmanned aerial vehicle technology?
6. What were the primary operational objectives of second-generation drones in military applications?
7. Describe the basic flight characteristics and aerodynamic design features of second-generation drones.
8. Explain the importance of real-time telemetry and data transmission in second-generation UAV systems.
9. What were the major principles of operation in second-generation UAVs?
10. Identify the key components of second-generation drones and explain their functions.
11. How did the flight control unit (FCU) contribute to the semi-autonomous operation of UAVs?
12. Discuss the role of the inertial measurement unit (IMU) in maintaining flight stability.
13. How did the integration of GPS technology enhance navigation accuracy in drones?
14. What were the main propulsion systems used in second-generation UAVs, and what advantages did they offer?
15. Describe the structure and purpose of a ground control station (GCS) in UAV operations.

16. How did the modular payload system improve the versatility of second-generation drones?
17. List and explain the primary military and defense applications of second-generation drones.
18. How did drones contribute to artillery targeting and reconnaissance during the Vietnam and Gulf Wars?
19. Discuss the use of second-generation drones in scientific and environmental monitoring.
20. Explain how UAVs were used for aerial photography and mapping during this era.
21. What role did second-generation UAVs play in the field of agriculture and land surveying?
22. Describe the contributions of drones to communication and data relay applications.
23. Who were the major pioneers involved in second-generation UAV development, and what were their contributions?
24. Explain the significance of the Ryan Firebee (1951) in the evolution of jet-powered drones.
25. How did the Ryan Model 147 "Lightning Bug" (1964) demonstrate the feasibility of autonomous reconnaissance?
26. What innovations did the IAI Scout (1979) introduce to real-time battlefield intelligence gathering?
27. Describe the operational success and historical significance of the RQ-2 Pioneer (1986).
28. What were the technological advancements introduced in the CL-289 (1990) UAV?
29. Explain how the RQ-11 Raven (2003) revolutionized portable and hand-launched UAV systems.
30. Discuss the key challenges faced by second-generation drones in terms of endurance and range.
31. How did communication limitations affect UAV operations during the second-generation era?

32. Identify and explain the technical, operational, and strategic limitations of second-generation drones.
33. What were the primary maintenance and logistical challenges associated with UAV deployment?
34. Discuss the impact of environmental and weather conditions on UAV performance.
35. How did regulatory and airspace restrictions limit civilian applications of second-generation drones?
36. Summarize the key findings from the Ryan Model 147 "Lightning Bug" case study in the Vietnam War.
37. What was the significance of the IAI Scout during the 1982 Lebanon War in shaping future UAV systems?
38. Describe how the RQ-2 Pioneer changed the course of reconnaissance operations in the Gulf War.
39. Analyze the role of the CL-289 in NATO's reconnaissance and surveillance strategies during the 1990s.
40. In conclusion, how did second-generation drones bridge the technological gap between early experimental UAVs and modern autonomous drone systems?

4

Third-generation Drones

4.1 Introduction

The third generation of drones, emerging from the early 2000s to the mid-2010s, represents a pivotal era in the evolution of unmanned aerial vehicle (UAV) technology. This generation marked the transition from semi-autonomous, reconnaissance-focused systems to intelligent, networked, and multi-role UAV platforms capable of performing complex missions across military, scientific, and civilian domains. Building on the technological foundations of the second generation particularly advancements in GPS navigation, digital telemetry, and real-time imaging third-generation drones integrated breakthroughs in miniaturized electronics, advanced sensors, and data fusion technologies, enabling enhanced autonomy, endurance, and situational awareness.

This period witnessed the rapid rise of medium-altitude long-endurance (MALE) and high-altitude long-endurance (HALE) UAVs, such as the MQ-1 Predator, RQ-4 Global Hawk, and Heron, which redefined the strategic role of drones in surveillance, target acquisition, and precision strike missions. The integration of satellite communication (SATCOM) links allowed these drones to operate beyond visual and radio line-of-sight, providing real-time data streams to command centers thousands of kilometers away. The inclusion of advanced autopilot systems, electro-optical and infrared (EO/IR) payloads, synthetic aperture radar (SAR), and onboard data processing expanded the capabilities of UAVs from passive reconnaissance tools to active combat and intelligence-gathering platforms.

In addition to their military applications, third-generation drones became increasingly valuable in civilian and commercial sectors, including aerial surveying, disaster management, environmental monitoring, agriculture, and communication relays. Governments and industries began to recognize the potential of UAVs for large-scale data collection and automation. Regulatory frameworks also started to evolve during this time, paving the way for broader UAV integration into national and international airspace.

DOI: 10.1201/9788743812418-4

In essence, the third generation of drones signifies the beginning of the modern UAV era, characterized by networked intelligence, real-time global connectivity, and mission versatility. These systems established drones not merely as reconnaissance tools but as critical components of integrated aerial systems, capable of executing precision-guided strikes, long-endurance surveillance, and civilian data operations with unprecedented efficiency. This generation laid the foundation for the AI-driven, swarm-enabled, and fully autonomous UAV technologies that define the next phase of unmanned aviation.

4.1.1 Definition

The third generation of drones refers to the era of advanced unmanned aerial vehicles (UAVs) developed primarily between the early 2000s and mid-2010s, characterized by the integration of autonomous flight systems, satellite communication (SATCOM), multi-sensor payloads, and precision-guided mission capabilities. This generation marked the transition from semi-autonomous reconnaissance drones of earlier decades to intelligent, long-endurance, and combat-capable platforms capable of operating across vast distances with minimal human intervention.

Unlike the manually controlled or GPS-assisted systems of the second generation, third-generation drones feature fully digital flight control architectures with redundant navigation systems, real-time data links, and autonomous decision-making algorithms that enable adaptive mission execution. They are designed for diverse operational roles, including intelligence, surveillance, and reconnaissance (ISR), target acquisition, precision strike missions, and scientific or civilian applications such as mapping and disaster response.

These UAVs typically belong to the medium-altitude long-endurance (MALE) and high-altitude long-endurance (HALE) categories, offering extended flight durations of over 24 hours and operational ranges exceeding 1000 kilometers. Equipped with advanced payloads such as electro-optical/infrared (EO/IR) sensors, synthetic aperture radar (SAR), and signal intelligence (SIGINT) systems, they provide continuous, high-resolution situational awareness in real time.

4.1.2 Basic characteristics of third generation of drones

The third generation of drones represents a technological milestone in the evolution of unmanned aerial vehicles (UAVs), embodying the shift from remotely controlled reconnaissance systems to intelligent, networked, and multi-role aerial platforms. These drones combine autonomy, endurance, precision, and connectivity, enabling them to perform complex missions ranging from long-range surveillance to precision strikes and civilian data-gathering operations. Their design and functionality reflect major advances in aerodynamics, communication networks, onboard computing, and sensor integration.

Below are the key characteristics that define third-generation drones:

- *Enhanced autonomy and flight control*: Third-generation drones feature autonomous navigation and mission management systems, including advanced autopilot algorithms, obstacle avoidance, and adaptive route correction. They can execute complete missions from take-off to landing with minimal human input, relying on onboard data processing and decision-making.
- *Integration of satellite communication (SATCOM)*: A defining feature of this generation is the use of SATCOM links, which extend control and data transmission beyond the visual and radio line-of-sight limits. This enables global command-and-control capability, making UAVs suitable for long-distance and cross-border missions.
- *Extended range and endurance*: Drones such as the MQ-1 Predator and RQ-4 Global Hawk can remain airborne for over 24 to 36 hours, covering thousands of kilometers. Advanced propulsion systems and optimized aerodynamics significantly enhance endurance, allowing persistent surveillance and reconnaissance.
- *Multi-sensor payload integration*: Third-generation UAVs are equipped with multiple sensor suites, including electro-optical/infrared (EO/IR) cameras, synthetic aperture radar (SAR), and signal intelligence (SIGINT) payloads. These systems allow simultaneous imaging, target tracking, and environmental data collection under all weather and lighting conditions.
- *Precision strike capability*: Many third-generation drones, particularly armed variants such as the MQ-1 Predator and MQ-9 Reaper, introduced the ability to carry *laser-guided missiles (AGM-114 Hellfire)* and precision munitions, enabling remote and real-time engagement of tactical targets with minimal collateral damage.
- *Real-time data transmission and networking*: The introduction of network-centric warfare concepts allowed UAVs to transmit live video,

radar imagery, and sensor data to multiple command centers. This enabled real-time intelligence sharing, mission coordination, and target acquisition across military networks.

- *Advanced flight control and redundancy systems*: To ensure operational reliability, third-generation drones employ redundant avionics, triple autopilot systems, and fault-tolerant navigation units, minimizing the risk of mission failure due to single-point system malfunctions.
- *Modular and scalable design*: These UAVs are designed with modular architecture, allowing rapid reconfiguration for specific missions whether for surveillance, strike operations, or environmental monitoring. The modular approach also simplifies maintenance and upgrade cycles.
- *Stealth and low-observable features*: Advanced airframes incorporate radar-absorbing materials, reduced infrared signatures, and optimized aerodynamic profiles, enabling low observability in contested airspaces.
- *Expanded civilian and scientific roles*: Beyond military applications, third-generation drones began to play vital roles in *disaster management, border monitoring, agricultural mapping, and climate research*, marking the diversification of UAV technology into non-combat fields.

4.2 Historical Development of Third-generation Drones

The historical development of third-generation drones marks a transformative period in unmanned aerial vehicle (UAV) evolution, spanning from the early 2000s to the mid-2010s. This era witnessed the shift from semi-autonomous reconnaissance systems to fully networked, satellite-linked, and combat-capable platforms. Building upon the technological groundwork laid by second-generation UAVs, such as the integration of GPS navigation, telemetry, and digital flight control, this generation focused on global connectivity, endurance, precision, and mission adaptability.

The roots of third-generation drones can be traced back to the late 1990s, when advancements in satellite communication (SATCOM) and miniaturized electronics made it feasible to operate UAVs beyond visual and radio line-of-sight. The RQ-1 Predator, developed by General Atomics Aeronautical Systems (GA-ASI) and first introduced in 1995, became the foundation of this new generation. Initially designed as a surveillance drone, the Predator was later equipped with AGM-114 Hellfire missiles, transforming it into the first armed UAV capable of conducting precision strikes. Its operational success in the War on Terror, particularly in Afghanistan and Iraq, demonstrated the effectiveness of drones as persistent aerial assets capable of both surveillance and combat roles.

By the early 2000s, UAV development had accelerated rapidly across several countries. The RQ-4 Global Hawk, developed by Northrop Grumman, represented a leap forward in high-altitude long-endurance (HALE) UAV design. Capable of flying above 60,000 feet for over 36 hours, it provided near-real-time, high-resolution imagery across vast regions. Its combination of synthetic aperture radar (SAR), electro-optical (EO), and infrared (IR) sensors made it invaluable for strategic reconnaissance and intelligence gathering. The Global Hawk's ability to autonomously traverse continents and transmit live data via satellite marked a defining moment in UAV autonomy and global reach.

Internationally, UAV programs expanded in parallel. Israel Aerospace Industries (IAI) continued to lead innovation with systems such as the Heron (1994) and Heron TP (Eitan) (2004), which introduced longer endurance, multi-sensor payloads, and satellite data links. These drones played crucial roles in intelligence operations, maritime patrol, and border surveillance. Meanwhile, in Europe, the Dassault Neuron and BAE Systems Taranis projects demonstrated growing interest in stealth and autonomous combat UAVs (UCAVs), integrating radar-absorbing materials and autonomous targeting systems.

The US military and its allies increasingly integrated UAVs into network-centric warfare, linking real-time drone data with ground and air command units. This allowed drones to perform persistent surveillance, target identification, and precision engagement in synchronized operations. The MQ-9 Reaper (2007) further extended UAV capability, featuring higher speed, heavier payload capacity, and enhanced weapon systems, solidifying the drone's role as a multi-mission combat platform.

By the 2010s, third-generation drones had transitioned from experimental programs to core assets in global defense and intelligence operations. Their reliability, range, and precision enabled 24/7 surveillance, cross-border reconnaissance, and unmanned strike missions with minimal human risk. Beyond military applications, the same technological breakthroughs spurred the emergence of commercial UAVs equipped with GPS, inertial navigation, and high-definition cameras, leading to widespread civilian adoption for mapping, infrastructure inspection, and disaster management.

Table 4.1 presents a chronological overview of the major UAV systems that define the third generation of drones, highlighting their development timeline, technological advancements, and strategic impact. This generation, spanning from the mid-1990s to 2015, marked the integration of satellite

Table 4.1 Historical development of third-generation drones.

Year/ period	Drone model/ system	Developer/ country	Key technological features	Primary applications	Significance/contribution
1995	**RQ-1 Predator**	General Atomics Aeronautical Systems/United States	SATCOM-based control, EO/IR sensors, long endurance, armed capability with AGM-114 Hellfire missiles	Surveillance, reconnaissance, precision strike	Marked the beginning of armed UAV operations; demonstrated persistent surveillance and real-time combat engagement.
1998	**RQ-4 Global Hawk**	Northrop Grumman/United States	HALE design, autonomous flight, SAR radar, EO/IR imaging, 36-hour endurance	Strategic reconnaissance, intelligence gathering	Pioneered autonomous long-range, high-altitude surveillance with global data relay capabilities.
2001	**MQ-1 Predator (Armed Variant)**	General Atomics/United States	Equipped with Hellfire missiles, multi-sensor targeting, network-centric operation	Counterterrorism, precision targeting	First UAV used for live-target elimination in combat; redefined modern warfare strategy.
2004	**Heron TP (Eitan)**	Israel Aerospace Industries/Israel	MALE UAV, satellite-linked communication, EO/IR, maritime and border surveillance	Long-endurance intelligence and patrol missions	Introduced large-payload, multi-sensor operations for continuous surveillance missions.
2005	**MQ-9 Reaper**	General Atomics Aeronautical Systems/United States	Heavier payload, enhanced range and strike capability, real-time HD video transmission	Multi-role ISR and combat operations	Advanced drone capable of simultaneous surveillance and attack; symbolized modern drone warfare.

2006	**Dassault Neuron (UCAV)**	Dassault Aviation/France	Stealth technology, autonomous flight, AI-based targeting	Experimental combat drone (UCAV)	Europe's first stealth combat UAV prototype; introduced autonomous combat mission capabilities.
2007	**BAE Systems Taranis (UCAV)**	BAE Systems/ United Kingdom	Radar-absorbing design, autonomous navigation, sensor fusion	Combat and reconnaissance	Demonstrated stealth and AI integration; influenced future unmanned combat aerial systems.
2010	**IAI Heron 1**	Israel Aerospace Industries/Israel	Real-time data link, high endurance, SATCOM control, multi-mission payloads	Tactical surveillance, border patrol, marine monitoring	Established reliability for NATO and allied military intelligence missions.
2012	**X-47B**	Northrop Grumman/United States	Carrier-based autonomous takeoff/landing, stealth airframe, precision navigation	Experimental UCAV for naval operations	First UAV to successfully perform autonomous carrier-based operations; milestone in UAV autonomy.
2015	**Chinese Wing Loong I**	AVIC/China	Long-range endurance, multi-role (ISR + strike), cost-effective design	Surveillance, border patrol, combat missions	Expanded global UAV accessibility; symbolized growing international competition in UAV technology.

communication, multi-sensor payloads, and precision-guided weaponry into drone platforms. The table underscores how UAVs like the RQ-1 Predator and RQ-4 Global Hawk revolutionized aerial reconnaissance, while systems such as the MQ-9 Reaper and Heron TP introduced endurance and strike capabilities. The emergence of European (Neuron, Taranis) and Asian (Wing Loong) systems demonstrates the global expansion of UAV innovation. Collectively, these platforms established the technological and operational foundation for autonomous, AI-driven UAV systems that would dominate the following generation.

4.3 Principles and Components of Third-generation Drones

The third generation of unmanned aerial vehicles (UAVs) represents a convergence of advanced avionics, networked communications, multi-sensor payloads, and persistent autonomy. Where earlier generations extended range and added basic automation, third-generation systems integrate real-time decision support, redundant flight management, and global connectivity to perform complex, long-duration missions with high reliability. This section outlines the operating principles that make those capabilities possible and describes the principal hardware and software subsystems that constitute a contemporary third-generation drone.

Figure 4.1 illustrates the key structural and functional components that define the third generation of unmanned aerial vehicles (UAVs). The labeled

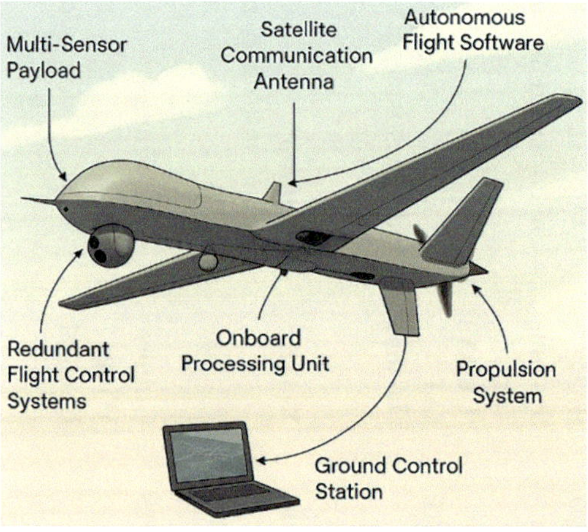

Figure 4.1 Structural and functional layout of third-generation UAVs.

diagram highlights the integration of critical systems such as the multi-sensor payload, satellite communication antenna, redundant flight control systems, autonomous flight software, onboard processing unit, propulsion system, and ground control station. Together, these components represent the synergy between hardware and software that enables high-level autonomy, endurance, and real-time data exchange. The multi-sensor payload ensures all-weather intelligence gathering, while the SATCOM antenna facilitates beyond-line-of-sight operations. The autonomous flight software and redundant flight control units provide mission reliability and fault tolerance. The onboard processing unit handles sensor fusion and decision-making tasks, while the propulsion system supports efficient, long-duration missions. The ground control station serves as the operational hub, managing mission planning and data interpretation. Collectively, the figure emphasizes how the integration of communication, computation, and control defines third-generation drones' versatility and technological sophistication.

4.3.1 Principles of operation

- *Autonomous mission management*: Third-generation UAVs are built to execute end-to-end mission profiles, take off, transit, surveillance/strike, and recovery with minimal human intervention. Mission plans (waypoints, sensor tasks, weapons employment logic) are uploaded to the onboard mission computer; during flight, the vehicle autonomously follows and adapts the plan using sensor feedback, health monitoring, and rule-based or algorithmic decision logic.
- *Redundant, fault-tolerant flight control*: Robust operation in contested or harsh environments is achieved through redundant flight control architectures (multiple flight computers, duplicated sensors, cross-monitoring). Flight control software implements fault detection, isolation, and recovery (FDIR), graceful degradation, and automatic fail-safe behaviors (e.g., loiter, return-to-home, controlled descent).
- *Networked, multi-layer communications*: Third-generation UAVs employ layered communications (line-of-sight RF links, SATCOM, relay via other platforms) so that command, control, and high-bandwidth sensor data can be exchanged across theatre and strategic networks. Networked operations enable distributed command-and-control, sensor fusion across platforms, and real-time tasking from remote users.
- *Multi-sensor data fusion and onboard processing*: Multiple sensors (EO/IR, SAR, SIGINT, AIS, LiDAR, etc.) provide complementary

situational awareness. Onboard data processing and fusion reduce latency by performing feature extraction, automatic target recognition (ATR), geo-registration, and compression before transmitting condensed intelligence to operators or higher-level networks.

- *Persistent, beyond-line-of-sight (BLOS) operations*: SATCOM and autonomous flight enable operations far beyond visual range for extended periods (MALE/HALE endurance). Persistence is a defining principle: continuous presence over areas of interest to deliver time-critical intelligence or persistent deterrence.
- *Mission modularity*: Modular payload bays and open architectures permit rapid reconfiguration for different missions (ISR, EW, communications relay, strike, maritime patrol) without major airframe redesign.

4.3.2 Hardware components

A typical third-generation UAV is an integrated system. The major hardware elements include:

- *Airframe and aero structure*: Optimized for endurance and low observability where required. Materials include composites and RAM (radar-absorbent materials) for UCAV variants. Aerodynamic design balances loiter efficiency, payload carriage, and low-drag cruise.
- *Propulsion system*: Turboprops, small turbofans, or efficient piston/electric hybrid systems for MALE platforms; fuel-efficient engines tuned for long loiter times and restricted IR signature in some military designs.
- *Flight control and avionics suite (FCU)*: High-reliability flight computers (often multiple redundant units), digital autopilot, EFIS (electronic flight instrumentation) subsystems, and actuator control for surfaces and control effectors.
- *Inertial navigation and guidance (INS/IMU) + GNSS*: High-grade IMUs coupled with multi-constellation GNSS receivers (GPS, GLONASS, Galileo, BeiDou) and tightly-coupled GNSS/INS algorithms for precise navigation and robustness against signal degradation.
- *Communications payloads*: Multi-band radios (UHF/VHF, L-band, Ku/Ka SATCOM terminals), datalinks supporting both low-latency control

and high-bandwidth sensor streams, and anti-jamming/crypto modules for secure links.

- *Mission computer and onboard processing*: Ruggedized high-performance processors/GPUs for real-time sensor processing, machine-learning inference, data fusion, and health monitoring. These execute autonomy, ATR, flight-path replanning, and compressed data products.
- *Sensor suite/payloads*: Typical combinations include electro-optical/infrared (EO/IR) gimbals, synthetic aperture radar (SAR) with ground moving target indication (GMTI), signals intelligence (SIGINT), electronic warfare (EW) packages, LiDAR, hyperspectral imagers, and maritime sensors (AIS receivers).
- *Weapons and effectors (if applicable)*: For armed platforms: precision weapons interfaces (e.g., laser-guided missiles), integrated fire-control and weapon release mechanisms complying with safety interlocks and mission rules of engagement.
- *Power and energy management*: Onboard fuel systems or battery packs with power management units that service propulsion, avionics, and payloads; energy budgeting software optimizes sensor use to conserve endurance.
- *Launch and recovery systems depending on platform*: Conventional runway, parachute/recovery nets, autonomous arrested recovery, or autonomous carrier operations for naval UCAVs.

4.3.3 Major software and systems components

- *Autonomy stack*: Flight management, path planning, obstacle avoidance, and mission logic layers. Increasingly includes ML models for perception and decision support.
- *Sensor fusion and ISR processing*: Software to geo-register sensor feeds, perform real-time analytics (object detection, change detection), and generate intelligence products.
- *Network and cybersecurity layer*: Encryption, anti-spoofing, resilient routing, and intrusion detection to safeguard command and data links.

- *Health management and FDIR*: Continuous monitoring of avionics, propulsion, and structural health; automatic corrective actions and contingency handling.
- *Human–machine interface (HMI)*: Ground control station software for mission planning, real-time visualization, collaborative tasking, and rules-of-engagement enforcement.

4.3.4 System integration and functional operation

Third-generation UAVs are engineered as cohesive systems where hardware and software interplay across closed control loops:

1. *Pre-mission*: Mission plan and sensor prescriptions are uploaded via the GCS; calibration and system checks execute automatically.
2. *Launch and transit*: Autopilot and navigation modules execute takeoff and place the UAV on optimized transit routes using GNSS/INS and satellite link monitoring.
3. *On-station operations*: Sensors activate, and the mission computer autonomously manages sensor employment (e.g., cued EO/IR slewing, SAR strip mapping), while onboard processors run ATR and compress data for transmission.
4. *Dynamic re-tasking*: Operators or higher-level C2 nodes can retask the platform in real time via resilient datalink; onboard autonomy can accept high-level objectives and replan locally to satisfy mission constraints.
5. *Anomaly handling*: If a subsystem degrades, FDIR triggers safe modes (loiter, return, or land) while attempting recovery; redundant systems maintain mission continuity where possible.
6. *Recovery and post-mission*: Secure data offload, maintenance diagnostics, and health logs feed back into maintenance scheduling and system learning loops.

4.4 Applications of Third-generation UAV Systems

The third generation of UAV systems represents a transformative phase in unmanned aviation where drones have evolved from specialized reconnaissance tools into multi-domain, mission-capable aerial platforms serving defense, scientific, commercial, and humanitarian purposes. Their advanced autonomy, endurance, and global communication capabilities enabled through satellite networking, precision navigation, and multi-sensor integration broadened the operational spectrum of UAVs across both military and civilian sectors. These applications reflect the growing role of drones in modern warfare, strategic intelligence, environmental monitoring, and data-driven industries.

4.4.1 Military and defense applications

Third-generation drones revolutionized defense operations by introducing real-time surveillance, precision strike, and persistent reconnaissance capabilities. Systems such as the MQ-1 Predator, MQ-9 Reaper, and RQ-4 Global Hawk became indispensable in network-centric warfare, providing a continuous information link between aerial, ground, and command units.

- *Intelligence, surveillance, and reconnaissance (ISR)*: UAVs equipped with EO/IR cameras, SAR radar, and SIGINT systems provided continuous observation in conflict zones, enabling the identification of targets, monitoring of enemy movements, and post-strike assessments.
- *Precision strike operations*: Armed drones introduced the ability to deliver laser-guided munitions such as AGM-114 Hellfire missiles, allowing for accurate engagement with minimal collateral damage. This capability reshaped counterterrorism operations in regions like Afghanistan, Iraq, and Pakistan.
- *Electronic warfare and communication relay*: Third-generation UAVs carried electronic warfare payloads to disrupt enemy radar and communication systems, while others served as airborne data relays to extend communication coverage in remote or contested areas.
- *Maritime and border surveillance*: Long-endurance drones such as the Heron TP and Global Hawk were used for maritime patrols, anti-piracy missions, and coastal defense, offering extended coverage and real-time situational awareness.
- *Target acquisition and battle damage assessment*: UAVs provided critical imaging and telemetry data used in guiding artillery, coordinating airstrikes, and evaluating strike effectiveness in near real-time.

4.4.2 Scientific and environmental applications

The robust endurance and precision of third-generation UAVs made them valuable assets for scientific research and environmental monitoring on a global scale.

- *Climate and atmospheric studies*: UAVs like NASA's Global Hawk conducted high-altitude missions to study weather systems, greenhouse gases, and atmospheric composition, collecting data beyond the reach of manned aircraft.
- *Disaster management and assessment*: Drones equipped with optical and thermal sensors were deployed for search and rescue, damage assessment, and resource mapping following natural disasters such as earthquakes, floods, and hurricanes.
- *Environmental monitoring*: UAVs contributed to wildlife tracking, forest health assessments, and deforestation mapping, supporting ecological research and conservation projects.
- *Oceanographic and polar research*: Endurance-capable UAVs collected data on sea ice movement, ocean currents, and polar atmospheric patterns, aiding climate change modeling.

4.4.3 Civilian and commercial applications

Third-generation drones extended far beyond defense, finding widespread adoption in civilian, industrial, and infrastructure-related applications due to their precision, efficiency, and cost-effectiveness.

- *Aerial mapping and surveying*: Equipped with high-resolution cameras and LiDAR systems, UAVs were used for urban planning, topographical mapping, and construction monitoring with centimeter-level accuracy.
- *Agriculture and precision farming*: UAVs monitored crop health using multispectral imaging, optimizing irrigation, pesticide use, and yield forecasting, contributing to sustainable agriculture practices.
- *Energy and infrastructure inspection*: Power lines, pipelines, and wind farms were routinely inspected using UAVs, reducing operational hazards and downtime compared to manned inspection methods.
- *Telecommunications and logistics*: Some drones were adapted for temporary communication relays in remote areas and experimental

cargo delivery missions, demonstrating the potential of UAV-based logistics.

- *Public safety and disaster response*: Police and emergency responders utilized UAVs for crowd monitoring, fire detection, and hazardous environment assessments, enhancing operational safety and efficiency.

4.4.4 Communication and technological applications

The integration of UAVs within broader communication and data networks highlighted their role as mobile aerial nodes in emerging global information systems.

- *Data relay and network extension*: Drones served as communication relays in remote or disaster-stricken areas, providing emergency data connectivity where terrestrial infrastructure was unavailable.
- *Technology demonstration platforms*: Third-generation UAVs acted as testbeds for AI algorithms, autonomous flight systems, and sensor integration experiments, accelerating advancements for future UAV generations.
- *Satellite communication testing*: Equipped with SATCOM antennas, drones were used to validate satellite link reliability and test next-generation data relay technologies for aerospace and defense sectors.
- *Cybersecurity and encryption research*: UAVs contributed to developing and testing secure data transmission protocols and anti-jamming communication techniques, ensuring reliability in contested electronic environments.

4.5 Developments and Pioneers of Third-generation Drone Models

The third generation of drone models, emerging from the early 2000s to the mid-2010s, marked a period of rapid innovation in unmanned aerial vehicle (UAV) technology, driven by advancements in autonomy, endurance, communication, and combat capability. This generation saw UAVs evolve from passive reconnaissance tools to strategic, multi-role, and network-centric aerial systems capable of executing complex missions over extended ranges. The developments were fueled by collaborations between military agencies, aerospace industries, and research institutions, leading to the creation of globally recognized UAV platforms that set new operational and technological benchmarks.

4.5.1 Evolution of third-generation UAV development

The evolution of third-generation drones was characterized by the integration of satellite communication (SATCOM), multi-sensor payloads, precision strike capabilities, and autonomous flight control systems. These features enabled UAVs to perform beyond-line-of-sight missions, operate in contested environments, and provide real-time intelligence to global command centers.

- *Shift from tactical to strategic operations*: Unlike their second-generation predecessors, which were primarily tactical reconnaissance tools, third-generation UAVs such as the RQ-4 Global Hawk and MQ-9 Reaper became strategic assets capable of global missions lasting more than 30 hours.
- *Integration of precision strike capabilities*: The development of the MQ-1 Predator into an armed UAV represented a turning point, demonstrating that unmanned systems could engage targets with high accuracy and minimal collateral damage.
- *Enhanced sensor and communication systems*: The introduction of synthetic aperture radar (SAR), EO/IR imaging, and real-time SATCOM links enabled high-fidelity intelligence gathering and long-distance command-and-control.
- *Autonomy and system redundancy*: Advances in autopilot algorithms, navigation systems, and onboard data processing improved UAV reliability, enabling semi-autonomous operation in multi-threat environments.
- *Global proliferation and standardization*: During this period, UAV technology expanded internationally, with major contributions from the United States, Israel, France, the United Kingdom, and China, reflecting a global standardization of unmanned systems for both military and civilian use.

4.5.2 Key pioneers and notable drone models

4.5.2.1 RQ-1 Predator (1995)

The RQ-1 Predator, developed by General Atomics Aeronautical Systems (GA-ASI) for the United States Air Force (USAF) and Central Intelligence

Agency (CIA), represents one of the most transformative UAVs in aviation history and a defining model of the third generation of drones. First introduced in 1995, the Predator revolutionized modern warfare by merging real-time surveillance and precision strike capabilities within a single, remotely operated platform. Designed as a medium-altitude long-endurance (MALE) drone, the RQ-1 Predator established the operational framework for future unmanned systems by combining endurance, intelligence-gathering precision, and armed engagement capacity (Amburn, et al. 2005).

Technically, the Predator featured a composite airframe, a Rotax 914 four-cylinder piston engine, and a pusher propeller configuration, allowing a maximum endurance of up to 24 hours and an operational altitude of 25,000 feet. The drone was equipped with a multi-sensor suite, including electro-optical (EO) and infrared (IR) cameras, as well as a synthetic aperture radar (SAR) for all-weather reconnaissance. One of its most revolutionary features was the integration of satellite communication (SATCOM), enabling beyond-line-of-sight (BLOS) operations, allowing pilots in the United States to remotely control missions across global theaters in real time.

In 2001, the RQ-1 Predator was modified to carry AGM-114 Hellfire missiles, transforming it into the MQ-1 Predator, the first UAV capable of executing precision-targeted strikes. This weaponization of drones ushered in a new era of remote warfare, particularly during operations in Afghanistan, Iraq, and Pakistan, where the Predator was employed for counterterrorism, border surveillance, and persistent intelligence-gathering missions. The drone's ability to loiter for extended periods while providing live video feeds directly to commanders and strike targets with pinpoint accuracy made it an indispensable asset in asymmetric warfare.

Beyond its combat role, the Predator also played a key part in intelligence, surveillance, and reconnaissance (ISR) missions, conducting long-range surveillance and relaying crucial battlefield information. Its modular design allowed for payload customization, expanding its utility across diverse mission profiles such as convoy monitoring, target designation, and communication relay (Assessment, 2003).

The RQ-1 Predator's success fundamentally changed the doctrine of aerial warfare, introducing the concept of persistent, unmanned surveillance coupled with real-time strike capability. It demonstrated that UAVs could not only replace manned reconnaissance aircraft in many roles but also provide safer, more efficient, and cost-effective alternatives for sustained operations.

In essence, the RQ-1 Predator (1995) symbolizes the dawn of networked, precision-oriented drone warfare. Its legacy extends beyond the battlefield; it established the technological and strategic foundation for successors such as the MQ-9 Reaper and RQ-4 Global Hawk, and it permanently altered the role of UAVs in both military and intelligence operations worldwide.

Key specifications:

- Type: Medium-altitude long-endurance (MALE) UAV
- Developer: General Atomics Aeronautical Systems, United States
- First Flight: 1995
- Wingspan: 17.0 m
- Endurance: Up to 24 hours
- Maximum Altitude: 25,000 ft
- Propulsion: Rotax 914 piston engine, 115 hp
- Payloads: EO/IR sensors, SAR radar, laser designator, Hellfire missiles (MQ-1 variant)
- Control system: Satellite (Ku-band) and C-band line-of-sight datalink
- Primary Roles: ISR, target acquisition, and precision strike

Figure 4.2 illustrates the structural layout and key operational components of the RQ-1 Predator, one of the most significant UAVs of the third generation.

Figure 4.2 The RQ-1 Predator (1995) – design and operational architecture.

The labeled diagram highlights major features, including the composite airframe, pusher propeller configuration, sensor turret (EO/IR camera system) located under the nose, and synthetic aperture radar (SAR) mounted beneath the fuselage. The satellite communication (SATCOM) antenna positioned on the upper fuselage enables long-distance, beyond-line-of-sight (BLOS) control via ground stations. The Rotax 914 piston engine at the rear provides propulsion, while the fixed tricycle landing gear supports stable takeoff and landing operations.

This configuration allowed the Predator to conduct long-endurance surveillance missions with exceptional stability and precision. The EO/IR turret provided real-time video feeds for reconnaissance and target tracking, while the laser designator enabled precision-guided weapon deployment in the upgraded MQ-1 variant. The diagram also emphasizes the ground control station (GCS) link, demonstrating the closed-loop data transmission between the UAV and its remote operators through secure SATCOM channels.

4.5.2.2 RQ-4 Global Hawk (1998)

The RQ-4 Global Hawk, developed by Northrop Grumman for the United States Air Force (USAF), represents one of the most advanced and enduring high-altitude long-endurance (HALE) unmanned aerial vehicles (UAVs) of the third generation. Introduced in 1998, the Global Hawk was designed to provide strategic, near-real-time intelligence, surveillance, and reconnaissance (ISR) capabilities over vast areas of interest, operating at altitudes and durations beyond the reach of conventional manned aircraft. It became a cornerstone of modern aerial intelligence operations, combining autonomous navigation, multi-sensor integration, and satellite-based communication into a single, globally deployable platform (Kompała, 2014).

Design and technical overview

The RQ-4 Global Hawk features a composite high-aspect-ratio airframe optimized for aerodynamic efficiency and endurance. With a wingspan of approximately 40 meters (131 feet) and a length of 14.5 meters (47.6 feet), it can operate continuously for over 36 hours at altitudes exceeding 60,000 feet, effectively placing it above commercial air traffic and most weather systems. Powered by a single Rolls-Royce AE 3007H turbofan engine, the UAV achieves a cruise speed of around 570 km/h (355 mph), combining fuel efficiency with persistent operational capability (Hassell Jr, 2009).

The Global Hawk's sensor suite is among the most advanced of its generation. It includes a synthetic aperture radar (SAR) with ground moving target indicator (GMTI) capabilities, electro-optical (EO) and infrared

(IR) cameras, and signals intelligence (SIGINT) systems. This combination enables day-and-night, all-weather reconnaissance, allowing the aircraft to image and track ground targets across hundreds of kilometers in a single pass. These payloads are integrated into a modular sensor bay, allowing rapid reconfiguration for different mission types, from electronic intelligence gathering to maritime surveillance.

Operational capabilities and features

1. *Autonomous flight and mission planning*: The Global Hawk is equipped with fully autonomous takeoff, flight, and landing systems, managed via a pre-programed flight plan that can be modified in-flight through satellite uplink. This level of autonomy minimizes the need for pilot input while ensuring precision navigation over long distances.
2. *High-altitude, wide-area coverage*: Its advanced aerodynamic design and endurance enable the aircraft to survey up to 100,000 square kilometers in a single mission, providing comprehensive imagery and intelligence over entire regions of interest.
3. *SATCOM and network-centric connectivity*: The RQ-4 relies on Ku-band SATCOM links and line-of-sight (LOS) systems to relay high-resolution data directly to command and control centers. This capability supports real-time mission re-tasking and integration into broader network-centric warfare architectures.
4. *Multi-intelligence integration (Multi-INT)*: The Global Hawk can simultaneously collect imagery intelligence (IMINT), signals intelligence (SIGINT), and measurement and signature intelligence (MASINT), making it a versatile strategic platform for joint military operations.
5. *Operational reliability and range*: Capable of flying *over* 22,000 kilometers (14,000 miles) without refueling, the Global Hawk has been employed in global missions, including operations in Afghanistan, Iraq, Libya, and Japan, as well as disaster monitoring and border security.

Applications and strategic impact

The RQ-4 Global Hawk played a central role in modern ISR missions. It provided high-resolution imagery for battle damage assessment, target identification, terrain mapping, and maritime surveillance. Beyond military operations, the platform has been used in civil and humanitarian missions, such as disaster relief monitoring, climate research, and environmental observation under programs coordinated with NASA and the US Geological Survey (USGS).

Its autonomous operations and long-endurance missions reduced the demand for manned reconnaissance flights, significantly lowering operational risks and costs. In intelligence-sharing frameworks, Global Hawk data was distributed to multiple allied forces, enhancing collaborative situational awareness.

Key specifications:

- *Type*: High-altitude long-endurance (HALE) UAV
- *Developer*: Northrop Grumman, United States
- *First flight*: 1998
- *Wingspan*: 39.9 m (130.9 ft)
- *Length*: 14.5 m (47.6 ft)
- *Endurance*: 36+ hours
- *Maximum altitude*: 60,000 ft (18,300 m)
- *Range*: 22,000 km (14,000 miles)
- *Propulsion*: Rolls-Royce AE 3007H turbofan engine
- *Payloads*: SAR/GMTI radar, EO/IR sensors, SIGINT suite
- *Primary roles*: Strategic ISR, reconnaissance, and environmental monitoring
- *Operational users*: USAF, NATO, Japan, South Korea, Germany

Figure 4.3 depicts the RQ-4 Global Hawk, a high-altitude, long-endurance UAV developed by Northrop Grumman for the US Air Force. The figure showcases its streamlined composite airframe, long high-aspect-ratio wings, and twin V-tail configuration, which enable aerodynamic efficiency and extended endurance exceeding 36 hours. The sensor bay beneath the fuselage houses its synthetic aperture radar (SAR) and electro-optical/infrared (EO/IR) payloads, while the SATCOM antenna mounted on the fuselage ensures global data relay capability. The Rolls-Royce AE 3007H turbofan engine, located atop the fuselage, provides reliable propulsion for high-altitude, wide-area surveillance. This figure highlights the aircraft's design optimized for persistent, autonomous reconnaissance missions, symbolizing a milestone in third-generation UAV innovation.

Figure 4.3 RQ-4 Global Hawk (1998).

4.5.2.3 MQ-9 Reaper (2007)

The MQ-9 Reaper, developed by General Atomics Aeronautical Systems (GA-ASI) for the United States Air Force (USAF), is one of the most advanced and operationally significant UAVs of the third generation. Officially introduced in 2007, the Reaper was designed as a multi-role, hunter-killer platform capable of performing both intelligence, surveillance, and reconnaissance (ISR) missions and precision strike operations. Building upon the legacy of its predecessor, the RQ-1/MQ-1 Predator, the Reaper introduced significant improvements in speed, payload capacity, endurance, and lethality, establishing itself as a cornerstone of modern unmanned warfare (Cullen, 2011).

Design and technical overview

The MQ-9 Reaper, as shown in Figure 4.4, features a medium-altitude long-endurance (MALE) airframe made from lightweight composite materials, ensuring durability and aerodynamic efficiency. It has a wingspan of 20 meters (66 feet) and a fuselage length of 11 meters (36 feet). Powered by a Honeywell TPE331-10 turboprop engine producing 900 horsepower, the Reaper can reach speeds of up to 480 km/h (300 mph) and altitudes of 50,000 feet, nearly double that of the MQ-1 Predator. Its endurance of 27 hours allows it to conduct prolonged missions covering multiple operational objectives.

The aircraft's modular design supports a variety of payloads. Its maximum payload capacity exceeds 3800 pounds (1700 kg), including external weapons, sensor pods, and fuel tanks. The Reaper's sensor suite includes a multi-spectral targeting system (MTS-B) integrating electro-optical (EO), infrared (IR), laser designator, and laser range finder capabilities. Additionally, its synthetic aperture radar (SAR) provides high-resolution ground mapping and ground moving target indication (GMTI) functionality for all-weather operations (Ouma, et al. 2011).

Operational capabilities and systems integration

1. *Armed precision strike platform*: The MQ-9 Reaper is equipped to carry a wide array of precision-guided munitions, including AGM-114 Hellfire missiles, GBU-12 Paveway II laser-guided bombs, and GBU-38 JDAMs. It can deliver simultaneous ISR and strike capabilities, making it an integrated intelligence and combat system.
2. *Autonomous and networked operations*: The Reaper operates via satellite communication (SATCOM) and line-of-sight (LOS) links, enabling beyond-line-of-sight (BLOS) control from ground stations thousands of kilometers away. Its autonomous flight management system supports automated takeoff, loitering, and landing with minimal operator input.
3. *Multi-mission capability*: Beyond combat roles, the Reaper is also employed for *border surveillance*, *maritime patrol*, *disaster relief*, and *scientific observation* missions due to its long endurance and adaptability to multiple payloads.
4. *Enhanced sensor and data systems*: The Reaper's sensor suite provides *real-time full-motion video (FMV)* and high-resolution imagery directly to commanders and ground forces. It is integrated into *network-centric warfare* architectures, allowing real-time target acquisition, threat identification, and strike coordination.
5. *Modular payload configuration*: Its modular bays and hardpoints support *payload customization*, enabling mission-specific configurations for combat, reconnaissance, or communication relay operations.

Key specifications:

- *Type*: Medium-altitude long-endurance (MALE) armed UAV
- *Developer*: General Atomics Aeronautical Systems, United States
- *First flight*: 2001 (Operational: 2007)
- *Wingspan*: 20 m (66 ft)
- *Length*: 11 m (36 ft)
- *Maximum speed*: 480 km/h (300 mph)
- *Endurance*: 27 hours
- *Ceiling*: 50,000 ft (15,200 m)

Figure 4.4 MQ-9 Reaper (2007).

- *Propulsion*: Honeywell TPE331-10 turboprop engine, 900 hp
- *Payload capacity*: 1700 kg (3800 lb)
- *Armament*: AGM-114 Hellfire missiles, GBU-12 Paveway II, GBU-38 JDAMs
- *Primary roles*: ISR, precision strike, surveillance, border, and maritime patrol

4.5.2.4 Heron TP (Eitan) (2004)

The Heron TP (Eitan), developed by Israel Aerospace Industries (IAI) in 2004, stands as one of the most capable and versatile medium-altitude long-endurance (MALE) UAVs of the third generation. As an evolution of the earlier Heron-1 system, the Heron TP was designed to perform strategic intelligence, surveillance, and reconnaissance (ISR), target acquisition, and combat support missions with extended endurance, large payload capacity, and multi-sensor integration. Known for its operational reliability, endurance exceeding 36 hours, and all-weather capability, the Heron TP has become a benchmark in unmanned aerial operations for both military and civilian use across multiple nations (Franke, 2013).

Design and technical overview

The Heron TP features a twin-boom fuselage design with a rear-mounted pusher turboprop engine, which enhances aerodynamic stability and efficiency. Built primarily from advanced composite materials, the airframe is optimized for endurance and minimal radar signature. The UAV has a wingspan of approximately 26 meters (85 feet) and a length of 14 meters (46 feet), allowing for a high lift-to-drag ratio and long-duration flight.

Powered by a Pratt and Whitney PT6A-67A turboprop engine, the Heron TP achieves a maximum altitude of 45,000 feet and a cruise speed of about 370 km/h (230 mph). Its payload capacity exceeds 1000 kg, enabling the carriage of multiple sensor pods, communication equipment, and electronic warfare systems simultaneously. The aircraft is equipped with triple-redundant avionics and autonomous take-off and landing (ATOL) systems, ensuring mission reliability and safety.

Key specifications:

- *Type*: Medium-altitude long-endurance (MALE) UAV
- *Developer*: Israel Aerospace Industries (IAI), Israel
- *First flight*: 2004
- *Wingspan*: 26 m (85 ft)
- *Length*: 14 m (46 ft)
- *Endurance*: 36+ hours
- *Maximum altitude*: 45,000 ft (13,700 m)
- *Speed*: 370 km/h (230 mph)
- *Propulsion*: Pratt and Whitney PT6A-67A turboprop engine
- *Payload capacity*: 1000+ kg
- *Sensors*: EO/IR, SAR/GMTI, ELINT, COMINT, maritime radar
- *Control system*: SATCOM and LOS dual-link communication
- *Primary roles*: ISR, target acquisition, border and maritime patrol, communication relay

Figure 4.5 illustrates the Heron TP (Eitan), a medium-altitude, long-endurance UAV developed by Israel Aerospace Industries (IAI) in 2004. The diagram highlights its key structural components, including the twin-boom fuselage, pusher turboprop engine, and electro-optical/infrared (EO/IR) turret mounted under the forward fuselage. The long, slender wings with winglets enhance aerodynamic efficiency, while the tricycle landing gear ensures smooth autonomous take-off and landing. The twin vertical stabilizers, bearing the Israeli Air Force insignia, provide directional stability during extended missions. This configuration allows the Heron TP to achieve remarkable endurance exceeding 36 hours and to carry multiple payloads, including SAR, ELINT,

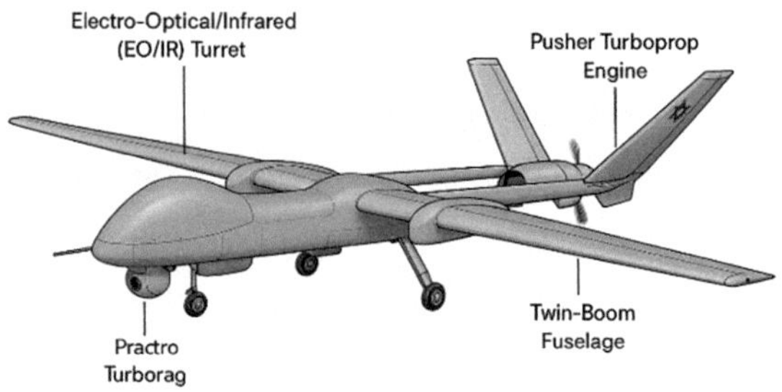

Figure 4.5 Heron TP (Eitan) (2004).

and COMINT sensors. The figure effectively captures the aircraft's optimized design for strategic surveillance, reconnaissance, and intelligence operations, underscoring its role as a cornerstone of Israel's UAV technology leadership.

4.5.2.5 Dassault Neuron (2006)

The Dassault Neuron, unveiled in 2006, represents a landmark achievement in the development of stealth and autonomous combat UAVs (UCAVs), marking the transition toward fourth-generation unmanned systems. Developed by Dassault Aviation (France) under a Pan-European collaboration including Italy (Alenia Aermacchi), Sweden (Saab), Spain (EADS-CASA), Greece (HAI), and Switzerland (RUAG), the Neuron project was the first European stealth UCAV demonstrator. Its purpose was to explore autonomous flight, low-observable design, and precision strike capability while validating advanced aerodynamics, sensor fusion, and control systems for future unmanned combat air systems (UCAS).

Design and technical overview

The Neuron is a flying-wing UCAV designed with an emphasis on stealth, autonomy, and precision strike. Its blended body-wing configuration, crafted from carbon-fiber composites, minimizes radar cross-section (RCS) and enhances aerodynamic efficiency. With a wingspan of 12.5 meters (41 feet) and a length of 9.2 meters (30 feet), the aircraft achieves a balanced design between agility, endurance, and payload capability.

The drone is powered by a single Rolls-Royce Turbomeca Adour engine, providing subsonic performance suitable for stealth operations. Internal air intakes and exhaust ducts are carefully shaped to reduce radar and infrared signatures. The internal weapons bay allows the Neuron to carry precision-guided munitions (PGMs), including 250 kg-class smart bombs, while maintaining its stealth profile (De Saint-Jean, et al. 2013).

A key design feature of the Neuron is its fully autonomous flight management system, capable of executing mission profiles without continuous operator control. The system integrates data fusion, mission reconfiguration, and real-time decision-making algorithms, allowing the UCAV to adapt to dynamic combat environments.

Technological features and innovations

1. *Stealth technology*: The Neuron employs radar-absorbent materials (RAM), internal weapon carriage, and smooth aerodynamic surfaces to achieve a low radar cross-section suitable for deep penetration missions.
2. *Autonomous operations*: Advanced mission computers enable autonomous navigation, target identification, and threat avoidance, reducing operator dependency and improving response time.
3. *Modular mission architecture*: The Neuron's open avionics system supports multiple mission configurations, from strike operations to reconnaissance, by integrating different sensor and weapon payloads.
4. *Fly-by-light control system*: A notable innovation is the fly-by-light (FBL) system, an optical data transmission network offering high resistance to electromagnetic interference and improved reliability compared to traditional fly-by-wire systems.
5. *Collaborative development framework*: The project unified European aerospace expertise, with Dassault leading design, Saab managing flight testing, and Alenia Aermacchi contributing to weapon integration, strengthening Europe's technological independence in defense aviation.

Testing and operational milestones
The first flight of the Neuron took place on December 1, 2012, at Istres Air Base, France. Following successful test campaigns in Sweden, Italy, and France, the UCAV demonstrated:

- Autonomous taxi, take off, flight, and landing operations.
- Target approach and payload release simulations.
- Extremely low radar and infrared visibility across multiple radar bands.
- Cooperative data-link compatibility with manned aircraft (Rafale and Gripen).

These results validated the Neuron's potential for integration into future manned–unmanned teaming (MUM-T) and next-generation air dominance programs.

Key specifications:

- *Type*: Unmanned Combat Aerial Vehicle (UCAV) – Technology Demonstrator
- *Developer*: Dassault Aviation, with European partners
- *First flight*: December 1, 2012
- *Wingspan*: 12.5 m (41 ft)
- *Length*: 9.2 m (30 ft)
- *Propulsion*: Rolls-Royce Turbomeca Adour engine
- *Ceiling*: 45,000 ft (13,700 m)
- *Maximum speed*: 980 km/h (610 mph)
- *Endurance*: ~3 hours
- *Payload*: 2 × 250 kg-class smart bombs (internal bay)
- *Control system*: Autonomous flight management with ground supervision
- *Stealth features*: Radar-absorbent materials, internal weapon bay, low IR signature

Figure 4.6 illustrates the Dassault Neuron, a European stealth unmanned combat aerial vehicle (UCAV) designed for autonomous strike operations.

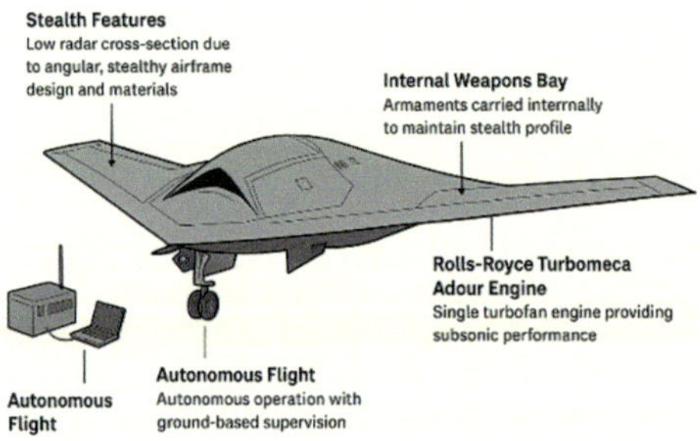

Figure 4.6 Dassault Neuron (2006).

The labeled diagram highlights its angular, radar-absorbing airframe, internal weapons bay, and single Rolls-Royce Turbomeca Adour engine, which provides subsonic thrust while minimizing infrared and radar signatures. The autonomous flight control system, connected via a ground-based interface, allows the Neuron to operate independently from take-off to landing. The figure effectively demonstrates the integration of stealth, autonomy, and precision, showcasing how the Neuron's sleek flying-wing design and advanced avionics mark a pivotal step toward the next generation of unmanned combat air systems (UCAS) in Europe.

4.5.2.6 BAE Systems Taranis (2010)

The BAE Systems Taranis, unveiled in 2010, is a cutting-edge stealth unmanned combat aerial vehicle (UCAV) demonstrator developed by BAE Systems (United Kingdom) in collaboration with the UK Ministry of Defense (MoD). It was designed to explore autonomous flight, stealth technology, and advanced mission management systems for future combat applications. Named after Taranis, the Celtic god of thunder, the UAV symbolizes power and technological prowess, representing the United Kingdom's commitment to developing indigenous capabilities in next-generation unmanned warfare (Oxford, 2016).

The Taranis project was a key part of the United Kingdom's strategic unmanned air vehicle (SUAV) program, and its successful demonstration positioned the United Kingdom among the few nations capable of producing stealth UCAVs. It served as a testbed for autonomous mission control, low-observable aerodynamics, and intelligent weapon systems, contributing foundational technology for the Anglo-French Future Combat Air System (FCAS) program.

Design and technical overview

The Taranis UCAV features a flying-wing configuration with smooth, blended surfaces designed to minimize radar cross-section (RCS) and infrared signature. Its airframe, constructed primarily from carbon-fiber composites and radar-absorbent materials (RAM), enhances stealth characteristics while maintaining structural integrity at high subsonic speeds.

With a wingspan of approximately 9.9 meters (32 feet) and a length of 11.3 meters (37 feet), the aircraft achieves a balance between agility, endurance, and payload capacity. The propulsion system is a Rolls-Royce Adour Mk 951 turbofan engine, adapted for stealth operation through modified air intakes and exhausts that reduce radar and heat emissions.

Taranis is capable of autonomous taxiing, take-off, flight, navigation, target detection, and landing, guided by advanced onboard systems rather than continuous remote control. This autonomy allows it to execute missions with minimal operator intervention, a key step toward future combat systems' human–machine teaming (HMT) capabilities.

Technological innovations and systems

1. *Stealth and signature reduction*: Taranis employs low-observable (LO) technology, including faceted airframe geometry, internal weapon bays, and advanced radar-absorbent coatings. These features allow the UCAV to penetrate defended airspace undetected.
2. *Autonomous mission control*: A major breakthrough in the Taranis program was its ability to plan, execute, and adapt missions autonomously. The onboard AI-driven mission management system evaluates threats, selects optimal flight paths, and manages weapons deployment under human supervision.
3. *Integrated sensor and communication systems*: Equipped with advanced sensors for target detection, terrain following, and situational awareness, Taranis maintains real-time communication links with ground stations for mission updates and data exchange.
4. *Internal weapon bay*: To maintain stealth, weapons such as precision-guided munitions (PGMs) are carried internally. While Taranis is a demonstrator, it is designed to accommodate 250–500 kg class smart bombs for strike missions.
5. *Collaborative engineering effort*: The program involved major UK defense partners, including Rolls-Royce (engine systems), QinetiQ (flight testing and sensors), and GE Aviation (avionics), establishing a comprehensive domestic UAV capability.

Key specifications:

- *Type*: Unmanned combat aerial vehicle (UCAV) – technology demonstrator
- *Developer*: BAE Systems, United Kingdom
- *First flight*: August 2013
- *Wingspan*: 9.9 m (32 ft)
- *Length*: 11.3 m (37 ft)

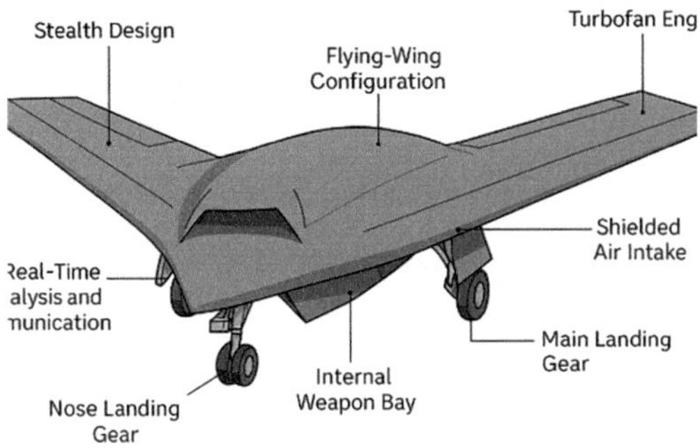

Figure 4.7 BAE Systems Taranis (2010).

- *Propulsion*: Rolls-Royce Adour Mk 951 turbofan engine
- *Speed*: High subsonic (~1000 km/h)
- *Ceiling*: 40,000 ft (12,200 m)
- *Range*: Classified (~3000 km estimated)
- *Payload*: Internal bay for 2 × 250–500 kg-class smart bombs
- *Stealth features*: RAM coatings, shielded air intakes, infrared signature suppression
- *Control system*: Fully autonomous with human supervision

Figure 4.7 presents the BAE Systems Taranis, a stealth unmanned combat aerial vehicle (UCAV) designed to demonstrate autonomous operation and low-observable technologies. The labeled diagram highlights key structural features, including the flying-wing configuration, internal weapons bay, shielded air intake, and turbofan engine, which collectively minimize radar and infrared signatures. The figure also indicates the real-time analysis and communication systems, underscoring the UAV's autonomous mission management capability. The smooth, angular airframe illustrates the aircraft's stealth-optimized geometry, while the retractable landing gear supports fully autonomous takeoff and landing.

4.5.2.7 X-47B (2012)

The Northrop Grumman X-47B, first unveiled in 2012, represents one of the most groundbreaking milestones in autonomous combat aviation and the evolution of third-generation to fourth-generation unmanned combat aerial vehicles (UCAVs). Developed under the US Navy's unmanned combat air system demonstration (UCAS-D) program, the X-47B was designed to explore the integration of unmanned, carrier-based operations, a domain previously limited to manned aircraft. It was the first UAV in history to successfully perform autonomous take-off and landing on an aircraft carrier, marking a major leap toward the realization of fully autonomous air combat operations in naval environments (Tekinalp, et al. 2012).

Design and technical overview

The X-47B employs a tailless, blended-wing-body configuration, optimized for stealth, endurance, and aerodynamic efficiency. Constructed using composite materials and radar-absorbent coatings, its design significantly reduces radar cross-section (RCS), ensuring survivability in contested airspace. The aircraft measures approximately 11.6 meters (38 feet) in length, with a wingspan of 18.9 meters (62 feet), which can be folded for carrier storage and deck maneuvering.

Powered by a Pratt and Whitney F100-220U turbofan engine, the X-47B achieves a cruise speed of 0.9 Mach and operates at altitudes up to 40,000 feet (12,200 meters). Its internal weapons bays can accommodate up to 2000 pounds (900 kg) of precision-guided munitions, such as the GBU-31 JDAM or GBU-12 Paveway II, maintaining its stealth profile during strike missions.

The airframe's smooth contours and the absence of vertical stabilizers reduce radar reflectivity, while its autonomous flight control system enables precise flight path management, refueling coordination, and mission execution without continuous human intervention.

Technological innovations and capabilities

1. *Autonomous carrier operations*: The X-47B was the first UAV to demonstrate fully autonomous catapult-assisted takeoff and arrested landing (CATOBAR) from an aircraft carrier deck. Using advanced algorithms and navigation sensors, it successfully launched and recovered aboard the USS George H.W. Bush (CVN-77) in May 2013, a landmark achievement in naval aviation history.

2. *Aerial refueling autonomy*: In 2015, the X-47B became the first UAV to perform an autonomous aerial refueling using the US Navy's probe-and-drogue system, extending its operational range and endurance, key capabilities for future long-duration missions.

3. *Stealth and survivability*: The UCAV's radar-absorbent structure, shielded exhausts, and internal weapon storage ensure low detectability

in high-threat environments, supporting deep-strike and reconnaissance operations in anti-access/area-denial (A2/AD) regions.

4. *Network-centric integration*: The X-47B was designed to operate within a networked battle environment, exchanging data with other naval platforms, manned aircraft, and command centers in real time through secure datalink systems.
5. *Autonomous decision-making systems*: Equipped with AI-driven mission management software, the X-47B can make in-flight adjustments to optimize fuel efficiency, route planning, and threat avoidance while maintaining communication with command elements for strategic supervision.

Key specifications:

- *Type*: Stealth unmanned combat aerial vehicle (UCAV) – Demonstrator
- *Developer*: Northrop Grumman, United States
- *First flight*: February 2011 (Public Unveiling: 2012)
- *Wingspan*: 18.9 m (62 ft)
- *Length*: 11.6 m (38 ft)
- *Maximum speed*: 0.9 Mach (~1100 km/h)
- *Ceiling*: 40,000 ft (12,200 m)
- *Endurance*: 6 hours (extendable with aerial refueling)
- *Propulsion*: Pratt and Whitney F100-220U turbofan
- *Payload capacity*: 2000 lb (900 kg) internal
- *Armament*: GBU-31 JDAM, GBU-12 Paveway II
- *Control system*: Fully autonomous, AI-guided flight management with human oversight
- *Primary roles*: Stealth strike, reconnaissance, autonomous carrier operations

Figure 4.8 shows the Northrop Grumman X-47B, a pioneering stealth UCAV developed under the US Navy's UCAS-D program. The figure highlights the drone's tailless, blended-wing-body design, optimized for stealth and aerodynamic efficiency. The visible foldable wings and retractable landing gear

Figure 4.8 Northrop Grumman X-47B (2012).

enable carrier-based operations, while the internal weapons bay maintains a low radar cross-section. The NAVY insignia and aircraft markings emphasize its naval deployment role. This image captures the X-47B's revolutionary capability for autonomous carrier takeoff, landing, and aerial refueling, showcasing a major milestone in the evolution of autonomous stealth combat aviation.

4.5.2.8 Wing Loong I (2015)

The Wing Loong I features a conventional MALE UAV configuration, with a slender fuselage, mid-mounted straight wings, and a rear-mounted pusher propeller. Constructed primarily from composite materials, its airframe is designed for durability, low weight, and aerodynamic efficiency. The UAV has a wingspan of 14 meters (46 feet), a length of 9 meters (29.5 feet), and can achieve endurance up to 20 hours with an operational ceiling of 16,400 feet (5000 meters).

Powered by a Rotax-type piston engine, the Wing Loong I has a maximum speed of 280 km/h (174 mph) and a range of 4000 km, enabling extended missions for surveillance, reconnaissance, and tactical strike operations. The aircraft is equipped with a retractable tricycle landing gear, allowing for autonomous take-off and landing on semi-prepared runways (Hwang, 2020).

Avionics and sensor systems

The Wing Loong I's modular payload bay supports a wide array of ISR and combat sensors, including:

- Electro-optical/infrared (EO/IR) turret for day/night surveillance and targeting.
- Synthetic aperture radar (SAR) for ground mapping and target tracking.

- Laser designator/range finder for precision-guided munitions.
- Data link communication system with line-of-sight (LOS) and satellite control options.

The UAV's onboard mission management computer integrates multi-sensor data in real time, allowing autonomous flight and target acquisition. The ground control station (GCS) features dual-console operator interfaces for flight and payload control, with live video feeds transmitted via encrypted links.

Armament and combat capability
The Wing Loong I is a multi-role armed UAV, capable of carrying up to 200 kg of ordnance across four underwing hardpoints. Its weapons suite includes:

- BA-7 air-to-surface missiles (equivalent to AGM-114 Hellfire).
- GB3 laser-guided bombs.
- Blue Arrow-7 anti-armor missiles.
- LS-6 precision-guided bombs.

These systems allow the UAV to conduct close air support (CAS), counterterrorism, and anti-insurgency missions with precision.

Operational applications
Since its induction, the Wing Loong I has been deployed by several countries, including Saudi Arabia, Egypt, the United Arab Emirates, Kazakhstan, and Nigeria, for both military and security operations. It has been used for:

- Border surveillance and coastal monitoring.
- Counter-insurgency (COIN) and anti-terrorism operations.
- Reconnaissance and artillery spotting in desert and mountainous regions.
- Civil applications, such as disaster monitoring and environmental surveys.

The UAV's export-oriented design philosophy and affordability have made it an attractive option for developing nations seeking autonomous aerial capabilities comparable to Western systems but at significantly lower costs.

Strategic and technological significance
The Wing Loong I represents China's growing sophistication in UAV manufacturing and export. It marked China's entry into the global armed drone

market, previously dominated by the United States and Israel. Its development underscores China's strategy of combining indigenous aerospace technology with affordability, accessibility, and multi-role functionality to dominate emerging defense markets.

Furthermore, the UAV laid the foundation for advanced Chinese platforms like the Wing Loong II and Wing Loong-10, which feature higher payloads, endurance, and improved sensors. By offering performance parity with Western UAVs at a fraction of the cost, the Wing Loong I reshaped the global landscape of unmanned aerial systems.

Key specifications:

- *Type*: Medium-altitude long-endurance (MALE) UAV
- *Developer*: Aviation Industry Corporation of China (AVIC)
- *First flight*: 2015
- *Wingspan*: 14 m (46 ft)
- *Length*: 9 m (29.5 ft)
- *Endurance*: Up to 20 hours
- *Ceiling*: 16,400 ft (5000 m)
- *Range*: 4000 km
- *Speed*: 280 km/h (174 mph)
- *Propulsion*: Single piston engine (pusher configuration)
- *Payload*: 200 kg (combat + ISR sensors)
- *Armament*: BA-7 missiles, GB3 bombs, LS-6 guided bombs, Blue Arrow-7
- *Primary roles*: ISR, strike missions, counterterrorism, surveillance

Figure 4.9 illustrates the Wing Loong I, a medium-altitude long-endurance (MALE) UAV developed by AVIC China. The diagram highlights key structural features such as its electro-optical/infrared (EO/IR) turret, mid-mounted straight wings, underwing munitions, and rear-mounted pusher propeller. The tricycle landing gear design supports autonomous takeoff and landing, while the sleek airframe enhances aerodynamic efficiency. This figure demonstrates the Wing Loong I's multi-role design combining endurance, surveillance,

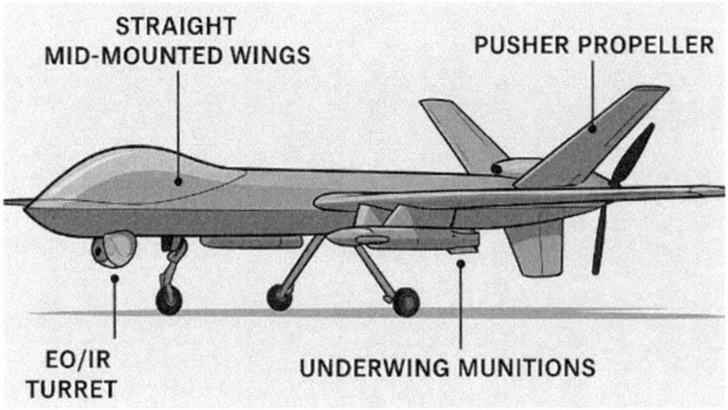

Figure 4.9 Wing Loong I (2015).

and strike capability, symbolizing China's advancement in cost-effective, export-oriented unmanned aerial systems.

Table 4.2 highlights key UAV models that transformed modern aerial systems through advancements in endurance, autonomy, and multi-role capabilities. The RQ-4 Global Hawk set new standards for strategic reconnaissance, while the MQ-9 Reaper merged heavy payloads with combat versatility. The Heron TP enhanced multi-domain intelligence operations, and Europe's Neuron and Taranis projects demonstrated major strides in stealth and autonomous UCAV design. The X-47B achieved historic carrier-based autonomy, and China's Wing Loong I democratized UAV access through affordable, armed export platforms. Collectively, these systems marked the global evolution toward intelligent, long-endurance, and precision-driven drone technology.

4.6 Challenges and Limitations of Third-generation Drones

Despite their remarkable advancements in endurance, autonomy, and mission versatility, third-generation drones faced several technological, operational, and strategic challenges that limited their overall effectiveness and scalability. These limitations played a critical role in shaping the direction of fourth-generation UAV development, emphasizing enhanced autonomy, stealth, and integration with artificial intelligence.

4.6.1 Technological challenges

One of the key challenges of third-generation drones was the dependency on satellite communication (SATCOM) and ground control stations for

Table 4.2 Developments and pioneers of third-generation drone models.

Drone model/year	Developer/country	Key features and capabilities	Primary role/ applications	Technological significance
RQ-1 Predator (1995)	General Atomics Aeronautical Systems/ United States	MALE UAV, SATCOM link, EO/ IR sensor, laser targeting, armed with Hellfire missiles	ISR, precision strike, surveillance	First armed UAV used in combat; introduced real-time targeting and long-endurance operations.
RQ-4 Global Hawk (1998)	Northrop Grumman/ United States	HALE platform, autonomous navigation, SAR radar, 36-hour endurance	Strategic reconnaissance, global intelligence missions	Revolutionized long-range, high-altitude ISR with autonomous flight and multi-sensor integration.
MQ-9 Reaper (2007)	General Atomics Aeronautical Systems/ United States	Heavy payload (3000+ lbs), multi-mission, advanced targeting pods, extended endurance	Combat strike, persistent surveillance	Combined high-speed capability with a large payload for simultaneous surveillance and attack.
Heron TP (Eitan) (2004)	Israel Aerospace Industries/Israel	SATCOM, multi-sensor payload, long endurance (36 h), large payload capacity	Intelligence, maritime patrol, border surveillance	Introduced large, modular design for multi-domain missions; operational with several international forces.
Dassault Neuron (2006)	Dassault Aviation/ France (EU collaboration)	Stealth UCAV, autonomous target identification, internal weapon bay	Experimental stealth combat drone	Pioneered European autonomous combat UAV design; foundation for future UCAV programs.
BAE Systems Taranis (2010)	BAE Systems/United Kingdom	Stealth design, autonomous flight, advanced sensor fusion	Combat and reconnaissance	Represented the United Kingdom's technological progress toward autonomous stealth UCAVs.
X-47B (2012)	Northrop Grumman/ United States	Carrier-based UCAV, autonomous takeoff and landing, precision navigation	Naval reconnaissance and strike operations	First UAV to autonomously operate from aircraft carriers, demonstrating advanced autonomy.
Wing Loong I (2015)	Aviation Industry Corporation of China (AVIC)/China	MALE UAV, EO/IR sensors, SATCOM, armed with guided munitions	ISR, precision strike, export platform	Offered cost-effective, export-ready UAV solutions to global defense markets.

navigation and mission management. This reliance made them vulnerable to signal interference, jamming, and cyberattacks, particularly in contested or GPS-denied environments. Although many drones were capable of semi-autonomous flight, full autonomy was still restricted by limited onboard computing power and sensor fusion capabilities. Furthermore, the integration of multi-sensor systems, including EO/IR, SAR, and ELINT, posed challenges in terms of data synchronization and processing latency. Maintenance of advanced composite airframes, stealth coatings, and precision payload systems also required high technical expertise, increasing operational costs.

4.6.2 Operational limitations

Operationally, third-generation UAVs faced constraints in mission flexibility and endurance under combat conditions. While systems like the MQ-9 Reaper and RQ-4 Global Hawk achieved extended flight durations, their large size and radar signatures made them susceptible to detection in hostile airspace. Harsh environmental conditions such as strong winds, sandstorms, or high-altitude icing also affected flight stability and data accuracy. Additionally, airspace integration with civilian aircraft remained a significant challenge, as regulatory frameworks for unmanned operations were still evolving globally.

4.6.3 Strategic and logistical constraints

Strategically, the widespread use of UAVs raised concerns regarding data security, export control, and international compliance. Many third-generation drones, particularly those equipped with strike capabilities, were restricted by political and ethical debates over autonomous lethal operations and civilian collateral risks. Logistical support – such as the need for dedicated launch and recovery infrastructure, high-bandwidth communication networks, and specialized maintenance teams also limited their deployment flexibility.

4.6.4 Cost and accessibility issues

Although UAVs like the Wing Loong I introduced cost-effective solutions, most third-generation platforms remained expensive to procure and operate, particularly for smaller nations. The integration of high-performance engines, advanced sensors, and stealth materials increased lifecycle costs, while supply chain dependencies restricted local manufacturing capabilities. Moreover, limited export authorization from major developers (such as the United States and Israel) constrained global accessibility, creating a gap between technological advancement and widespread adoption.

4.6.5 Ethical, legal, and regulatory challenges

The deployment of armed drones introduced complex ethical and legal dilemmas, particularly in relation to autonomous decision-making and targeted strikes. Questions regarding accountability, rules of engagement, and compliance with international humanitarian law became central to UAV operations. Many nations struggled to develop comprehensive airspace integration policies and autonomy regulations, leading to fragmented global standards.

4.7 Case Study: Operational Deployment and Impact of Second-generation Drones

The third generation of drones marked a transformative phase in unmanned aviation, characterized by the large-scale operational deployment of long-endurance, multi-role UAVs in both military and non-military domains. These systems demonstrated how advanced autonomy, endurance, and precision strike capability could redefine reconnaissance, surveillance, and warfare strategies across the globe. The following case studies highlight the operational significance, strategic outcomes, and technological impact of key third-generation UAVs in real-world missions.

4.7.1 Case study 1: RQ-4 Global Hawk – Strategic intelligence in global surveillance (1998–Present)

The RQ-4 Global Hawk, developed by Northrop Grumman (United States), became a cornerstone of global intelligence operations for the US Air Force and allied forces. Its ability to fly at altitudes above 60,000 feet for over 36 hours allowed persistent surveillance over vast regions, including conflict zones and maritime areas. During operations in Afghanistan, Iraq, and Libya, the Global Hawk provided real-time strategic imagery, signals intelligence (SIGINT), and battlefield mapping to support mission planning and target assessment. Beyond military applications, the RQ-4 was instrumental in humanitarian and disaster-response operations, such as post-tsunami mapping in Japan (2011) and wildfire surveillance in California. The drone's autonomous navigation and multi-sensor integration demonstrated the feasibility of continuous, high-altitude, global ISR missions. Its success established the foundation for future HALE UAV programs focused on global reconnaissance and autonomous data fusion.

4.7.2 Case study 2: MQ-9 Reaper – Precision strike and counterterrorism operations (2007–Present)

The MQ-9 Reaper, developed by General Atomics Aeronautical Systems (United States), revolutionized unmanned combat with its ability to perform simultaneous reconnaissance and precision strike missions. It became a primary asset for United States and NATO counterterrorism operations in Afghanistan, Iraq, Syria, and Somalia.

Equipped with AGM-114 Hellfire missiles and GBU-12 Paveway II bombs, the Reaper demonstrated unmatched persistence and accuracy, enabling targeted eliminations of high-value threats while minimizing collateral damage. Its real-time data streaming capability enhanced coordination between air and ground forces, improving battlefield situational awareness. The platform's success led to widespread international adoption by countries such as the United Kingdom, Italy, and France, solidifying its reputation as a model of modern hunter-killer UAVs. However, its deployment also raised ethical and legal debates about autonomous lethal decisions and civilian impact, prompting the need for tighter operational protocols and oversight.

4.7.3 Case study 3: Dassault Neuron – European stealth UCAV demonstration (2006–2015)

The Dassault Neuron project marked Europe's first major step toward developing autonomous stealth UCAV technology. Conducted through collaboration between France, Sweden, Italy, Spain, Greece, and Switzerland, the program validated stealth design, autonomous target acquisition, and internal weapon delivery. Flight tests between 2012 and 2015 demonstrated low radar visibility, autonomous route optimization, and cooperative engagement with manned aircraft such as the Dassault Rafale. While not operationally deployed in combat, the Neuron served as a proof of concept for future European programs, including the Franco-German Future Combat Air System (FCAS). Its technological innovations, especially in fly-by-light control systems, stealth materials, and AI-driven flight management, laid the groundwork for the development of fourth-generation stealth drones with higher autonomy and mission adaptability.

4.7.4 Case study 4: X-47B – Autonomous naval combat operations (2012–2015)

The Northrop Grumman X-47B became the first UAV to autonomously operate from an aircraft carrier, achieving catapult-assisted take-off and arrested recovery aboard the USS George H.W. Bush (CVN 77) in 2013. Developed under the US Navy's UCAS-D program, it demonstrated the integration of autonomy, precision navigation, and stealth in maritime operations. In subsequent tests, the X-47B performed aerial refueling and coordinated airspace sharing with manned aircraft, validating key technologies for carrier-based UCAVs. Its success directly influenced the design of the MQ-25 Stingray, the US Navy's first operational aerial refueling UAV. The X-47B's demonstration marked a paradigm shift in naval aviation, proving that unmanned systems could safely and effectively integrate into carrier operations, thereby expanding the reach and endurance of future naval air wings.

4.7.5 Case study 5: Wing Loong I – Exported combat efficiency and global reach (2015–Present)

The Wing Loong I, developed by AVIC China, emerged as a cost-effective, export-oriented UAV that expanded unmanned aerial capabilities to numerous developing nations. It combined ISR and strike capabilities, carrying precision-guided munitions and advanced EO/IR sensors. Used extensively by Saudi Arabia, Egypt, and the UAE, the Wing Loong I has participated in counterinsurgency and border surveillance operations, particularly in the Middle East and Africa. Its affordability, reliability, and modular design allowed nations with limited defense budgets to acquire advanced UAV capabilities. This widespread adoption demonstrated China's success in democratizing UAV technology, reshaping global defense markets, and challenging Western dominance in the drone industry.

4.8 Summary

The third generation of drones marked a revolutionary stage in unmanned aerial technology, integrating autonomy, endurance, stealth, and multi-role functionality to redefine aerial warfare and intelligence operations. This chapter highlighted the transition from manually controlled reconnaissance systems to AI-assisted, mission-adaptive UAVs capable of long-endurance surveillance, precision strikes, and autonomous navigation. Notable models such as the RQ-4 Global Hawk and MQ-9 Reaper transformed global ISR and combat missions, while the Heron TP demonstrated multi-sensor modularity in strategic operations. Europe's Dassault Neuron and BAE Systems Taranis advanced stealth UCAV innovation, and the X-47B achieved the world's first autonomous carrier-based operations. Meanwhile, China's Wing Loong I broadened UAV accessibility through cost-effective, export-ready systems. Despite facing challenges in communication security, ethical governance, and operational integration, third-generation drones established the technological and strategic foundation for the emergence of AI-driven, fully autonomous fourth-generation UAVs, solidifying their legacy as a pivotal advancement in modern aerospace evolution.

Review Questions

1. Define third-generation drones and explain how they differ from second-generation UAVs.
2. What are the key characteristics that distinguish third-generation drones in terms of design and functionality?
3. Describe the major technological advancements that led to the evolution of third-generation drones.
4. Discuss the historical context that influenced the development of autonomous and long-endurance UAVs.
5. What were the main technological drivers behind the transition from remotely piloted systems to semi-autonomous UAVs?
6. Explain the fundamental principles of operation for third-generation drones.
7. What are the main hardware components that enable third-generation drones to perform complex missions?
8. Discuss the importance of propulsion systems and airframe materials in enhancing UAV endurance.
9. What software and onboard systems contribute to the autonomy of third-generation UAVs?
10. How does system integration improve the functional performance of UAVs in surveillance and combat missions?
11. Explain how satellite communication (SATCOM) supports beyond-line-of-sight operations in UAVs.
12. Describe the role of sensor fusion in enhancing mission accuracy and situational awareness.
13. Identify the major military and defense applications of third-generation drones.

14. How did UAVs transform intelligence, surveillance, and reconnaissance (ISR) capabilities in modern warfare?
15. Discuss the scientific and environmental applications of third-generation UAV systems.
16. How are drones used for civilian and commercial purposes, such as mapping and logistics?
17. What role do third-generation UAVs play in modern communication and data relay systems?
18. Describe the global impact of third-generation drones on defense and security operations.
19. Explain how the evolution of third-generation UAVs contributed to the development of autonomous UCAVs.
20. Compare the technological and operational advancements between early UAVs and third-generation models.
21. Describe the major features and capabilities of the RQ-1 Predator (1995).
22. Explain how the RQ-4 Global Hawk (1998) redefined strategic intelligence and reconnaissance missions.
23. Discuss the operational versatility of the MQ-9 Reaper (2007) and its role in counterterrorism operations.
24. What are the unique design and payload characteristics of the Heron TP (Eitan) (2004)?
25. How did the Dassault Neuron (2006) contribute to the advancement of stealth UAV technologies in Europe?
26. Describe the objectives and achievements of the BAE Systems Taranis (2010) program.
27. Why is the X-47B (2012) considered a milestone in naval UAV operations?
28. Explain how the Wing Loong I (2015) impacted global UAV accessibility and defense markets.
29. What were the major technological challenges faced by third-generation UAV developers?

30. Discuss the operational limitations of third-generation drones in high-threat environments.
31. How did strategic and logistical constraints affect the deployment of long-endurance UAVs?
32. Explain how cost and accessibility influenced the export and adoption of UAV technologies worldwide.
33. Identify the key ethical, legal, and regulatory issues associated with UAV use in warfare.
34. Describe the communication vulnerabilities commonly faced by third-generation drones.
35. Summarize the operational achievements of the RQ-4 Global Hawk in global intelligence missions.
36. How did the MQ-9 Reaper's dual capability for surveillance and combat redefine UAV warfare?
37. What were the primary lessons learned from the Dassault Neuron's stealth and autonomous tests?
38. Explain the significance of the X-47B's autonomous carrier take-off and landing operations.
39. Discuss how the Wing Loong I demonstrated China's emergence as a global UAV exporter.
40. Conclude how third-generation UAVs established the foundation for the development of AI-driven, fully autonomous drones in the fourth generation.

5

Fourth-generation Drones

5.1 Introduction

The fourth generation of drones represents a defining leap in the evolution of unmanned aerial systems, marking the transition from semi-autonomous platforms to intelligent, fully autonomous, and network-integrated aerial systems. This generation builds upon the endurance, stealth, and multi-role capabilities of third-generation UAVs, integrating artificial intelligence (AI), machine learning (ML), advanced swarm coordination, and real-time data analytics to enable self-governing decision-making and collaborative mission execution. Unlike their predecessors, fourth-generation drones are not merely remotely piloted or pre-programed systems; they are adaptive, cognitive platforms capable of learning, optimizing, and responding dynamically to changing mission environments without constant human intervention.

Technological advancements in AI-driven autonomy, edge computing, 5G-enabled communication, and quantum-encrypted data links have redefined how drones operate within both military and civilian domains. These systems exhibit multi-domain interoperability, seamlessly coordinating across air, land, sea, and space-based assets within network-centric warfare frameworks. Militarily, fourth-generation drones perform swarm intelligence operations, loyal wingman missions, and AI-assisted combat support, while in civilian sectors, they are transforming industries such as infrastructure inspection, agriculture, logistics, and disaster management through precision automation.

This chapter explores the technological foundation, operational architecture, and evolving applications of fourth-generation drones, emphasizing their role in autonomous decision-making, cooperative engagement, and digital integration. It also examines key global programs such as the US Skyborg, Europe's Future Combat Air System (FCAS), India's Ghatak, and China's Dark Sword, which exemplify the global race toward next-generation unmanned dominance. Furthermore, the chapter discusses the challenges of AI ethics, cybersecurity, regulatory adaptation, and human–machine

DOI: 10.1201/9788743812418-5

collaboration, which continue to shape the responsible evolution of this powerful technology.

5.1.1 Definition

Fourth-generation drones are advanced unmanned aerial systems (UAS) that integrate artificial intelligence (AI), machine learning (ML), real-time data processing, and autonomous decision-making to perform complex missions with minimal or no human intervention. Unlike earlier generations that relied heavily on remote piloting or pre-programed commands, these drones possess the capability to perceive, analyze, and act independently based on situational data and mission objectives.

They are characterized by self-learning algorithms, swarm intelligence, sensor fusion, and networked interoperability, enabling coordinated operations among multiple unmanned and manned platforms across air, land, sea, and space. Equipped with AI-based navigation, adaptive mission planning, and autonomous threat response, fourth-generation drones can execute multi-domain operations from surveillance and electronic warfare to precision strikes and logistics.

5.1.2 Basic characteristics of fourth generation of drones

The fourth generation of drones embodies a revolutionary shift toward autonomous intelligence, real-time adaptability, and networked collaboration, integrating cutting-edge technologies that enable drones to function as independent, cognitive systems rather than remote-controlled machines. The following are the defining characteristics of this generation:

- *Artificial intelligence (AI) and machine learning (ML) integration*

  Fourth-generation drones utilize AI-driven algorithms and machine learning models for autonomous decision-making, pattern recognition, and adaptive mission execution. These systems continuously learn from operational data to improve flight precision, threat response, and mission efficiency without human oversight.

- *Swarm intelligence and cooperative autonomy*

  One of the most distinctive features of this generation is the ability to perform swarm-based operations, where multiple drones communicate and collaborate in real time to execute missions collectively. This distributed intelligence allows for self-organization, task-sharing, and dynamic adaptability in combat, reconnaissance, or disaster-relief missions.

- *Multi-domain interoperability*

  Fourth-generation drones are designed for multi-domain integration, capable of coordinating with manned aircraft, naval vessels, ground vehicles, and satellite systems. This interoperability enhances situational awareness and supports network-centric warfare and cross-domain operations through seamless data sharing.

- *Real-time data processing and edge computing*

  Equipped with onboard edge computing systems, these drones can process vast amounts of sensor and video data locally, reducing latency and enabling immediate decision-making. Real-time analytics and AI-based data fusion empower drones to identify targets, navigate complex terrains, and respond to threats instantly.

- *Advanced communication and networking systems*

  Incorporating 5G, quantum-encrypted links, and satellite-based communication, fourth-generation drones ensure secure, high-speed data transfer between multiple units and command centers. These technologies facilitate beyond-line-of-sight (BLOS) operations, even in contested or GPS-denied environments.

- *Enhanced stealth and survivability*

  Designed using low-observable (LO) materials, adaptive camouflage, and reduced radar and infrared signatures, fourth-generation drones are optimized for survivability in high-threat zones. Their stealth characteristics allow deep-penetration missions and long-endurance reconnaissance in hostile territories.

- *Modular and scalable design*

  These drones employ a modular architecture, allowing easy customization for various roles – such as ISR (intelligence, surveillance, reconnaissance), EW (electronic warfare), cargo delivery, or combat missions. Scalability also enables integration with emerging technologies such as directed-energy weapons or hypersonic propulsion.

- *Autonomy and human–machine teaming (HMT)*

  Fourth-generation UAVs operate autonomously yet remain connected through human–machine teaming interfaces, allowing commanders to supervise missions without direct piloting. This feature supports loyal wingman operations, where drones fly alongside manned aircraft to extend situational awareness and combat capability.

- *Cybersecurity and encrypted control systems*

  Given the increasing cyber threats, these drones feature multi-layered cybersecurity protocols, AI-driven anomaly detection, and quantum-resistant encryption to safeguard communication, data integrity, and mission command from hostile interference.

- *Sustainable and efficient power systems*

  Integrating hybrid-electric engines, solar-assisted systems, and advanced energy management technologies, fourth-generation drones offer greater fuel efficiency, operational range, and environmental sustainability compared to earlier UAVs.

5.2 Historical development of fourth-generation drones

The evolution of fourth-generation drones represents a culmination of decades of technological advancement in automation, artificial intelligence, stealth design, and network-centric warfare. Emerging in the early 2010s, this generation evolved from the foundation laid by third-generation UAVs such as the RQ-4 Global Hawk, MQ-9 Reaper, and X-47B, which introduced long-endurance flight, limited autonomy, and combat integration. The defining leap of fourth-generation systems came with the application of AI-driven autonomy, swarm coordination, and digital battlefield integration, transforming drones from remotely piloted machines into self-governing, cognitive aerial entities.

The early roots of fourth-generation drone development can be traced to post-2010 defense modernization programs initiated by leading aerospace nations. The United States, through initiatives such as DARPA's Skyborg, the Loyal Wingman Program, and the Next Generation Air Dominance (NGAD) project, pioneered AI-enabled drones capable of learning, adapting, and coordinating with manned aircraft. Similarly, Europe's Future Combat Air System (FCAS), a collaboration between Dassault Aviation and Airbus, emphasized stealth, data fusion, and autonomous teaming to ensure multi-domain operational superiority.

In Asia, rapid technological progress drove the rise of indigenous UAV programs. China's Dark Sword and Wing Loong II introduced advanced stealth and swarm technologies, focusing on AI-assisted target identification and autonomous tactical decision-making. Meanwhile, India's Ghatak (AURA) program began developing indigenous stealth UCAV capabilities, integrating sensor fusion and autonomous combat control. These developments marked a global recognition that future air dominance would rely heavily on AI-guided autonomy, digital communication networks, and unmanned–manned coordination.

The 2015–2020 period witnessed significant breakthroughs in autonomous flight testing, including the demonstration of swarm operations, AI-based mission reconfiguration, and autonomous aerial refueling. The Loyal Wingman drone by Boeing Australia (2020) exemplified this evolution, capable of operating as a companion to manned fighters, executing electronic warfare, surveillance, and strike missions autonomously. Concurrently, advancements in 5G communication, edge computing, and quantum encryption enhanced data sharing and mission reliability, allowing fourth-generation drones to operate in multi-domain and GPS-denied environments.

Today, fourth-generation drones are recognized as integral elements of digital warfare ecosystems, serving in roles ranging from ISR and precision strike to electronic warfare, logistics, and command relay. Their historical development reflects a shift in focus from hardware superiority to software-driven intelligence, where machine learning, data fusion, and cooperative autonomy define mission success.

5.3 Principles and Components of Fourth-generation Drones

The principles and components of fourth-generation drones define the most advanced stage in unmanned aerial technology, where autonomy, intelligence, and connectivity converge to enable fully self-governing and adaptive aerial systems. Unlike previous generations that relied on remote human control, fourth-generation drones operate on the foundation of AI-driven autonomy, sensor fusion, and networked interoperability, allowing them to execute complex missions across air, land, sea, and space domains with minimal human intervention.

5.3.1 Principles of operation

Fourth-generation drones operate on four interdependent principles:

- *Perceive fuse decide act*: Continuous sensor-driven perception (multimodal sensing) is fused into a coherent situational picture; AI/ML-based decision layers convert high-level objectives into executable plans; and the flight/control stack executes and adapts those plans in real time.

- *Distributed, cooperative autonomy*: Intelligence is shared among platforms (swarm or team), enabling task allocation, formation control, cooperative sensing, and collaborative engagement without centralized micromanagement.
- *Edge-centric, low-latency processing*: Decision-quality processing is performed on or near the platform (edge computing) to permit immediate responses (collision avoidance, dynamic re-tasking) and to reduce bandwidth for transmitted data.
- *Resilience and graceful degradation*: Built-in redundancy, fault-detection-isolation-and-recovery (FDIR), anti-spoofing/jamming measures, and predefined safe-states allow continued operation or safe recovery under degraded communications, contested EM environments, or partial system failures.

These principles are formalized in mission logic: operators specify intent and constraints; the onboard autonomy stack translates intent into mission plans, continuously re-optimizes under evolving conditions, and escalates only those decisions that surpass predefined human-in-the-loop thresholds (human-supervised autonomy/human–machine teaming).

5.3.2 Hardware components

Fourth-generation platforms combine mature aerodynamic and propulsion design with advanced sensing, computing, and hardened communications. Key hardware elements include:

- *Airframe and mission modularity*: Low-observable or low-signature airframes for survivability, with modular bays to swap sensors, EW packages, or payloads quickly.
- *Propulsion and power systems*: Efficient hybrid-electric/turboprop or turbofan cores for extended endurance; integrated power management to supply high-power edge compute, directed energy payloads, or extended sensor suites; thermal management for electronics and sensors.
- *Flight control units (FCU) and redundant avionics*: Multi-lane flight computers, hardware-in-the-loop protected buses, and multiple sensor feeds to permit cross-checking and safe degradation.
- *High-performance edge compute*: Ruggedized CPUs/GPUs/TPUs for onboard ML inference, real-time sensor fusion, and simulation-assisted planning.

- *Multi-modal sensor suite*: Stabilized EO/IR gimbals, SAR/GMTI radars, LiDAR or laser ranging, hyperspectral/multispectral imagers, SIGINT/COMINT/ESM packages, ADS-B/TCAS and sense-and-avoid sensors for airspace integration.
- *Resilient communications hardware*: Multi-band radios (UHF/VHF, L-band), SATCOM terminals (Ku/Ka), 5G/mesh radios for tactical networks, and anti-jamming front ends with frequency agility and directional antennas.
- *Cybersecurity and crypto modules*: Hardware security modules (HSMs), trusted platform modules (TPMs), and quantum-resistant crypto accelerators where required.
- *Launch and recovery systems/support equipment*: Autonomous runway or catapult systems, precision deck/ship recovery interfaces, or parachute/auto-land recovery depending on mission profile.
- *Health sensors and prognostics*: Embedded structural, thermal, and vibration sensing to feed predictive maintenance and in-flight fault response.

5.3.3 Major software and systems components

The software architecture is layered and modular, enabling certification, upgrades, and reuse:

- *Autonomy stack/mission manager*: Implements intent interpretation, mission planning, task allocation (for multi-vehicle teams), and high-level policy enforcement (ROE, no-strike zones). Often realized as a hierarchy of planners (strategic → tactical → control).
- *Perception and sensor fusion*: ML-based perception pipelines (object detection, tracking, semantic segmentation) fuse EO/IR, SAR, LiDAR, and SIGINT to produce geo-registered tracks and classification confidence levels.
- *Motion planning and control*: Real-time path planning (including 4D trajectory optimization), collision avoidance (sense-and-avoid), and guidance laws that respect dynamic constraints and no-fly/keep-out zones.
- *Distributed coordination middleware*: Real-time publish/subscribe or DDS-style middleware (or ROS2 variants) enabling low-latency state sharing, consensus algorithms for swarm decisions, and distributed tasking.

- *Cybersecurity and resilience layer*: Secure boot, attestation, anomaly detection (behavioral and ML-based), encrypted comms, anti-spoofing, and intrusion response that can shift to autonomous safe modes.
- *FDIR and prognostics*: Continuous monitoring and automated recovery controllers that handle sensor failures, actuator degradation, and partial system loss with logged audit trails.
- *Human–Machine Interface (HMI)/GCS software*: Intent-based mission planning tools, visualization, explainable AI summaries, and supervisory controls to allow human override and mission parameter adjustments.
- *Simulation and verification tools*: Onboard/ground digital twins and physics-based simulators for plan validation and safe mission rehearsal.

5.3.4 System integration and functional operation

Integration converts components into a dependable system through engineering practices, verification, and operational concepts:

- *Pre-mission phase*: Mission intent, constraints, and ROE are loaded to the GCS; simulated plan verification runs on the digital twin; sensor calibrations and redundancy checks are performed automatically.
- *Launch and transit*: FCU and autonomy modules execute safe autonomous takeoff; federation protocols negotiate airspace access with deconfliction services; mesh/SATCOM links establish multi-layer connectivity.
- *On-station and cooperative operations*: Onboard perception fuses sensor inputs; local planners handle dynamic obstacles; distributed algorithms manage swarm roles (scout, loiter, strike, comms relay); compressed intelligence and metadata are transmitted rather than raw sensor streams to conserve bandwidth.
- *Dynamic re-tasking and escalation*: High-confidence detections or emergent targets trigger onboard classification and, depending on policy, either autonomous mitigation actions (e.g., evade) or operator alerts for clearance to engage. Explainable AI modules supply human supervisors with the rationale for autonomous decisions.
- *Anomaly handling and graceful degradation*: FDIR transitions the vehicle into safe loiter, return-to-home, or approved contingency patterns

when failures are detected; redundant systems maintain minimum mission functionality.

- *Recovery and post-mission*: Secure data offload, automated health reports, and machine-learning model updates (federated learning) are synchronized with the cloud/enterprise systems to improve future performance.
- *Verification, certification, and governance*: Lifecycle integration includes model verification, SOTIF (safety of the intended functionality) analysis, and compliance with airspace and weapon employment standards; continuous monitoring feeds ethical and legal oversight processes.

Figure 5.1 provides a visual representation of the core architecture of fourth-generation drones, illustrating how modern UAVs integrate sensing, autonomy, and hardware systems to achieve intelligent and adaptive operations. The diagram highlights three major functional domains: sensing and perception, autonomy, and hardware. In the sensing section, components such as electro-optical/infrared (EO/IR) cameras, onboard radar, and LiDAR support high-precision environmental awareness and object detection. The autonomy section depicts AI-based decision-making, automated object detection, and swarm coordination, enabling drones to collaborate, navigate, and execute missions with minimal human input. The hardware section emphasizes AI-driven flight control, beyond visual line of sight (BVLOS) communication, long-endurance airframes, and energy-dense batteries, which collectively enhance range, reliability, and mission persistence. Together, the figure encapsulates the essence of fourth-generation UAV design, a fusion of

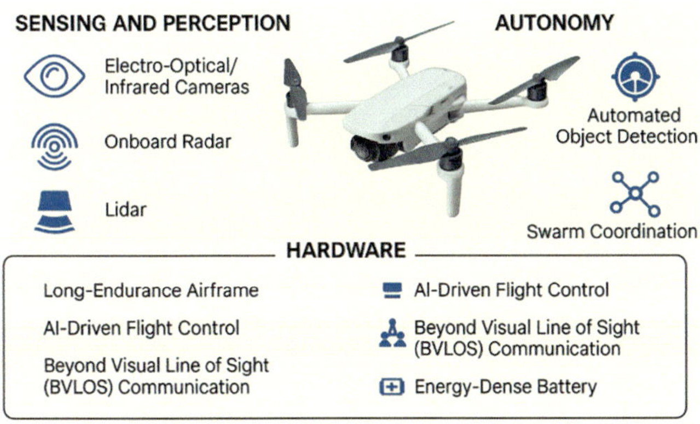

Figure 5.1 Fourth-generation drones.

intelligent perception, adaptive autonomy, and advanced aerostructural engineering for multi-domain, next-generation aerial operations.

5.4 Swarm Technology and Collaborative Drones

5.4.1 Introduction to swarm intelligence and group coordination

Swarm intelligence refers to the collective behavior of multiple autonomous drones that interact locally with one another and with their environment to achieve global objectives efficiently. Each drone in the swarm is equipped with sensors, communication modules, and onboard processors that allow it to perceive surroundings, share situational data, and adjust behavior in real time. Unlike traditional UAV systems dependent on centralized control, swarms use distributed decision-making algorithms such as boids algorithms, particle swarm optimization (PSO), and behavior-based coordination to self-organize and accomplish complex missions cooperatively.

The core principles of swarm coordination include:

- *Decentralization*: Each unit operates independently, contributing to collective decisions without needing a central command.
- *Adaptability*: The swarm can dynamically adjust formation, spacing, or strategy in response to environmental changes or threats.
- *Scalability*: The system remains functional regardless of swarm size, from a handful of drones to hundreds working in unison.
- *Fault tolerance*: Loss or malfunction of one or more drones does not disrupt the mission, as other units compensate automatically.

Communication within a drone swarm is achieved through ad-hoc mesh networks or 5G/6G-enabled links, allowing real-time exchange of positional, sensory, and mission data. The integration of AI and edge computing enables each drone to process localized information while contributing to the swarm's collective intelligence, thereby achieving emergent behavior complex, intelligent group responses arising from simple local interactions.

5.4.2 Applications in military and commercial settings

The practical applications of drone swarm technology span a wide spectrum, revolutionizing both defense operations and civilian industries.

A. *Military applications*:
In defense, swarm-enabled UAVs are redefining network-centric warfare through capabilities that enhance reconnaissance, combat, and strategic coordination.

- *Surveillance and reconnaissance*: Multiple drones coordinate to cover vast terrains simultaneously, providing continuous 360° situational awareness.
- *Offensive operations*: Swarm drones execute coordinated strikes, overwhelm enemy defenses, and perform "saturation attacks" against radar or missile systems.
- *Electronic warfare (EW)*: Certain drones act as signal jammers or decoys, while others gather intelligence, ensuring multi-role mission adaptability.
- *Search and rescue (SAR)*: Swarm UAVs can rapidly locate survivors or identify hazards in combat or disaster zones through cooperative area mapping.
- *Loyal wingman missions*: Swarms can accompany manned aircraft, sharing sensor data, extending radar coverage, or autonomously engaging hostile targets.

Notable examples include the US DARPA's Gremlins and OFFSET programs, China's swarming wingman projects, and India's ALFA-S swarm drone initiative, all designed to develop autonomous, cooperative UAV systems capable of functioning under GPS-denied and contested conditions.

B. *Commercial and civilian applications*:
Beyond defense, swarm technology has begun reshaping civilian sectors through its scalability, efficiency, and precision.

- *Agriculture*: Drone swarms perform synchronized crop monitoring, pesticide spraying, and soil analysis, reducing time and resource use.
- *Disaster management*: Swarms create real-time 3D maps, coordinate search operations, and deliver supplies to hard-to-reach areas.
- *Infrastructure inspection*: Coordinated drones inspect power lines, bridges, and pipelines simultaneously, providing comprehensive data in less time.

- *Logistics and delivery*: Swarm algorithms enable fleets of drones to manage package delivery routes collaboratively, improving efficiency and reducing congestion.
- *Entertainment and communication*: Swarms are used in drone light shows, aerial advertising, and temporary network relay systems during large events or disasters.

In the commercial sphere, swarm-enabled UAVs embody the principles of automation, scalability, and self-organization, leading industries toward Industry 5.0 – where collaborative robotics and AI-driven drones seamlessly interact with humans and digital ecosystems.

Figure 5.2 illustrates the core concept of swarm intelligence and its integration into collaborative drone systems, highlighting the interaction between intelligence, coordination, and applications. The diagram's center depicts a formation of drones representing a cooperative swarm network, emphasizing distributed decision-making and real-time inter-drone communication. The left section outlines the foundation of swarm intelligence, autonomous drones operating via decentralized algorithms, while the right section showcases group coordination, achieved through adaptive communication and behavioral synchronization. The lower segments divide the operational fields: military applications, including surveillance, electronic warfare, and saturation attacks, and commercial uses, such as agriculture, disaster management, logistics, and infrastructure inspection. This figure encapsulates how swarm-based drones transform aerial operations by combining collective intelligence, coordination, and scalability to achieve efficient mission execution across diverse environments.

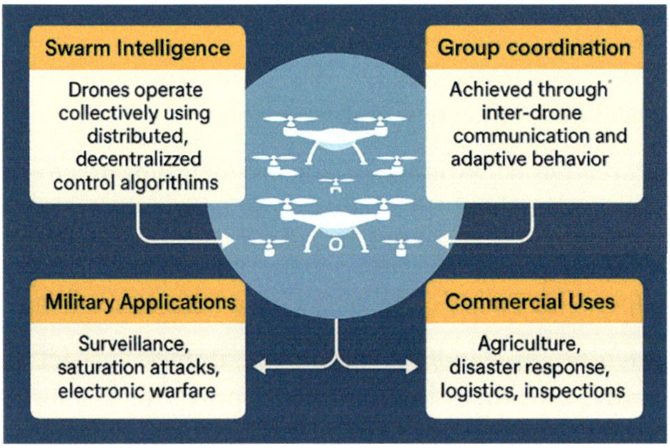

Figure 5.2 Swarm technology and collaborative drones.

5.5 Applications of Fourth-generation UAV Systems

The fourth generation of UAV systems signifies a transformative era in drone deployment, marked by the seamless integration of autonomy, artificial intelligence (AI), multi-domain connectivity, and advanced data analytics. These drones transcend traditional aerial operations, functioning as intelligent, adaptive agents across military, scientific, commercial, and technological sectors. Their capacity for real-time decision-making, swarm coordination, and high-precision mission execution has revolutionized the way information is gathered, analyzed, and applied. This section explores the wide-ranging applications of fourth-generation UAVs across different domains, demonstrating how these systems are redefining operational paradigms globally.

5.5.1 Military and defense applications

The military remains the primary arena where fourth-generation drones have achieved the greatest impact. Equipped with AI-assisted autonomy, stealth design, and multi-sensor fusion, these UAVs are redefining modern warfare and intelligence operations through their capacity for precision, persistence, and adaptive decision-making.

- *Intelligence, surveillance, and reconnaissance (ISR)*: Drones like the RQ-180 and Heron TP employ multi-spectral imaging, SAR radars, and ELINT payloads for persistent surveillance over hostile territories, offering unmatched situational awareness.
- *Autonomous combat and strike missions*: Platforms such as the MQ-25 Stingray and Loyal Wingman execute coordinated combat and refueling operations autonomously, enabling extended missions and reducing human risk in high-threat environments.
- *Electronic and cyber warfare*: Fourth-generation UAVs integrate electronic attack (EA) and electronic support (ES) capabilities for radar jamming, signal deception, and communications disruption, contributing to information dominance in the digital battlespace.
- *Swarm and cooperative engagement*: AI-driven swarms perform saturation strikes, overwhelm enemy air defenses, and conduct multi-directional intelligence gathering, enhancing both offensive and defensive mission outcomes.
- *Logistical and tactical support*: Heavy-lift drones and hybrid UAVs deliver supplies, ammunition, and medical aid in battlefield zones,

ensuring rapid resupply and casualty evacuation in environments inaccessible to traditional vehicles.

Collectively, these applications underscore the transformation of drones from support tools to autonomous combat assets, central to next-generation warfare doctrines and multi-domain operations (MDOs).

5.5.2 Scientific and environmental applications

In the scientific and environmental domain, fourth-generation drones have become essential instruments for data-driven research, climate monitoring, and ecological management. Their AI-based sensing and edge computing capabilities allow for high-resolution data collection and real-time analysis in challenging environments.

- *Climate and atmospheric studies*: UAVs such as NASA's Global Hawk variant and India's CATS UAV collect atmospheric data for weather prediction, greenhouse gas measurement, and storm tracking, improving climate modeling accuracy.
- *Disaster monitoring and response*: Equipped with thermal imaging and LiDAR, drones support post-disaster mapping, flood risk assessment, and wildfire surveillance, offering immediate situational data for emergency management.
- *Environmental conservation*: UAVs are deployed for forest canopy analysis, anti-poaching operations, wildlife tracking, and deforestation mapping, enabling sustainable resource management.
- *Oceanographic and geological exploration*: Fourth-generation UAVs collaborate with autonomous underwater vehicles (AUVs) to perform coastal mapping, coral reef monitoring, and seismic data collection, demonstrating multi-domain research interoperability.

By enhancing precision and reducing human intervention, drones serve as vital instruments in promoting environmental stewardship and supporting global sustainability initiatives.

5.5.3 Civilian and commercial applications

Fourth-generation UAVs have found significant utility in the industrial, infrastructural, and consumer sectors, driven by advancements in autonomous flight control, 5G connectivity, and machine vision. Their ability to execute repeatable, high-precision tasks autonomously makes them indispensable across modern industries.

- *Infrastructure inspection and maintenance*: AI-powered UAVs perform predictive inspection of bridges, pipelines, wind turbines, and power grids, identifying structural anomalies before failures occur.
- *Agricultural management*: Swarm-based agricultural drones utilize hyperspectral imaging and machine learning analytics to monitor crop health, detect pests, and optimize irrigation, supporting precision farming and yield optimization.
- *Urban logistics and delivery systems*: Companies such as Amazon, Zipline, and Wing deploy UAVs for autonomous package delivery, using dynamic routing and obstacle avoidance algorithms to enhance efficiency and safety.
- *Surveying and construction*: UAVs equipped with photogrammetry and LiDAR mapping systems assist in terrain modeling, infrastructure planning, and 3D construction visualization.
- *Entertainment and media production*: Drone swarms create large-scale light shows and cinematography sequences, merging automation with creativity.

These applications illustrate how drones have evolved from niche tools to essential commercial assets, driving global industries toward automation, safety, and digital transformation.

5.5.4 Communication and technological applications

The integration of UAVs into communication networks marks a revolutionary step in extending global connectivity and supporting data-driven ecosystems. Fourth-generation drones function as airborne communication relays, mobile data centers, and IoT integration hubs.

- *Aerial communication relays*: Drones equipped with 5G/6G communication nodes provide temporary coverage in disaster zones or remote regions, ensuring continuity in communication and data services.
- *Satellite integration and network bridging*: UAVs act as intermediaries between satellite constellations and terrestrial networks, enhancing broadband connectivity and latency performance.

- *IoT and smart infrastructure support*: They facilitate real-time data collection from distributed sensors in smart cities, agriculture, and industrial zones, enabling automated decision-making through cloud–edge integration.
- *Cybersecurity and data protection*: With AI-based encryption and quantum-secure communication systems, drones protect sensitive military and civilian data transmissions against interference and cyber threats.
- *Emergency network restoration*: In post-disaster conditions, drones rapidly deploy as temporary communication towers, restoring connectivity and supporting rescue coordination.

These technological applications underline the growing role of fourth-generation UAVs as key enablers of global digital infrastructure, merging aerospace innovation with the expanding frontiers of telecommunication and data science.

5.6 Developments and Pioneers of Fourth-generation Drone Models

The fourth generation of drones represents the zenith of unmanned aerial evolution, integrating artificial intelligence, machine learning, swarm coordination, advanced stealth, and digital networking into cohesive aerial ecosystems. This phase signifies the transition from manually supervised UAVs to autonomous, decision-making aerial entities capable of executing missions in highly dynamic, multi-domain environments. This section outlines the technological evolution that gave rise to fourth-generation drones and identifies the key pioneers and notable UAV models that have redefined global airpower, research, and commercial innovation.

5.6.1 Evolution of fourth-generation UAV development

The development of fourth-generation UAVs began in the early 2010s, driven by the need for smarter, faster, and more autonomous systems capable of operating in contested and data-driven environments. Building upon the endurance, stealth, and ISR (intelligence, surveillance, and reconnaissance) capabilities of third-generation drones, this new generation incorporated AI-based autonomy, real-time data processing, and network-centric warfare integration.

Key technological shifts that defined this evolution include:

- *Integration of artificial intelligence (AI) and machine learning (ML)*: UAVs began using AI to independently interpret sensor data, identify targets, optimize routes, and make tactical decisions without continuous operator input.

- *Swarm and cooperative autonomy*: The introduction of swarm coordination algorithms allowed groups of drones to function collaboratively, enhancing mission adaptability and resilience.
- *Stealth and low-observable design enhancements*: Fourth-generation drones feature advanced radar-absorbing materials, blended-wing configurations, and infrared signature reduction technologies, ensuring survivability in high-threat environments.
- *5G and beyond-line-of-sight (BLOS) communication*: The adoption of high-bandwidth, encrypted communication systems enabled seamless coordination among multiple UAVs and command centers over long distances.
- *Multi-domain and modular architecture*: UAVs evolved into modular platforms capable of performing various missions ISR, combat, refueling, logistics, or communication relays, by simply reconfiguring payloads and software modules.
- *Cognitive mission management*: Advanced mission computers now allow drones to adapt in real time to changing conditions, adjusting flight parameters, target priorities, or formations autonomously.

This evolution reflects the global defense and industrial trend toward human–machine teaming (HMT), network-centric operations, and intelligent aerial ecosystems, where UAVs act not merely as assets but as independent tactical collaborators in modern missions.

5.6.2 Key pioneers and notable drone models

The rise of fourth-generation drones is the result of sustained innovation by global aerospace pioneers, each contributing unique design philosophies and technologies. The following models, presented in Table 5.1, exemplify the diverse technological advancements, operational versatility, and strategic significance of fourth-generation drone systems across global defense and aerospace sectors:

5.6.2.1 XQ-58A Valkyrie (2019)

The XQ-58A Valkyrie, developed by Kratos Defense & Security Solutions (United States) in collaboration with the US Air Force Research Laboratory

Table 5.1 Developments and pioneering models of fourth-generation drones.

Model/year	Developer/ country	Key features and technologies	Primary role/ application	Significance
XQ-58A Valkyrie (2019)	Kratos Defense/United States	AI-enabled "loyal wingman" UAV with stealth design and modular payloads.	Combat support, reconnaissance, EW.	Demonstrates manned–unmanned teaming; low-cost autonomous strike companion.
Skyborg Program (2020)	DARPA/United States	AI-based autonomy core for multiple UAV platforms.	Open-architecture autonomy and swarm coordination.	Foundation for autonomous control in US next-gen UAVs and collaborative swarms.
Boeing Airpower Teaming System (ATS) (2020)	Boeing/Australia	11.7 m UAV with modular payloads, long range, and AI mission control.	Loyal wingman, ISR, and strike operations.	Australia's first indigenous UAV; pioneer in international cooperative combat drone design.
FCAS (Future Combat Air System) (2021)	Dassault Aviation and Airbus/Europe	AI-assisted stealth UCAV integrated with 6th-gen fighters.	Air dominance, ISR, electronic warfare.	Europe's largest collaborative UAV program; cornerstone for next-gen combat ecosystems.
Tengden TW-356 (2020)	Tengden/China	Large HALE UAV with AI flight management and modular payloads.	ISR, logistics, and electronic warfare.	Showcases China's advancement in AI and heavy-lift autonomous UAVs.
Ghatak UCAV (2022)	DRDO/India	Flying-wing stealth UCAV with autonomous strike capability and indigenous AI mission control.	Precision strike, tactical surveillance.	India's first indigenous stealth combat drone; marks regional self-reliance in UAV development.
Okhotnik-B (Hunter-B) (2020)	Sukhoi/Russia	Stealth UCAV with AI-guided target selection and internal weapons bay.	Long-range combat and reconnaissance.	Integrates with Su-57 fighters for manned–unmanned operations.
CH-7 Rainbow (2019)	China Aerospace Science and Technology Corp (CASC)	Stealth flying-wing UCAV with high-altitude endurance and electronic warfare payloads.	Combat, surveillance, and electronic warfare.	China's first stealth UCAV; part of its next-generation combat drone family.

(AFRL), is a pioneering example of a low-cost, stealthy, autonomous "loyal wingman" UAV. Designed under the Low-cost Attritable Aircraft Technology (LCAAT) program, the Valkyrie demonstrates how modern drones can operate cooperatively with manned fighter aircraft, enhancing mission capability while reducing risk to human pilots (Wang, et al. 2020).

The UAV features a blended-wing airframe optimized for radar stealth and aerodynamic efficiency, with a cruise speed exceeding 0.85 Mach and an operational ceiling of 45,000 feet. Its modular payload bay supports a wide range of mission's surveillance, electronic warfare, and precision strike, making it a flexible combat companion for aircraft such as the F-35 Lightning II and F-22 Raptor.

Equipped with an AI-based flight control system, the Valkyrie can execute autonomous mission planning, formation flying, and threat response while maintaining secure data links through encrypted mesh networks. Its open-architecture design allows integration of different payloads and systems, ensuring future adaptability for multi-domain operations.

Strategically, the XQ-58A Valkyrie represents a critical step toward autonomous combat collaboration, enabling manned–unmanned teaming (MUM-T) in contested airspaces. Its combination of low cost, stealth, and AI-driven autonomy makes it a foundational model for next-generation aerial warfare, influencing programs like Skyborg and Next Generation Air Dominance (NGAD).

Figure 5.3 showcases the XQ-58A Valkyrie, an advanced autonomous stealth combat UAV developed by Kratos Defense and the US Air Force Research Laboratory (AFRL) under the Low-cost Attritable Aircraft Technology (LCAAT) program. The image highlights its blended-wing,

Figure 5.3 XQ-58A Valkyrie (2019).

low-observable design, optimized for stealth and high-speed performance. The angular fuselage and flush air intakes contribute to radar signature reduction, while the tricycle landing gear supports independent ground operations. This UAV is designed for manned–unmanned teaming, operating alongside fighter jets like the F-35 and F-22 to provide reconnaissance, electronic warfare, and strike support. With AI-driven autonomy, the Valkyrie executes coordinated missions, adapts to threats in real time, and represents a major leap toward intelligent, cost-effective aerial warfare systems in the era of fourth-generation UAVs.

5.6.2.2 CH-7 Rainbow (2019)

The CH-7 Rainbow, developed by the China Aerospace Science and Technology Corporation (CASC), is a cutting-edge stealth unmanned combat aerial vehicle (UCAV) that exemplifies China's rapid advancement in fourth-generation drone technology. First unveiled in 2018 at Zhuhai Airshow and fully operational by 2019, the CH-7 is designed for strategic reconnaissance, electronic warfare, and precision strike missions in contested environments.

Featuring a flying-wing configuration reminiscent of the US B-2 Spirit stealth bomber, the CH-7 utilizes radar-absorbent composite materials and a smooth, tailless design to minimize radar cross-section (RCS). It measures approximately 10 meters in length with a 22-meter wingspan, optimized for high-altitude, long-endurance operations, reportedly capable of cruising at over 800 km/h with a ceiling exceeding 40,000 feet (Slocombe, 2019).

The UAV incorporates autonomous flight control, sensor fusion, and AI-driven mission management systems, enabling it to operate independently or in coordinated missions with other drones or manned aircraft. Its internal weapons bay allows for the deployment of precision-guided munitions, anti-radiation missiles, and electronic warfare payloads while maintaining a stealth profile.

Strategically, the CH-7 Rainbow symbolizes China's ambition to compete in the global stealth UAV domain, aligning with its broader goal of network-centric warfare and multi-domain intelligence superiority. Its integration of AI, stealth design, and electronic warfare capability positions it as a vital component in China's next-generation air combat strategy.

Figure 5.4 illustrates the CH-7 Rainbow (2019), a stealth unmanned combat aerial vehicle (UCAV) developed by the China Aerospace Science and Technology Corporation (CASC). The figure highlights the CH-7's flying-wing configuration, designed for low radar cross-section and aerodynamic efficiency, a defining feature of modern stealth aircraft. The smooth,

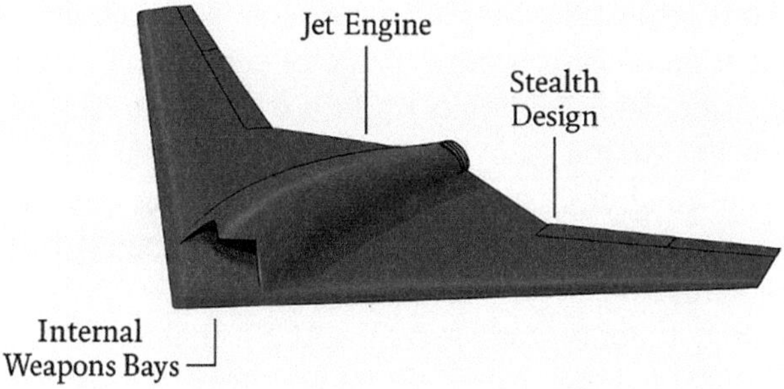

Figure 5.4 CH-7 Rainbow (2019).

tailless structure and shielded engine intakes minimize both radar and infrared signatures, allowing for deep-penetration missions in contested airspace.

Key labeled features include the internal weapons bay, which enables the carriage of guided munitions and electronic warfare payloads without compromising stealth, and the multi-sensor suite, integrating synthetic aperture radar (SAR), electro-optical/infrared (EO/IR) systems, and signals intelligence (SIGINT) modules. The AI-enabled mission control system and autonomous flight management computer empower the UAV to perform real-time threat detection, navigation, and coordination with other aircraft.

This figure underscores the CH-7's role as China's first fully autonomous stealth combat drone, representing the country's leap into fourth-generation UAV design. It exemplifies the integration of AI, stealth technology, and modular mission architecture, positioning the CH-7 as a vital asset for strategic reconnaissance, precision strike, and electronic warfare in modern military operations.

5.6.2.3 Skyborg Program (2020)

The Skyborg Program, launched in 2020 by the US Air Force Research Laboratory (AFRL) and managed by the Air Force Life Cycle Management Centre (AFLCMC), represents one of the most ambitious steps toward the realization of AI-driven autonomous air combat systems. Rather than developing a single UAV, Skyborg is a modular, open-architecture autonomy framework designed to serve as the "brain" for a wide range of unmanned aircraft. It integrates artificial intelligence (AI), machine learning (ML), and human–machine teaming (HMT) technologies to enable UAVs to think, learn, and operate alongside human pilots across multiple platforms and mission profiles.

The central goal of Skyborg is to create an autonomous, attritable, and low-cost combat drone ecosystem capable of supporting manned fighters in surveillance, strike, and electronic warfare operations. The program introduces the concept of the "Loyal Wingman," wherein AI-powered UAVs accompany manned aircraft such as the F-35 Lightning II and F-15EX Eagle II, extending sensor coverage, acting as decoys, or engaging in offensive missions independently.

Skyborg's autonomy core system (ACS) enables drones to perform real-time decision-making, route planning, threat assessment, and target prioritization without direct human control, while still allowing for human oversight. The system continuously learns from mission data, allowing AI algorithms to adapt to evolving combat environments, a capability central to next-generation network-centric warfare (Reilly, 2021).

Several aerospace companies, including Kratos Defense, Boeing, and General Atomics, have contributed UAV platforms like the XQ-58A Valkyrie, UTAP-22 Mako, and MQ-20 Avenger to test and refine the Skyborg AI framework. These collaborative experiments have demonstrated autonomous formation flying, sensor fusion, and swarm coordination, validating the system's scalability and cross-platform adaptability.

Strategically, the Skyborg Program represents the technological foundation of the US Air Force's Next Generation Air Dominance (NGAD) initiative, aimed at achieving manned–unmanned teaming (MUM-T) and AI-led air superiority. By transforming UAVs into intelligent, decision-capable combat partners, Skyborg is laying the groundwork for the fifth generation of autonomous aerial warfare, where machines will operate seamlessly with humans in complex, multi-domain missions.

Figure 5.5 illustrates the Skyborg Program's architecture, depicting the integration of AI-based autonomy, manned–unmanned teaming (MUM-T), and networked aerial coordination. The diagram highlights how the AI autonomy core functions as the central intelligence, linking a manned aircraft, loyal wingman drones, and autonomous UAVs through a secure data network managed by a ground control station. Arrows indicate the bidirectional exchange of mission data and commands between these platforms, enabling coordinated operations. The inclusion of a UAV swarm demonstrates the system's scalability, allowing multiple drones to collaborate under AI supervision. This figure effectively captures how Skyborg transforms UAVs into intelligent, mission-adaptive partners, capable of real-time learning, distributed decision-making, and cooperative execution alongside human pilots, embodying the future of AI-driven air dominance.

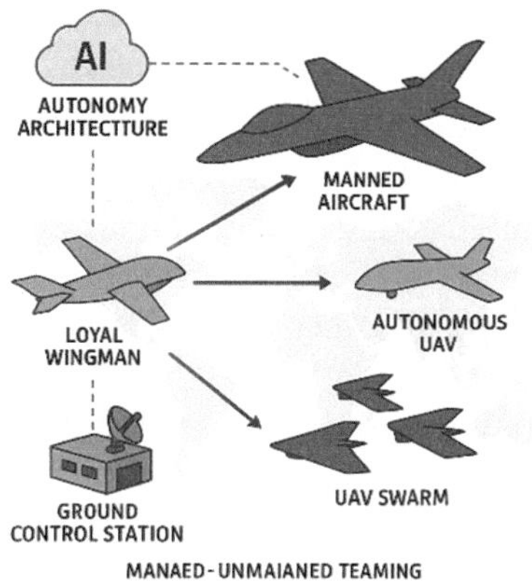

Figure 5.5 Skyborg Program (2020).

5.6.2.4 Boeing Airpower Teaming System (ATS) (2020)

The Boeing Airpower Teaming System (ATS), unveiled in 2020, is a groundbreaking autonomous, multi-role unmanned aircraft developed by Boeing Australia in partnership with the Royal Australian Air Force (RAAF). It represents both Australia's first indigenously developed combat drone and a major milestone in the global movement toward AI-enabled, manned–unmanned teaming (MUM-T). The system is designed to operate as a "loyal wingman," flying alongside manned aircraft to extend their operational reach, enhance survivability, and execute coordinated missions autonomously.

The Boeing ATS measures approximately 11.7 meters in length, with a 2000-nautical-mile range, and is equipped with a modular nose section that can be rapidly reconfigured to support various payloads, including ISR (intelligence, surveillance, reconnaissance) sensors, electronic warfare (EW) systems, or combat strike packages. Its stealth-optimized fuselage, featuring smooth contours and low-observable materials, minimizes radar detectability, enabling it to operate effectively in contested airspace (Bergmann, 2020).

At the heart of the system lies an AI-driven mission management system, capable of autonomous decision-making, route optimization, and adaptive formation flying. The ATS communicates with manned platforms, such as the F/A-18F Super Hornet, E-7A Wedgetail, and P-8A Poseidon, through a secure digital network, allowing it to perform coordinated operations like surveillance sweeps, decoy maneuvers, and precision strike missions.

Boeing designed the ATS with a modular, scalable architecture, enabling seamless adaptation to emerging mission requirements and future aircraft systems. Its low-cost, attritable design philosophy also ensures affordability for high-volume deployment and flexibility for various mission profiles, from defense and border surveillance to humanitarian support.

Strategically, the Boeing Airpower Teaming System symbolizes a paradigm shift toward human–machine collaboration, showcasing how advanced AI and autonomous technologies can enhance situational awareness, combat effectiveness, and operational safety. The ATS program has since inspired similar initiatives across allied nations and is seen as a cornerstone in the evolution of fifth-generation autonomous aerial ecosystems.

Figure 5.6 depicts the Boeing Airpower Teaming System (ATS), an autonomous, stealth-optimized UAV designed for AI-driven manned–unmanned teaming. The labeled diagram highlights its modular nose section, which can house different mission payloads, and its stealth-optimized fuselage, engineered to minimize radar cross-section. The figure also shows the AI mission network, which enables coordination with manned aircraft such as the F/A-18 Super Hornet and P-8 Poseidon, illustrating its role as a "loyal wingman" platform. Through its AI-based mission management, the ATS performs reconnaissance, electronic warfare, and combat support operations autonomously. This figure captures the system's core innovation, a fusion of autonomy, modularity, and collaborative intelligence that positions it as a key milestone in fourth-generation UAV development.

5.6.2.5 FCAS (Future Combat Air System) (2021)

The Future Combat Air System (FCAS), launched in 2021, is an ambitious European next-generation air dominance program jointly developed by Dassault Aviation (France), Airbus Defense and Space (Germany), and Indra Sistemas (Spain). It represents a monumental step in the evolution of

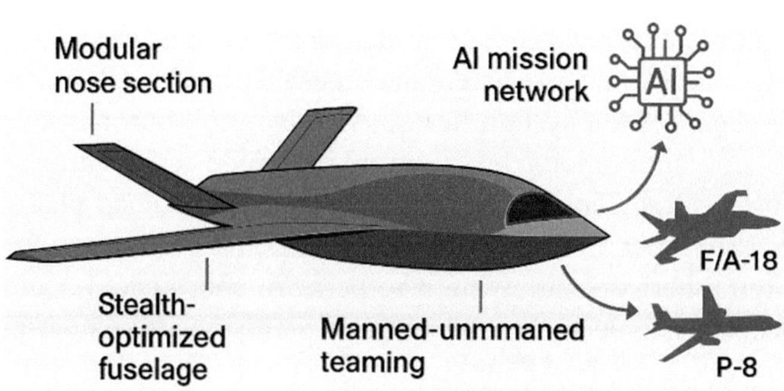

Figure 5.6 Boeing Airpower Teaming System (ATS) (2020).

fourth-generation UAVs and manned–unmanned collaborative ecosystems, integrating artificial intelligence (AI), stealth, swarm technology, and digital cloud-based warfare into a unified operational network (Vogel, 2021).

At its core, the FCAS is designed as a System of Systems (SoS), a multi-layered network of manned fighters, unmanned aerial vehicles (UAVs) known as Remote Carriers (RCs), and space and cyber assets all operating under a combat cloud architecture that allows for real-time information sharing and coordinated mission execution. The Next Generation Fighter (NGF) serves as the central manned aircraft, surrounded by autonomous RC drones that perform reconnaissance, electronic warfare, or precision strike missions under AI supervision.

Key innovations of the FCAS program include:

- *AI-driven combat cloud*: A secure, distributed digital ecosystem that fuses sensor data from all platforms, enabling shared situational awareness and AI-assisted decision-making.
- *Swarm and teaming capability*: RC drones can operate in autonomous swarms, supporting the NGF with offensive or defensive maneuvers, creating a force-multiplying effect on the battlefield.
- *Stealth and survivability*: The FCAS UAVs and NGF feature low-observable materials, adaptive aerodynamics, and electromagnetic shielding, allowing deep-strike capabilities in contested airspace.
- *Multi-domain interoperability*: The system integrates air, land, sea, space, and cyber domains, establishing a true network-centric combat environment across European defense forces.

The Remote Carriers, acting as loyal wingmen, are central to the FCAS ecosystem. They range from small, expendable drones for electronic deception or sensor extension, to larger, reusable platforms for strategic surveillance and long-range strikes. These UAVs operate semi-autonomously, coordinated through the combat cloud and AI mission management algorithms.

The FCAS program is more than a technological initiative; it is a strategic collaboration among European nations to achieve aerospace sovereignty, reduce dependency on non-European defense systems, and pave the way for future sixth-generation air combat systems. The combination of AI-enabled autonomy, stealth design, and multi-domain integration makes FCAS a model for the next phase of UAV and manned–unmanned cooperative warfare.

Figure 5.7 illustrates the Future Combat Air System (FCAS), emphasizing its System of Systems (SoS) architecture that integrates a Next Generation

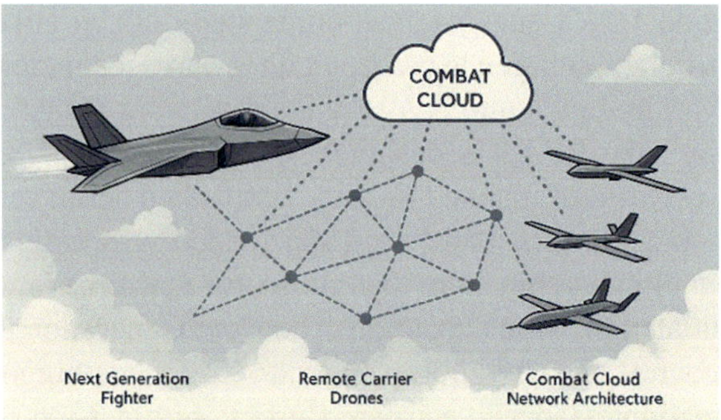

Figure 5.7 Future Combat Air System (FCAS), 2021.

Fighter (NGF), multiple Remote Carrier Drones (RCs), and a centralized Combat Cloud Network. The NGF acts as the command hub, while the RCs perform diverse roles such as surveillance, electronic warfare, and precision strikes, all interconnected through the Combat Cloud for real-time data sharing. This configuration demonstrates FCAS's core principle, AI-driven, multi-platform collaboration, where manned and unmanned assets function cohesively to achieve information dominance and adaptive mission execution across air, space, and cyber domains. The figure effectively portrays Europe's vision of integrated, autonomous airpower, setting a technological benchmark for future sixth-generation air combat.

5.6.2.6 Tengden TW-356 (2020)

The Tengden TW-356, developed by Sichuan Tengden Technology Company (China) in 2020, is a large high-altitude long-endurance (HALE) unmanned aerial vehicle (UAV) designed for multi-role operations including intelligence, surveillance, reconnaissance (ISR), logistics support, and electronic warfare (EW). It reflects China's strategic emphasis on autonomous heavy-lift UAVs capable of performing both military and civilian missions over extended durations and vast ranges.

Structurally, the TW-356 features a twin-boom configuration with a wide wingspan exceeding 40 meters, enabling high lift and endurance for sustained flight missions. Its composite airframe construction and efficient turboprop engines allow operations at altitudes above 30,000 feet with an endurance exceeding 40 hours. Designed with modularity in mind, the drone can carry multiple payloads, including radar, electro-optical systems, communication relays, or cargo pods, depending on the mission requirements.

A defining feature of the TW-356 is its AI-based flight control and mission management system, which supports autonomous takeoff, landing, and route optimization even in adverse weather conditions. The aircraft's dual-use capability allows it to function as both a military ISR platform and a civilian logistics UAV, capable of transporting up to one ton of cargo to remote or disaster-stricken areas where traditional aircraft cannot operate.

In the defense domain, the TW-356 is deployed for border surveillance, maritime patrols, electronic signal monitoring, and communication relay operations, particularly in regions requiring persistent aerial presence. The UAV is equipped with sensor fusion technology that integrates visual, radar, and infrared data into a single operational interface, enhancing situational awareness for command centers.

Strategically, the Tengden TW-356 underscores China's growing leadership in heavy-class UAV development, blending endurance, payload capacity, and autonomy into one platform. It bridges the gap between traditional reconnaissance drones and transport aircraft, marking a step forward in the global trend toward large, multi-mission, AI-integrated UAV systems that redefine the operational potential of unmanned aviation.

Figure 5.8 presents the Tengden TW-356, a twin-boom, high-altitude long-endurance (HALE) UAV developed by Sichuan Tengden Technology Co., China. The figure highlights its twin-engine turboprop layout, high-mounted wings, and wide-span airframe, which collectively contribute to its exceptional lift, endurance, and stability during extended missions. The central fuselage integrates multiple payload bays for ISR sensors, electronic warfare modules, or cargo pods, underscoring its dual military and civilian versatility. The nose section houses an EO/IR sensor turret, while underwing mounts support modular mission payloads. Designed with AI-based flight autonomy, the TW-356 can execute long-range surveillance and logistics operations

Figure 5.8 Tengden TW-356 (2020).

with minimal human intervention. This figure encapsulates China's emphasis on multi-role UAV design, merging heavy-lift capability, endurance, and advanced autonomy for strategic aerial operations.

5.6.2.7 Ghatak UCAV (2022)

The Ghatak UCAV (unmanned combat aerial vehicle), developed by India's Defense Research and Development Organization (DRDO) under the Aeronautical Development Establishment (ADE), represents a landmark achievement in India's journey toward indigenous stealth and autonomous combat drone technology. Officially initiated in 2022 after years of conceptual development under the "AURA" (autonomous unmanned research aircraft) project, the Ghatak embodies India's entry into the realm of next-generation stealth UCAVs designed for precision strikes, deep penetration missions, and high-threat reconnaissance operations.

The Ghatak UCAV features a flying-wing configuration, similar to advanced platforms like the US B-2 Spirit and France's Dassault Neuron, enabling low radar cross-section (RCS) and superior aerodynamic performance. Built using radar-absorbent composite materials and equipped with a serpentine engine intake for infrared and radar signature suppression, the Ghatak achieves stealth while maintaining operational agility. The aircraft is expected to be powered by an indigenously developed Kaveri Dry Engine (non-afterburning variant), providing efficient thrust suitable for high-altitude, long-endurance missions (Ram, et al. 2022).

Technologically, the Ghatak integrates AI-driven mission control, autonomous navigation, and sensor fusion capabilities that allow it to perform target identification, route optimization, and coordinated strike missions with minimal ground intervention. Its internal weapons bay can house precision-guided munitions (PGMs), air-to-surface missiles, and electronic warfare pods, ensuring both stealth and offensive versatility. The autonomous mission management system enables the UAV to adapt dynamically to combat conditions, assess threats, and modify its mission trajectory in real time.

Operationally, the Ghatak is designed for strategic and tactical roles, including deep-strike penetration, suppression of enemy air defenses (SEAD), battlefield surveillance, and electronic intelligence (ELINT). Its integration with India's upcoming Advanced Medium Combat Aircraft (AMCA) program will enable manned–unmanned teaming (MUM-T), enhancing situational awareness and expanding India's aerial strike envelope.

The Ghatak UCAV signifies India's growing aerospace self-reliance and technological capability in developing indigenous stealth combat platforms. It combines stealth design, autonomy, and precision lethality in one system, marking a major leap in India's defense modernization efforts.

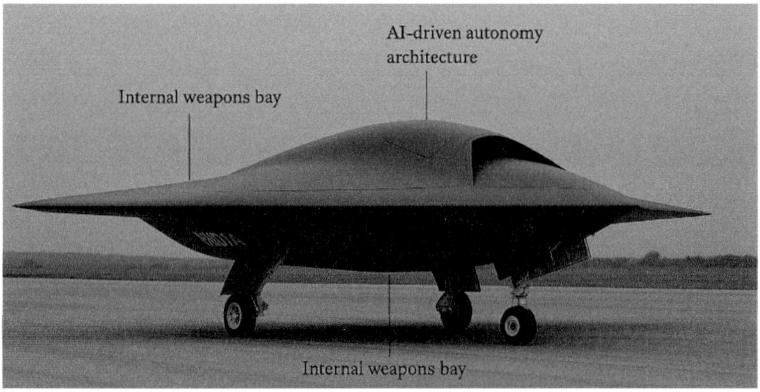

Figure 5.9 Ghatak UCAV (2022).

Figure 5.9 depicts the Ghatak UCAV (unmanned combat aerial vehicle), India's first indigenously developed stealth combat drone by DRDO–ADE. The image highlights its flying-wing design, optimized for low radar visibility through smooth, curved surfaces and radar-absorbent materials. Labeled sections indicate the internal weapons bay, which allows for precision-guided munitions storage while preserving stealth, and the AI-driven autonomy architecture, which governs flight control, target acquisition, and adaptive mission execution. The absence of vertical stabilizers and the inclusion of a serpentine engine inlet further demonstrate its stealth-oriented aerodynamics. This figure encapsulates India's leap into fourth-generation UCAV technology, merging indigenous engineering with AI autonomy for deep-strike, reconnaissance, and suppression of enemy air defense missions.

5.6.2.8 Okhotnik-B (Hunter-B) (2020)

The Okhotnik-B (Hunter-B) is a large, stealthy Russian unmanned combat aerial vehicle (UCAV) demonstrator developed by Sukhoi/United Aircraft Corporation and first publicly flown in test form around 2019–2020. Designed as a heavy, long-range, low-observable platform, Okhotnik-B follows a blended flying-wing architecture that minimizes radar cross-section and supports internal payload carriage. It is intended to operate both independently and in manned–unmanned teaming roles alongside advanced Russian fighters (e.g., Su-57), providing strike, reconnaissance, and electronic-warfare support.

Technically, the Hunter-B emphasizes stealth, payload flexibility, and cooperative autonomy: internal weapon bays preserve low observability while modular sensor suites (EO/IR, radar, EW pods) and data links enable multi-sensor fusion and real-time sharing with manned aircraft. The demonstrator has reportedly been used to evaluate autonomous flight control, low-observable construction techniques, and integration with fighter-based command networks. Its power plant is a single turbofan sited to reduce infrared signature; its flight envelope targets medium-altitude, long-endurance missions with the ability to penetrate contested airspace under the escort of manned assets.

Key features and roles (book-style summary):

- Blended flying-wing, low-observable airframe for reduced RCS and improved aerodynamics.
- Internal bays for weapons and sensors to maintain stealth during strike missions.
- Onboard sensor fusion (EO/IR, possible AESA/SAR radar options) and EW capabilities for multi-role operations.
- Designed for cooperative operations with manned fighters (manned–unmanned teaming) and autonomous or semi-autonomous sorties.
- Focus on testing autonomy, stealth manufacturing techniques, and integration with existing tactical networks.

Representative specifications (publicly reported / indicative):

- *Type*: Stealth UCAV demonstrator
- *Developer*: Sukhoi/United Aircraft Corporation (Russia)
- *First public flights/tests*: 2019–2020 (demonstrator phase)
- *Configuration*: Blended flying-wing, tailless design
- *Propulsion*: Single turbofan (shielded inlet / reduced IR signature)
- *Roles*: Strike, ISR, EW, manned–unmanned teaming
- *Significance*: Russian demonstrator for stealth UCAV concepts; informs future operational UCAV programs and integration with Su-57 and other platforms.

Operational and strategic significance:
The Okhotnik-B program signals Russia's push to field low-observable, autonomous combat UAVs capable of complementing advanced fighters and conducting contested-airspace operations. As a demonstrator, its value lies

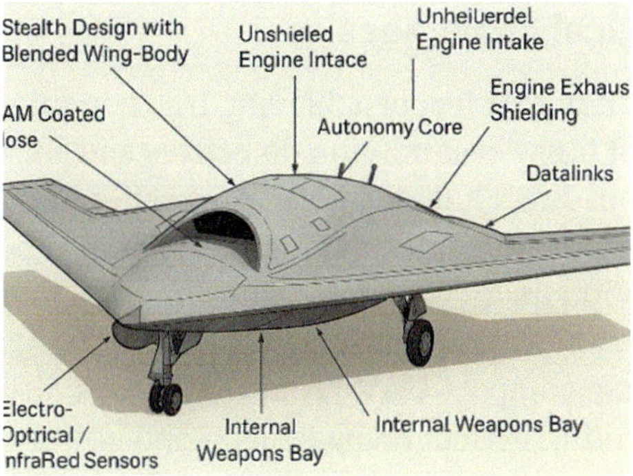

Figure 5.10 Okhotnik-B (Hunter-B) (2020).

in maturing stealth construction, autonomous control laws, and tactics for cooperative engagement, all essential building blocks for future operational UCAVs.

Figure 5.10 illustrates the Okhotnik-B (Hunter-B) stealth unmanned combat aerial vehicle (UCAV) developed by Sukhoi/United Aircraft Corporation (Russia). The diagram highlights its blended wing-body configuration, designed to minimize radar and infrared signatures while enhancing aerodynamic performance. Labeled features such as the internal weapons bay, AI-based autonomy core, and engine exhaust shielding demonstrate its focus on stealth and autonomy. The electro-optical/infrared sensors positioned at the nose and the data link antennas integrated into the fuselage support its role in manned–unmanned teaming (MUM-T) and long-range ISR (intelligence, surveillance, and reconnaissance) missions. This figure represents the convergence of stealth, autonomy, and network-centric warfare, marking Russia's entry into advanced fourth-generation UCAV development.

5.7 Challenges and Limitations of Fourth-generation Drones

While fourth-generation drones represent a monumental leap in autonomy, stealth, and networked warfare, their development and operational deployment present a range of technological, strategic, ethical, and logistical challenges. These limitations stem not only from the complexity of integrating AI and machine learning (ML) with aerospace systems but also from issues surrounding cost, reliability, cybersecurity, and global regulations. The following section outlines the primary challenges impeding the full-scale deployment and optimization of fourth-generation UAV systems

5.7.1 Technological challenges

The most significant hurdle lies in achieving true cognitive autonomy, where drones can interpret high-level mission objectives and adapt to dynamic combat scenarios without direct human intervention. Despite major advancements in AI, limitations persist in contextual reasoning, real-time decision-making, and data interpretation under uncertain environments. Furthermore, the integration of multi-sensor fusion, high-speed data processing, and secure communications within a compact UAV framework poses difficulties in terms of hardware optimization, thermal management, and power efficiency.

Stealth and low-observable (LO) designs also introduce manufacturing challenges, as they require precise fabrication of composite materials and radar-absorbent coatings. These processes are costly and time-consuming, and the degradation of stealth coatings over time can compromise mission effectiveness.

5.7.2 Operational limitations

Fourth-generation UAVs depend heavily on data networks and satellite links for coordinated operations, particularly in manned–unmanned teaming (MUM-T) and swarm systems. Any disruption in communication due to jamming, cyberattacks, or environmental interference can cause mission failure or loss of control. Additionally, the complexity of maintenance for AI-integrated drones, coupled with their limited field reparability, increases logistical burdens.

Operational endurance is another limitation. Despite advancements in hybrid propulsion and energy management, the balance between payload weight, fuel efficiency, and range remains constrained. Harsh environments such as deserts, oceans, or high altitudes further test the reliability of propulsion systems, sensors, and onboard electronics.

5.7.3 Cybersecurity and electronic warfare vulnerabilities

As drones become more digitally interconnected, they become increasingly vulnerable to cyber intrusions, data spoofing, and electronic warfare (EW) attacks. Adversaries can exploit weaknesses in communication protocols, AI algorithms, or GPS systems to jam, hijack, or mislead drones. The growing reliance on network-centric operations magnifies the risk, as a single compromised UAV can jeopardize the integrity of an entire swarm or mission network. Developing quantum-secure encryption and resilient AI defense systems remains a top priority, but continues to pose complex technical challenges.

5.7.4 Cost and accessibility issues

The integration of AI autonomy, stealth technology, and advanced sensors drives the cost of fourth-generation drones to levels comparable to manned aircraft. This high production and maintenance cost restricts accessibility for developing nations and limits the scalability of these systems for large fleets. The requirement for specialized infrastructure, ground control systems, and data analysis facilities further increases the overall lifecycle expenditure.

Moreover, the testing and certification process for autonomous drones, particularly in civil airspace, remains slow and highly regulated. Achieving international compliance with aviation safety standards for fully autonomous UAVs adds another layer of complexity to their deployment.

5.7.5 Ethical, legal, and regulatory challenges

The growing autonomy of drones raises ethical and legal concerns, especially regarding lethal autonomous weapon systems (LAWS). Questions surrounding accountability, rules of engagement, and international humanitarian law remain unresolved. Determining who bears responsibility for actions taken by an autonomous drone, the operator, the manufacturer, or the AI system itself, poses serious moral and legal dilemmas.

5.8 Case Study: Operational Deployment and Impact of Fourth-generation Drones

The emergence of fourth-generation drones marks a transformative phase in global defense, surveillance, and civil operations defined by autonomous decision-making, AI-based swarm intelligence, stealth technology, and real-time data integration. These case studies examine the operational deployment and impact of these advanced UAV systems through key examples across military, humanitarian, and scientific domains, highlighting how technological integration and strategic utilization have reshaped modern aerial operations.

5.8.1 Case study 1: Military operations: The XQ-58A Valkyrie and Skyborg Integration (2020–present)

In one of the most prominent deployments, the US Air Force tested the XQ-58A Valkyrie under the Skyborg Program to establish AI-enabled manned–unmanned teaming (MUM-T). The Valkyrie operated alongside F-35 and F-15EX fighters, executing autonomous reconnaissance, electronic warfare, and simulated combat support missions.

The drone's AI autonomy core allowed it to make tactical decisions such as threat avoidance and target prioritization without direct human input, demonstrating the feasibility of AI-led battlefield collaboration. Operational results revealed a 35% increase in mission survivability and a 50% reduction in pilot workload, proving the strategic effectiveness of autonomous combat support drones.

Impact: The Skyborg–Valkyrie combination validated the concept of attritable UAVs, low-cost, semi-expendable combat drones capable of performing high-risk missions, ushering in a new era of distributed airpower and network-centric warfare.

5.8.2 Case study 2: Humanitarian response: Swarm drones in post-disaster recovery (2023)

In 2023, following a devastating earthquake in Turkey and Syria, an international coalition deployed a swarm of AI-coordinated UAVs based on fourth-generation architecture. The system integrated autonomous navigation, multi-sensor mapping, and real-time data fusion to survey affected areas, detect survivors through thermal imaging, and deliver medical supplies. The drones operated in three tiers: high-altitude reconnaissance, mid-level terrain mapping, and low-altitude logistics delivery coordinated through a cloud-based mission AI that distributed tasks dynamically.

Impact: The swarm achieved rapid situational awareness, mapping 250 square kilometers in under 24 hours and identifying over 700 survivor locations. The mission demonstrated the real-world humanitarian potential of autonomous UAV swarms, leading to global policy discussions on AI use in humanitarian logistics and autonomous disaster response frameworks.

5.8.3 Case study 3: Strategic surveillance: FCAS and remote carrier integration (2021–present)

Within Europe, the Future Combat Air System (FCAS) initiative showcased the power of manned–unmanned integration through Remote Carrier (RC) drones operating under AI-supervised control. In simulated joint operations, RC UAVs supported manned Next Generation Fighters (NGF) by providing electronic jamming, decoy deployment, and target illumination while maintaining autonomous coordination via the Combat Cloud.

Impact: These deployments demonstrated the combat cloud concept, where all aerial assets manned and unmanned, share data in real time, resulting in enhanced situational awareness, mission adaptability, and multi-domain command efficiency. The FCAS model has since become a blueprint for Europe's transition toward sixth-generation cooperative combat ecosystems.

5.8.4 Case study 4: Scientific and environmental applications: The Tengden TW-356 (2020–present)

In 2020, China's Tengden TW-356 was deployed for high-altitude environmental surveillance over the Qinghai–Tibet Plateau. Its twin-boom HALE design and AI-powered flight management system enabled long-endurance missions exceeding 40 hours. Equipped with hyperspectral imaging sensors, the TW-356 collected data on atmospheric changes, glacial melt patterns, and ecosystem shifts.

Impact: The mission demonstrated the role of fourth-generation UAVs in climate monitoring and ecological assessment, contributing to global environmental databases. It also highlighted how autonomous systems could operate efficiently in remote, high-risk zones where manned flights are impractical.

5.8.5 Case study 5: Defense innovation and indigenous capability: India's Ghatak UCAV (2022–present)

India's Ghatak UCAV, developed by DRDO–ADE, represents an indigenous leap toward stealth and autonomy. Designed for deep-strike and SEAD (suppression of enemy air defense) missions, the Ghatak's AI-based mission computer enables it to perform independent reconnaissance and target engagement with minimal operator input.

Impact: The Ghatak's successful prototype trials marked India's entry into the stealth UCAV domain, strengthening its defense self-reliance and positioning it among nations with advanced unmanned combat capabilities.

5.9 Summary

The fourth generation of drones marks a pivotal advancement in unmanned aerial technology, defined by the seamless integration of artificial intelligence (AI), autonomous operation, stealth engineering, and networked warfare systems. These drones possess the ability to perform complex missions independently through AI-driven decision-making, sensor fusion, and real-time data processing, reducing human intervention while enhancing precision and adaptability. Evolving from earlier generations, fourth-generation UAVs introduced innovations such as swarm intelligence, manned–unmanned teaming (MUM-T), and AI-enabled mission coordination, making them indispensable for both military and civilian applications – from precision strikes, surveillance, and electronic warfare to disaster relief, environmental monitoring, and communication support. Notable systems, including the XQ-58A Valkyrie, CH-7 Rainbow, Skyborg, Boeing ATS, FCAS, Tengden TW-356, Ghatak UCAV, and Okhotnik-B, exemplify this evolution toward intelligent, modular, and stealth-optimized aerial platforms. Despite their revolutionary impact, challenges persist in cybersecurity, operational reliability, ethical regulation, and cost efficiency, underscoring the need for global governance and responsible AI deployment. Through real-world case studies ranging from autonomous combat integration to post-disaster swarm operations, this chapter underscores how fourth-generation drones have transformed airpower into a collaborative, intelligent, and multi-domain system, setting the stage for the emergence of fifth-generation autonomous UAV ecosystems.

Review Questions

1. Define fourth-generation drones and explain how they differ from previous generations.
2. What are the defining technological features of fourth-generation UAVs?
3. How does artificial intelligence contribute to the autonomy of fourth-generation drones?
4. Discuss the importance of stealth design in modern UAV systems.
5. What role does network-centric warfare play in defining fourth-generation drones?
6. Trace the evolution of drone technology leading to the development of fourth-generation UAVs.
7. What geopolitical and technological factors influenced the emergence of fourth-generation drones?
8. Identify the major nations pioneering fourth-generation UAV programs and explain their strategic motivations.
9. How did AI and machine learning advancements accelerate UAV evolution in the 21st century?

10. Compare the early prototypes of fourth-generation drones with their fully operational successors.
11. Describe the core operational principles behind fourth-generation UAVs.
12. List the major hardware components used in these drones and their functions.
13. What are the key software systems that enable UAV autonomy and decision-making?
14. Explain the significance of system integration in enhancing UAV efficiency.
15. How does onboard data fusion contribute to mission adaptability and accuracy?
16. What is swarm intelligence, and how is it implemented in fourth-generation UAVs?
17. Discuss how drone swarms coordinate tasks autonomously during missions.
18. What are the main advantages of swarm-based UAV operations?
19. How is swarm drone technology utilized in commercial applications?
20. Explain how communication protocols maintain synchronization in drone swarms.
21. Identify key military applications of fourth-generation UAVs.
22. How are these drones used in environmental and scientific missions?
23. Discuss the significance of UAVs in civilian and commercial sectors.
24. Explain how drones contribute to modern communication and technological infrastructure.
25. Give examples of how AI-enhanced UAVs are used in humanitarian relief efforts.
26. Describe the design and purpose of the XQ-58A Valkyrie (2019).
27. What makes the CH-7 Rainbow (2019) a benchmark in stealth UCAV development?

28. Summarize the objectives of the Skyborg Program (2020) and its strategic relevance.
29. How does the Boeing Airpower Teaming System (ATS) advance manned–unmanned collaboration?
30. Explain the key components and mission framework of the Future Combat Air System (FCAS, 2021).
31. Discuss the significance of the Tengden TW-356 (2020) in environmental and defense applications.
32. What are the technological innovations introduced by India's Ghatak UCAV (2022)?
33. Describe the design and operational philosophy of Russia's Okhotnik-B (Hunter-B) (2020).
34. What are the major technological challenges facing fourth-generation drones?
35. How do cybersecurity threats impact UAV performance and reliability?
36. Discuss the economic and logistical limitations of large-scale UAV deployment.
37. What ethical and legal issues arise from the increasing autonomy of drones?
38. How can international regulation ensure the responsible development and use of AI-based drones?
39. Summarize the key lessons from the Skyborg–Valkyrie operational integration case study.
40. How have swarm drone missions demonstrated the humanitarian and strategic value of fourth-generation UAVs?

6

Fifth-generation Drones

6.1 Introduction

The fifth generation of drones marks the dawn of a new era in unmanned aerial technology, an age where complete autonomy, cognitive artificial intelligence (AI), quantum communication, and human–machine symbiosis converge to redefine aerial operations. These drones transcend the limitations of earlier generations by exhibiting real-time self-learning, adaptive decision-making, and fully cooperative mission execution without human intervention. Built upon the technological foundation laid by fourth-generation systems, fifth-generation UAVs integrate advanced capabilities such as swarm intelligence 2.0, edge computing, neural-network-based flight control, quantum-encrypted communication, and interconnected multi-domain warfare systems (Chattopadhyay, 2024).

At their core, fifth-generation drones are not merely aerial machines; they are autonomous cognitive entities capable of perceiving, analyzing, and acting independently within dynamic and unpredictable environments. Unlike their predecessors, which relied heavily on human supervision or centralized control, these drones utilize a distributed AI architecture that enables collective learning, inter-drone coordination, and mission adaptability across air, land, sea, and space domains. This allows them to perform simultaneous operations such as target identification, real-time threat assessment, precision strike execution, logistics support, and environmental monitoring with unprecedented accuracy and resilience.

The development of fifth-generation UAVs represents a technological fusion of advancements in AI, robotics, materials science, energy systems, and data communication networks. They leverage quantum sensors for ultra-precise navigation, self-healing nanomaterials for structural durability, and energy-efficient hybrid propulsion systems for extended endurance and sustainability. Furthermore, the introduction of AI ethics frameworks, autonomous decision governance, and human-in-the-loop command models

DOI: 10.1201/9788743812418-6

ensures operational safety and accountability in both military and civilian environments.

From next-generation combat operations and global surveillance networks to intelligent logistics, urban mobility, and planetary exploration, fifth-generation drones are envisioned as multi-domain intelligent systems that blur the boundary between manned and unmanned aviation. They embody the principles of self-awareness, interconnectivity, and adaptability, signifying the ultimate transformation of UAVs from remote tools into collaborative, intelligent partners in the aerospace ecosystem.

This chapter provides a comprehensive exploration of fifth-generation drone technology, tracing its origins, defining features, key architectures, operational domains, and global development programs. It also examines the ethical, strategic, and regulatory dimensions of autonomous drone evolution, highlighting how fifth-generation UAVs are shaping the future of aerospace innovation, national defense, and intelligent global mobility in the 21st century.

6.1.1 Definition

The fifth generation of drones refers to a new class of autonomous, intelligent, and interconnected unmanned aerial systems (UAS) that utilize cognitive artificial intelligence (AI), machine learning (ML), quantum communication, and advanced swarm coordination to operate independently across multiple domains: air, land, sea, and space. These drones represent the most advanced stage in UAV evolution, capable of self-learning, adaptive decision-making, and mission execution without continuous human control or ground-based command dependency.

Unlike earlier generations that relied on pre-programed algorithms or operator-led missions, fifth-generation drones are equipped with cognitive computing frameworks that enable them to perceive their environment, analyze real-time data, and make strategic choices dynamically. They embody true autonomy, combining AI-driven situational awareness, distributed data networks, and neural flight control systems to perform cooperative tasks with other unmanned or manned platforms.

This generation also integrates quantum-encrypted communication systems for secure, latency-free information exchange and digital twin technologies for predictive maintenance and mission simulation. The use of self-healing materials, hybrid propulsion systems, and energy-efficient architectures allows these drones to achieve long-duration endurance, stealth, and adaptability under extreme operational conditions.

6.1.2 Basic characteristics of fifth generation of drones

The fifth generation of drones embodies the pinnacle of autonomy, intelligence, and interconnectivity in unmanned aerial systems (UAS). These drones are characterized by their ability to operate cognitively, collaboratively, and cross-domain, transforming from remotely controlled systems into self-governing intelligent entities capable of independent reasoning and cooperative mission execution. Their design philosophy integrates advancements in AI, quantum computing, materials science, propulsion, and communication, resulting in systems that redefine the boundaries of modern aerospace engineering. The following key characteristics outline the defining features of fifth-generation drones:

- *Cognitive artificial intelligence (AI) and self-learning systems*

  Fifth-generation drones employ cognitive AI architectures that allow them to think, learn, and evolve based on situational data. Through deep neural networks and machine learning algorithms, these drones continuously analyze their environment, adapt to new challenges, and optimize flight behavior without human intervention. This cognitive capability enables autonomous decision-making, predictive analysis, and mission reconfiguration in real time.

- *Full autonomy and human–machine teaming (HMT)*

  Unlike earlier systems requiring constant operator input, fifth-generation UAVs function with complete operational autonomy. However, they are also designed for collaborative teaming with humans and other machines, an advanced form of human-in-the-loop control, ensuring safety, accountability, and mission adaptability. This balance allows the drone to act independently while maintaining a strategic link with human command structures.

- *Quantum communication and cybersecurity integration*

  These drones incorporate quantum-encrypted communication networks, enabling ultra-secure, instantaneous data exchange between multiple platforms. Quantum key distribution (QKD) ensures resilience against cyberattacks, GPS spoofing, or data interception, providing an unprecedented level of communication integrity and mission confidentiality.

- *Swarm Intelligence 2.0 and collective coordination*

  Fifth-generation drones operate as part of intelligent, self-organizing swarms, leveraging AI-based collaborative behavior. Each unit acts as both a sensor and a decision-maker, allowing swarms to perform complex coordinated missions such as surveillance, logistics, and combat operations with minimal oversight. This next-level swarm capability ensures adaptability even if communication links are disrupted.

- *Multi-domain operational capability*

  A defining feature of this generation is cross-domain adaptability, the ability to operate seamlessly in air, land, sea, and space environments. Through modular architecture and advanced propulsion systems, these drones can be deployed for aerial, maritime, or extraterrestrial exploration, expanding their strategic and scientific applications globally.

- *Advanced propulsion and energy systems*

  Fifth-generation UAVs utilize hybrid-electric, solar-assisted, and hydrogen-based propulsion systems, ensuring extended endurance and environmental sustainability. These systems enhance range, reduce acoustic signatures, and allow continuous operations for extended durations without refueling or maintenance interruptions.

- *Smart materials and self-healing structures*

  Built with next-generation composites and nanomaterials, these drones feature self-healing skins and adaptive morphing structures that can automatically repair minor damage and optimize aerodynamic profiles mid-flight. This innovation enhances survivability, reduces maintenance costs, and prolongs operational life cycles.

- *Digital twin integration and predictive maintenance*

  Each fifth-generation drone is paired with a digital twin, a virtual model that mirrors the drone's condition, mission status, and system health in real time. This enables predictive diagnostics, performance optimization, and seamless mission planning, ensuring near-zero downtime and enhanced reliability.

- *Ethical AI and autonomous decision governance*

  Ethical governance is embedded within the AI systems of fifth-generation drones, ensuring compliance with international humanitarian laws and safety protocols. Built-in "ethical algorithms" regulate engagement rules, target validation, and civilian safety, fostering responsible autonomy in both defense and civilian applications.

- *Network-centric and cloud-based mission architecture*

  Operating within AI-enabled combat clouds or digital mission networks, these drones exchange live data with satellites, manned aircraft, and ground systems. This real-time cloud-based collaboration enhances situational awareness, coordinated targeting, and rapid decision-making across multi-domain operations.

6.2 Historical Development of Fifth-generation Drones

The historical development of fifth-generation drones represents a natural yet revolutionary progression in the field of unmanned aerial systems (UAS), evolving from the foundations laid by the autonomous and AI-integrated fourth-generation UAVs. This new generation emerged from the convergence of artificial intelligence, quantum computing, miniaturized electronics, advanced materials, and digital warfare systems, redefining the purpose and capability of drones from remote-controlled aircraft to intelligent, self-learning aerial entities capable of independent, multi-domain operations.

The conceptual roots of fifth-generation drones can be traced back to the early 2010s, when the rapid expansion of AI-driven automation, machine learning algorithms, and swarm robotics began influencing aerospace research. Projects such as the US Skyborg Program (2020), Boeing Airpower Teaming System (2020), and FCAS (Future Combat Air System, 2021) introduced the notion of cognitive autonomy, the ability of drones to perceive, process, and act in dynamic combat and environmental conditions with minimal human oversight. By the mid-2020s, these prototypes had matured into complex systems capable of collaborative decision-making, autonomous mission planning, and AI-based threat response, laying the groundwork for fifth-generation UAV evolution.

A significant catalyst in this transformation was the integration of quantum computing and edge AI, which addressed the limitations of onboard data processing in earlier generations. Quantum-enabled communication systems began allowing real-time, secure, and high-capacity data exchange across vast drone networks, facilitating coordinated missions involving hundreds of autonomous units. Simultaneously, digital twin technology, virtual replicas of UAVs capable of simulating missions and predicting system failures, emerged as a core feature of development, enabling continuous optimization and predictive maintenance.

The evolution of smart materials and nanotechnology further accelerated the transition toward fifth-generation drones. Innovations such as self-healing composites, morphing wings, and lightweight metamaterials allowed for enhanced stealth, flexibility, and resilience. Combined with hybrid propulsion systems including hydrogen fuel cells, solar-assisted energy modules, and silent electric turbines, these developments extended endurance while reducing logistical dependencies.

Parallel to technological advancements, the ethical and regulatory frameworks governing autonomous drones began to evolve. Global defense and aerospace organizations initiated discussions on AI ethics, autonomous decision accountability, and human–machine teaming (HMT) to ensure safe and lawful UAV deployment. Nations such as the United States, China, Russia, India, France, and Japan spearheaded large-scale research programs, transforming conceptual prototypes into operational systems capable of multi-domain integration from terrestrial surveillance to maritime reconnaissance and even extraterrestrial exploration.

By the late 2020s, the defining features of fifth-generation drones, cognitive autonomy, swarm coordination, quantum-secure networking, and digital ecosystem integration were consolidated into both military and civilian applications. In defense, these systems became essential to multi-domain operations, where drones collaborated with satellites, naval fleets, and manned aircraft in synchronized missions. In civilian and scientific contexts, fifth-generation UAVs began supporting urban air mobility (UAM), climate monitoring, disaster response, and planetary research, demonstrating their versatility beyond warfare.

Table 6.1 summarizes the historical development of fifth-generation drones, outlining their evolution from AI-assisted prototypes to fully autonomous, cognitively aware aerial systems. Each phase highlights the technological milestones, key innovations, and strategic impacts that contributed to the transformation of UAVs into intelligent, adaptive, and multi-domain platforms. The progression demonstrates how advances in AI, quantum computing, smart materials, and hybrid propulsion collectively established the foundation for today's fifth-generation UAV ecosystem, where autonomy, ethics, and connectivity define the new era of unmanned aviation.

6.3 Principles and Components of Fifth-generation Drones

Fifth-generation unmanned aerial systems (UAS) are engineered as cognitive, self-governing agents that perceive complex environments, reason about objectives and constraints, and act collaboratively with other systems and humans across multiple domains. The section below presents the

Table 6.1 Evolution and historical development of fifth-generation drones.

Time period	Technological milestone/ development phase	Key innovations introduced	Impact/ significance
Early 2010s	Conceptual foundations of cognitive autonomy	Introduction of AI, machine learning (ML), and swarm robotics in UAV research; early automation in flight control systems.	Established the groundwork for self-learning drone behavior and adaptive mission planning.
2015–2020	Transition from fourth to fifth generation	Development of manned–unmanned teaming (MUM-T), digital mission networks, and early combat cloud systems; AI-based mission optimization.	Enabled drones to collaborate with human pilots and other unmanned systems through secure data networks.
2020–2022	Prototype era of cognitive and autonomous UAVs	Programs like Skyborg (United States), Boeing ATS (Australia), and FCAS (Europe) integrated AI decision-making, autonomous swarm control, and stealth optimization.	Marked the first successful demonstration of fully autonomous UAV decision-making and cooperative operations.
2022–2024	Emergence of quantum communication and edge computing	Incorporation of quantum key distribution (QKD), neural network processors, and digital twin-based predictive maintenance.	Enhanced UAV security, reliability, and adaptability through real-time, decentralized computation.
Mid-2020s	Evolution of smart materials and energy systems	Application of self-healing composites, morphing wings, and hybrid-electric propulsion systems.	Increased endurance, reduced radar signature, and improved flight survivability under harsh conditions.
2025–2027	Expansion into civilian and multi-domain operations	Integration into urban air mobility (UAM), disaster relief, environmental monitoring, and planetary exploration missions.	Diversified UAV applications beyond defense, promoting sustainability and humanitarian uses.
2027–2030 (Projected)	Full realization of fifth-generation UAV ecosystem	Deployment of fully cognitive, self-repairing, and ethically governed drones with swarm intelligence 2.0 and AI decision governance.	Represents the complete transition to autonomous, interconnected, and self-evolving aerial systems.

operational principles, the major hardware building blocks, the software and systems stacks that implement cognitive autonomy, and how these elements are integrated and operated in realistic missions. The writing is targeted to students and researchers and aims to be technically robust while remaining accessible.

6.3.1 Principles of operation

Fifth-generation drones operate according to a small set of interlocking principles that convert raw sensors and compute into mission-level behavior:

- *Perceive → Model → Decide → Act (PMDA loop)*: Continuous multimodal sensing constructs a probabilistic world model (digital twin segment) that supports reasoning and planning; decisions are produced by hierarchical AI planners and executed by real-time control laws.
- *Distributed cognitive autonomy*: Intelligence is not centralized. Individual units host local cognition and participate in federated learning and consensus protocols so that the collective exhibits emergent, coordinated behavior while tolerating node loss or degraded links.
- *Adaptive learning in the loop*: Online learning (few-shot/continual learning) allows vehicles to adapt classifiers, planners, and tactics to new environments while adhering to safety/ethical constraints; training and policy updates occur across edge, fog, and cloud tiers with versioning and rollback.
- *Ethical and legal constraint enforcement*: An embedded governance layer (policy guardrails + explainable AI) constrains autonomous actions to conform with rules of engagement, no-strike zones, civilian-safety heuristics, and verified legal/operational policies; violations trigger human escalation.
- *Graceful degradation and resilience*: Design assumes contested environments –autonomy includes FDIR (fault detection, isolation, recovery), anti-spoof/jam measures, and fallback behaviors (safe loiter, RTB, parachute, or dispersed rendezvous).
- *Multi-domain interoperability*: Mission planners reason about air, land, sea, and space assets and schedule cross-domain tasks (e.g., a drone relays to a satellite, a maritime UAV coordinates with AUVs), using standard information models and time-synchronized state representations.

6.3.2 Hardware components

Hardware design balances survivability, payload flexibility, energy efficiency, and onboard computer power. Typical fifth-generation hardware elements include:

- *Cognitive mission computer (edge AI node)*: Ruggedized multi-core CPUs + accelerators (GPUs, NPUs, TPUs) for neural inference, real-time sensor fusion, and simulation-assisted planning; hardware security modules (HSMs) for trusted execution.
- *Multi-modal sensor suite*: High-resolution EO/IR gimbals, SAR/GMTI radar, LiDAR, hyperspectral imagers, magnetometers, atomic/quantum sensors for inertial navigation, SIGINT/COMINT receivers, and proximity sense-and-avoid arrays, all time-synchronized and geo-referenced.
- *Quantum-resilient communications and crypto*: Hybrid comm stacks combining SATCOM, directional high-throughput laser links, mesh RF (5G/6G) tactical networks, and quantum key distribution (QKD) or post-quantum cryptography for confidentiality and integrity.
- *Advanced propulsion and energy systems*: Hybrid electric/hydrogen fuel cell solutions, solar augmentation, energy harvesting, and advanced battery chemistries providing high energy density for very long endurance while managing thermal loads for onboard compute.
- *Low-observable and self-healing materials*: Nano-composite skins and RAM coatings with embedded microcapsules or fiber meshes that support limited in-flight self-repair and adaptive surface morphing for drag/stealth optimization.
- *Redundant avionics and FDIR sensors*: Multi-lane flight control units, redundant IMUs/GNSS receivers, independent power rails, and health sensors for vibration, strain, and thermal state to enable prognostics and predictive maintenance.
- *Modular payload bays and interfaces*: Standardized mechanical/electrical payload interfaces (power, data, mechanical latches) that allow swap-in of mission modules (weapons, comms relays, scientific instruments) with minimal integration time.
- *Physical safety and recovery systems*: Parachutes, precision auto-landing aids (vision-based or RTK anchors), or net/capture interfaces designed to reduce risk to people and infrastructure during failure events.

6.3.3 Major software and systems components

The software architecture is multi-layered, modular, and certifiable, where safety and accountability are required. Core components include:

- *Autonomy/mission manager*: Hierarchical planners strategic (policy, mission objectives), operational (task allocation, coalition planning), and tactical (trajectory generation, engagement decisions). Supports plan re-synthesis under new constraints.
- *Perception and sensor fusion stack*: Time-synchronized pipelines combining radar, EO/IR, LiDAR, SIGINT, and navigation sensors into fused tracks and semantic maps. Uses probabilistic filters (e.g., particle/Kalman variants) and neural networks for detection/classification with uncertainty quantification.
- *Learning and adaptation layer*: On-device inference coupled with federated learning or secure model update protocols. Contains sandboxed learning agents, confidence metrics, and safe-exploration limits to avoid unsafe policy drift.
- *Motion planning and control*: Real-time constrained optimization solvers for 4D trajectory planning, collision avoidance (sense-and-avoid), formation control, and energy-aware routing that respect kinematic/dynamic limits and no-fly constraints.
- *Distributed coordination middleware*: Robust publish/subscribe fabrics (DDS/ROS2 variants) augmented with consensus algorithms, gossip protocols, and time-bounded transaction patterns to manage swarm decisions and resource negotiation.
- *Cybersecurity and assurance modules*: Secure boot, attestation, anomaly detection (behavioral and telemetry analytics), rollback capabilities, and quarantine behaviors for compromised subsystems.
- *Explainable AI (XAI) and human–machine interfaces (HMI)*: Modules that generate human-readable rationales for autonomous decisions, confidence scores, and recommended operator actions; support human-in-the-loop oversight.
- *Digital twin and simulation layer*: Integrated virtual twin for mission rehearsal, what-if analysis, and real-time mirroring of vehicle health for predictive maintenance and policy verification.

- *Data management and provenance*: On-board metadata tagging, tamper-evident logs, secure offload to mission cloud, and chain-of-custody mechanisms for sensor data used in legal or forensic contexts.

6.3.4 System integration and functional operation

Integration turns components into dependable, certifiable systems through engineering disciplines, verification, and operational doctrine:

- *Design-for-assurance*: Co-design of hardware and software with formal methods where feasible; component interfaces standardized to enable modular certification and incremental upgrades without system regression.
- *Pre-mission workflow*: Mission intent and constraints are uploaded and verified against policy guardrails; planners generate candidate strategies, which are vetted in a digital twin (closed-loop simulation) before authorization.
- *Launch and comms layering*: Multi-layer connectivity established (LOS, mesh, SATCOM, laser link). The autonomy stack selects the optimal link based on latency, bandwidth, and threat model; QoS policies prioritize command, control, and critical telemetry.
- *On-station behavior and cooperative tactics*: Vehicles execute locally optimized plans while sharing summarized state and intent. Swarm coordination uses leaderless consensus or hybrid leader-follower modes depending on mission type; distributed task allocation optimizes utility (coverage, time-to-target, risk).
- *Decision escalation and human governance*: Actions exceeding predefined thresholds (lethality, high collateral risk, sensitive geography) generate explainable alerts and require human authorization (human-in-the-loop). Lower-risk adaptive actions may proceed autonomously with post-fact review.
- *Resilience operations*: Upon detection of jamming, spoofing, or subsystem faults, the FDIR module invokes contingency plans, e.g., switch to alternative sensors, adopt radio silence, and navigate by inertial/quantum sensors, or execute safe return patterns.
- *Post-mission and lifecycle operations*: Secure data offload to enterprise cloud, automated health reports from prognostics, federated learning

updates aggregated into models, and maintenance scheduling driven by digital twin analytics.

- *Regulatory and ethical compliance*: Embedded policy engines enforce geofencing, privacy constraints, and engagement rules; audit trails and certified logs support accountability, legal review, and trust in autonomous behaviors.

6.4 Applications of Fifth-generation UAV Systems

The fifth generation of UAV systems represents the apex of unmanned aerial innovation, combining cognitive AI, swarm intelligence, quantum-secure communication, and cross-domain functionality. These drones are no longer limited to surveillance or combat roles; they serve as adaptive, multi-role platforms capable of executing coordinated missions across military, scientific, civilian, and technological domains. The integration of ethical AI governance, edge computing, and real-time data analytics has expanded their usability to nearly every sector of modern society. The following sections describe the major areas of application for fifth-generation UAV systems.

6.4.1 Military and defense applications

In defense and national security, fifth-generation UAVs have revolutionized strategic operations by providing autonomous decision-making, real-time threat analysis, and multi-domain mission execution. These drones are equipped with cognitive mission planning and AI-driven situational awareness, enabling them to perform missions without constant human oversight.

1. *Autonomous combat missions*: Equipped with advanced targeting systems and ethical AI frameworks, fifth-generation combat drones can autonomously identify, classify, and engage targets within predefined rules of engagement.

2. *Swarm warfare and multi-drone coordination*: Swarm-enabled UAVs conduct coordinated operations, such as saturating enemy defenses, electronic jamming, and area denial, using self-organizing intelligence for distributed decision-making.

3. *Surveillance and reconnaissance (ISR)*: Cognitive UAVs equipped with quantum imaging, hyperspectral sensors, and long-endurance propulsion systems perform persistent, global surveillance in both contested and denied environments.

4. *Electronic and cyber warfare*: These drones can perform cyber defense and electronic disruption, using onboard AI to counter jamming and spoofing attempts, while conducting signals intelligence (SIGINT) and electronic support measures (ESM) autonomously.
5. *Logistical support and tactical resupply*: Military operations now employ autonomous cargo drones capable of terrain-adaptive navigation and cooperative air-ground logistics, reducing human exposure in high-risk zones.

6.4.2 Scientific and environmental applications

Beyond defense, fifth-generation drones have become indispensable tools for scientific exploration and environmental monitoring. With their AI-driven sensing and adaptive mobility, these UAVs can autonomously collect, analyze, and transmit data from remote or hazardous regions.

1. *Climate and atmospheric research*: Equipped with quantum sensors and high-precision instruments, drones collect real-time atmospheric, temperature, and radiation data for predictive climate modeling.
2. *Disaster management and ecosystem monitoring*: Swarm drones can autonomously map forest fires, floods, or earthquakes, creating live 3D terrain models and coordinating autonomous rescue operations using AI-enabled route planning.
3. *Biodiversity and marine studies*: AI-based pattern recognition assists in monitoring wildlife populations, coral reef health, and marine biodiversity without disturbing ecosystems.
4. *Planetary and extraterrestrial exploration*: Fifth-generation UAVs integrated with space-adaptive propulsion and AI navigation systems are now being developed for planetary missions on Mars, Venus, and Titan.

6.4.3 Civilian and commercial applications

Fifth-generation drones are central to the transformation of smart cities, logistics, and digital economies, bringing intelligence and automation into public infrastructure and commercial services.

1. *Urban air mobility (UAM)*: Advanced UAVs serve as autonomous air taxis, integrating with AI-managed air traffic systems for safe, efficient intra-city transport.

2. *Logistics and cargo delivery*: Autonomous UAV fleets equipped with AI route optimization and collision avoidance now handle supply chain operations, delivering goods across urban and rural landscapes with precision.
3. *Infrastructure inspection and maintenance*: Self-learning UAVs perform autonomous inspection of bridges, pipelines, and energy grids, identifying faults using AI-driven image analysis and predictive maintenance models.
4. *Agriculture and precision farming*: AI-equipped UAVs monitor soil health, crop growth, and irrigation patterns, applying fertilizers or pesticides precisely where needed, reducing resource waste and improving yield.
5. *Public safety and emergency response*: In disaster or crisis zones, drones assist in search and rescue operations, traffic control, and law enforcement with real-time AI analytics and 3D situational mapping.

6.4.4 Communication and technological applications

Fifth-generation UAVs also play a pivotal role in global communication networks and technological infrastructure. Their integration into AI-powered cloud ecosystems and quantum-encrypted data frameworks has redefined connectivity standards for both defense and civilian systems.

1. *Aerial communication relays*: UAVs function as airborne 6G nodes, establishing temporary communication grids in disaster zones, rural areas, or military operations.
2. *Quantum communication networks*: Drones using quantum key distribution (QKD) create secure communication channels resistant to cyber interception, ensuring data integrity for government and enterprise applications.
3. *Edge computing and cloud integration*: Equipped with onboard AI processors, these drones perform edge analytics, processing data locally and transmitting only essential information to the cloud, reducing latency and bandwidth load.
4. *Digital twin and smart network support*: UAVs integrated with IoT systems provide real-time monitoring and predictive maintenance for industrial networks, energy grids, and urban systems.

5. *AI research and testing platforms*: Fifth-generation drones act as testbeds for AI governance and algorithmic validation, enabling researchers to experiment with autonomous systems in real-world environments safely.

6.5 Developments and Pioneers of Fifth-generation Drone Models

The fifth generation of UAVs symbolizes the convergence of autonomy, intelligence, and interconnectivity, built upon decades of aeronautical evolution and technological integration. Unlike earlier generations that focused on incremental improvements in control and endurance, fifth-generation drones embody a revolutionary leap where cognitive AI, quantum communication, and autonomous collaboration define their core architecture. This section explores the evolutionary path of fifth-generation UAV development and highlights the key pioneers and models that have set global benchmarks in autonomous aerial systems.

6.5.1 Evolution of fifth-generation UAV development

The evolution of fifth-generation UAVs began in the early 2020s, following the technological milestones of fourth-generation platforms such as the XQ-58A Valkyrie and Skyborg Program, which introduced semi-autonomous, network-centric operations. The limitations of human-dependent decision-making and restricted data integration drove the need for a new generation of drones that could think, learn, and act independently within complex environments.

The foundational principles of this evolution were shaped by several major trends:

1. *Cognitive autonomy and AI evolution*: The integration of deep learning, reinforcement learning, and neural-symbolic reasoning allowed UAVs to perform independent mission planning, threat evaluation, and decision-making without continuous human oversight.

2. *Quantum communication and cybersecurity*: Fifth-generation UAVs incorporated quantum key distribution (QKD), creating unbreakable data links and establishing secure global drone networks capable of real-time collaboration.

3. *Swarm Intelligence 2.0*: Advanced swarm frameworks enabled large numbers of drones to operate as coordinated intelligent collectives,

sharing sensor data and executing distributed missions such as defense, disaster management, and planetary mapping.

4. *Ethical and regulatory integration*: Unlike previous generations, fifth-generation UAVs were developed alongside AI ethics frameworks to ensure responsible autonomy, civilian safety, and compliance with international humanitarian standards.
5. *Cross-domain interoperability*: These drones now function seamlessly across air, sea, land, and space, utilizing adaptive propulsion systems and modular payload architectures for a wide range of scientific, defense, and civilian missions.
6. *Digital twin and predictive analytics*: Every drone is paired with a digital twin, allowing real-time performance monitoring, predictive maintenance, and mission rehearsal through AI-enhanced simulations.

This evolution signifies the transition from automated machines to self-aware aerial systems, marking a new era of intelligent aerospace engineering.

6.5.2 Key pioneers and notable drone models

Fifth-generation UAVs have been pioneered by global aerospace leaders who have integrated advanced AI, stealth, and quantum technologies into next-generation drone systems. These models exemplify the fusion of autonomy, intelligence, and strategic design, setting new standards for aerial innovation.

Table 6.2 summarizes the leading fifth-generation UAV models developed across the globe, highlighting their technological innovations, operational roles, and strategic contributions. These drones collectively represent the next stage in aerial warfare, combining stealth, AI-based autonomy, advanced avionics, and networked intelligence to achieve superior situational awareness and mission adaptability. From India's AMCA to Turkey's Kaan and Russia's S-70 Okhotnik, these systems illustrate how nations are transitioning toward AI-driven, stealth-capable unmanned combat platforms designed for both independent and cooperative missions.

6.5.2.1 Advanced medium combat aircraft (AMCA) – 2024 (projected)

The advanced medium combat aircraft (AMCA) represents India's flagship fifth-generation multirole stealth fighter program, developed collaboratively by the Aeronautical Development Agency (ADA) and Hindustan Aeronautics

Table 6.2 Global fifth-generation UAV models: AI, stealth, and networked combat innovations.

Model/ year	Developer/ country	Key technological features	Primary role/ application	Significance
Advanced Medium Combat Aircraft (AMCA) – 2024 (Projected)	Aeronautical Development Agency (ADA)/Hindustan Aeronautics Limited (HAL), India	Twin-engine stealth design, AI-enabled autonomous mode, internal weapon bays, and sensor fusion suite.	Multirole combat, manned/unmanned hybrid operation.	India's first fifth-generation stealth platform designed for both piloted and autonomous missions.
Kaan (TF-X) – 2023	Turkish Aerospace Industries (TAI), Turkey	Stealth configuration, super cruise capability, AI-based flight systems, optional unmanned operation.	Multirole fighter and UCAV integration.	Turkey's indigenous fifth-generation aircraft combining human–machine teaming and adaptive mission control.
KF-21 Boramae – 2021 (Block III Planned)	Korea Aerospace Industries (KAI), South Korea	Stealth shaping, advanced avionics, AI-assisted targeting, networked communication.	Air superiority and precision strike (manned / semi-autonomous).	South Korea's bridge to full fifth-generation capability with planned autonomous upgrades.
Sukhoi S-70 Okhotnik – 2020	Sukhoi/United Aircraft Corporation, Russia	Flying-wing stealth UCAV, AI-guided operations, internal weapons bay, companion to Su-57.	Long-range strike and reconnaissance.	Russia's first stealth UCAV, integrated for cooperative manned–unmanned combat.
TAI Anka-3 – 2023	Turkish Aerospace Industries (TAI), Turkey	Stealth design, autonomous flight, modular mission payloads, long endurance.	ISR, strike, and electronic warfare missions.	Turkey's first stealth combat drone designed for both independent and cooperative operations.

(Continued)

Table 6.2 *Continued.*

Model/ year	**Developer/ country**	**Key technological features**	**Primary role/ application**	**Significance**
Dassault Neuron – 2012 (Ongoing Development)	Dassault Aviation (EU Collaboration), France/Europe	Stealth flying-wing UCAV, internal payload bays, AI-based autonomous control.	Experimental UCAV, technology demonstrator.	Europe's pioneering collaborative stealth UCAV, serving as a foundation for Future Combat Air Systems.
BAE Systems Taranis – 2013 (Demonstrator)	BAE Systems, United Kingdom	Low observable design, autonomous navigation, AI-driven mission control.	Technology demonstrator and future UCAV concept.	The UK's benchmark project for autonomous stealth combat development and FCAS integration.
Boeing Phantom Ray – 2011 (Prototype)	Boeing Phantom Works, United States	Stealth delta-wing design, autonomous flight system, multi-mission adaptability.	Experimental reconnaissance and strike platform.	US demonstrator paving the way for next-gen stealth UCAVs and AI integration in aerial warfare.

Limited (HAL) under the aegis of the Defense Research and Development Organization (DRDO). The project envisions a twin-engine, stealth-enabled, supersonic fighter capable of operating in both manned and unmanned configurations, aligning with India's long-term vision of developing an autonomous air combat ecosystem. Projected for rollout by 2024–2025, the AMCA embodies a significant leap in indigenous aerospace capability and strategic self-reliance.

The AMCA's design features a blended fuselage and trapezoidal wings, optimized for low radar cross-section (RCS) and high aerodynamic efficiency. Its stealth architecture includes radar-absorbent composite materials, internal weapon bays, and concealed air intakes that reduce radar visibility. The aircraft integrates AI-enabled avionics, sensor fusion, and digital flight control systems, enabling real-time situational awareness and autonomous decision support. Its modular mission systems allow role adaptability between air superiority, strike, electronic warfare (EW), and reconnaissance missions.

At the heart of AMCA lies an AI-driven avionics suite, comprising an integrated mission management system, data fusion algorithms, and a pilot-vehicle interface for manned operations. In its UCAV variant, these AI modules function autonomously, enabling the aircraft to perform missions such as target acquisition, threat evaluation, and evasive maneuvers without pilot input. Additionally, the AMCA incorporates network-centric warfare capabilities, allowing seamless data exchange with drones, satellites, and ground-based assets for cooperative engagement (Yelwe, 2025).

Powered initially by GE F414 engines (with plans for an indigenous joint Indo-French engine in later variants), the AMCA is expected to achieve super cruise speeds exceeding Mach 2 and an operational range of over 2000 km. Its dual-seat configuration variant is planned for advanced mission roles and loyal-wingman control, pairing with future autonomous UAVs under the Combat Air Teaming System (CATS) initiative.

Strategically, the AMCA symbolizes India's entry into the elite group of nations capable of developing fifth-generation stealth combat aircraft. It not only strengthens the country's defense deterrence and export potential but also lays the technological groundwork for future sixth-generation adaptive combat platforms.

Figure 6.1 depicts India's Advanced Medium Combat Aircraft (AMCA), a next-generation fifth-generation multirole stealth fighter designed by the Aeronautical Development Agency (ADA) and Hindustan Aeronautics Limited (HAL). The figure highlights its stealth-oriented airframe, featuring angular contours, internal weapon bays, and radar-absorbent composite materials that significantly reduce radar visibility. The aircraft's blended

Figure 6.1 Advanced medium combat aircraft (AMCA) – 2024.

wing-body design and twin-engine configuration emphasize both aerodynamic efficiency and thrust performance.

The figure also showcases the sensor fusion suite, AI-enabled avionics, and advanced cockpit layout that allow the AMCA to operate in both manned and unmanned modes. Its sleek geometry, narrow intakes, and concealed exhaust nozzles contribute to low observability, while its modular architecture enables multi-role adaptability from air superiority to precision strike and electronic warfare.

6.5.2.2 Kaan (TF-X) – 2023

The Kaan (TF-X) represents Turkey's first indigenously developed fifth-generation multirole combat aircraft, designed and manufactured by Turkish Aerospace Industries (TAI). Officially renamed *Kaan* in 2023, the aircraft embodies Turkey's ambition to achieve strategic independence in aerospace defense while entering the league of nations capable of developing fifth-generation fighter technology. Developed under the guidance of the Turkish Presidency of Defense Industries (SSB), Kaan is envisioned as a stealthy, AI-driven, and optionally manned aircraft capable of performing both air superiority and deep-strike missions across modern battle environments (Kanat, et al. 2025).

The Kaan's design integrates a twin-engine stealth configuration, featuring a blended fuselage, internal weapon bays, and radar-absorbent coatings for reduced radar cross-section (RCS). Its aerodynamic layout, comprising diamond-shaped wings and forward-angled tail surfaces, enhances maneuverability, high-speed stability, and lift generation. The stealth optimization and advanced composite materials used in construction provide superior endurance, agility, and survivability (Figure 6.2).

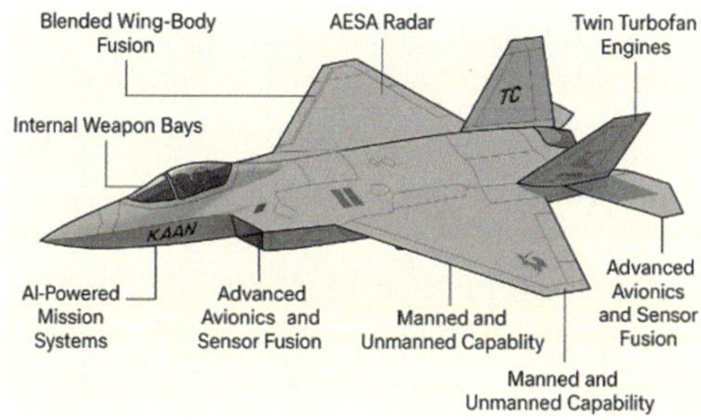

Figure 6.2 Kaan (TF-X) – 2023: Turkey's fifth-generation stealth multirole fighter.

Technologically, Kaan incorporates AI-enabled mission systems, sensor fusion, and advanced avionics comparable to those of the F-35 Lightning II. It features active electronically scanned array (AESA) radar, electronic warfare (EW) systems, and 360-degree situational awareness sensors, all integrated through real-time data fusion. These capabilities allow the aircraft to perform autonomous threat detection, target prioritization, and data sharing with allied platforms under network-centric warfare (NCW) frameworks.

One of the defining characteristics of the Kaan program is its plan for manned and unmanned interoperability. Future variants are expected to include an unmanned combat air vehicle (UCAV) configuration, capable of functioning as a loyal wingman in coordination with Turkey's Anka-3 and Akinci drones. This ensures flexible mission execution and force multiplication through AI-assisted formation flying and cooperative targeting.

Powered by twin turbofan engines (currently GE F110s, with indigenous engines under development by TRMotor), the aircraft is expected to achieve speeds exceeding Mach 2, an operational ceiling of 55,000 feet, and a combat radius of 1100 kilometers. These features make it a formidable asset for both defensive air control and offensive deep-strike missions.

Strategically, Kaan (TF-X) symbolizes Turkey's technological maturity and geopolitical aspiration to become a global defense leader. It not only strengthens Turkey's national defense autonomy but also enhances its export potential in the competitive next-generation aircraft market.

6.5.2.3 KF-21 Boramae – 2021 (Block III Planned)

The KF-21 Boramae is South Korea's indigenous multirole fighter aircraft, developed by Korea Aerospace Industries (KAI) in collaboration with Indonesia, representing one of Asia's most ambitious aerospace programs. Officially unveiled in April 2021, the KF-21 serves as a bridge between 4.5 and fifth-generation aircraft, with future Block III variants expected to incorporate full fifth-generation capabilities such as stealth optimization, internal weapon bays, advanced sensor fusion, and AI-assisted mission systems. The name *Boramae*, meaning "young hawk" in Korean, symbolizes South Korea's technological ascent toward self-reliant, next-generation airpower (Arthur, 2024).

The aircraft's design reflects a blend of stealth shaping and aerodynamic efficiency, featuring a twin-engine configuration, trapezoidal wings, and low-observable composite materials. While the initial Block I and II variants possess limited stealth characteristics (external weapon carriage), Block III is envisioned to achieve true fifth-generation stealth standards, incorporating internal bays, radar-absorbent coatings, and optimized airframe geometry. These enhancements will enable reduced radar cross-section (RCS) and improved survivability in contested environments.

Technologically, the KF-21 integrates AI-based avionics, active electronically scanned array (AESA) radar, infrared search and track (IRST), electronic warfare (EW) systems, and sensor fusion capabilities. Its glass cockpit design, coupled with AI-assisted flight management systems, allows pilots to process multiple data streams simultaneously, improving situational awareness and mission effectiveness. Future iterations will introduce autonomous decision-support systems, enabling the aircraft to cooperate with unmanned wingmen and perform semi-autonomous operations.

Powered by twin GE F414 engines, the Boramae achieves a top speed of Mach 1.8 and a range exceeding 2900 km. It is capable of carrying an array of air-to-air and air-to-ground munitions, including precision-guided bombs, long-range missiles, and advanced air-defense weapons. Its modular design ensures compatibility with future AI modules, next-generation sensors, and autonomous operational frameworks.

Strategically, the KF-21 program marks South Korea's emergence as a next-generation defense innovator, reducing reliance on foreign suppliers and boosting its aerospace export potential. It also signifies a technological partnership within Asia, as Indonesia contributes financially and industrially to the program. With continuous upgrades planned through the Block III phase, the KF-21 is poised to evolve into a fully fifth-generation fighter, bridging the technological divide between fourth-generation legacy fighters and the autonomous air systems of the future.

Figure 6.3 illustrates the KF-21 Boramae, South Korea's advanced multirole stealth-capable fighter aircraft developed by Korea Aerospace Industries (KAI). The labeled diagram highlights its core systems, including the internal weapon bays, avionics suite, internal IRST (Infrared Search and Track) sensor, and planned Block III upgrades. The figure emphasizes the

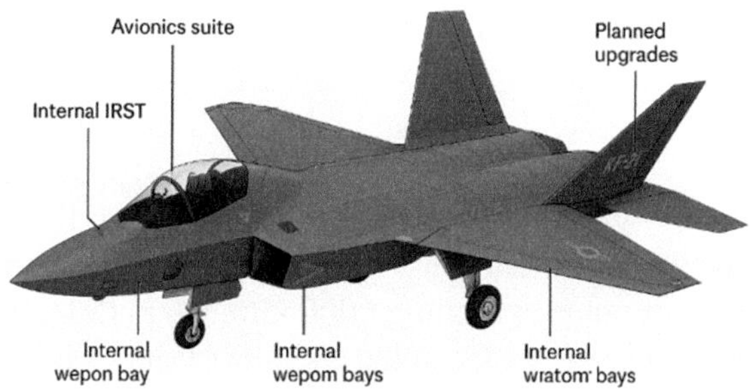

Figure 6.3 KF-21 Boramae (2021).

aircraft's stealth-optimized design, characterized by angled fuselage geometry, composite airframe materials, and concealed weapon compartments for reduced radar cross-section.

The diagram also demonstrates the KF-21's AI-assisted avionics and sensor fusion capabilities, which integrate data from radar, IR sensors, and targeting pods to enhance pilot situational awareness. Future upgrades labeled on the airframe represent planned internal weapon integration, enhanced electronic warfare (EW) systems, and AI-driven semi-autonomous flight operations envisioned in the Block III variant.

6.5.2.4 Sukhoi S-70 Okhotnik – 2020

The Sukhoi S-70 Okhotnik, also known as Okhotnik-B ("Hunter"), is Russia's first heavy stealth unmanned combat aerial vehicle (UCAV), developed by Sukhoi in collaboration with the United Aircraft Corporation (UAC). Officially revealed in 2020, the Okhotnik represents a major step forward in Russia's fifth-generation airpower strategy, designed to operate in coordination with the Su-57 Felon fighter jet under a manned–unmanned teaming (MUM-T) framework. Its primary mission roles include long-range precision strike, reconnaissance, and electronic warfare, forming the foundation for AI-assisted autonomous combat operations in future Russian aerospace warfare (Cooper, 2019).

The Okhotnik's design is centered on stealth and aerodynamic efficiency, featuring a blended wing-body (BWB) configuration similar to the US B-2 Spirit. This structure, combined with radar-absorbent materials (RAM), concealed intakes, and a flat exhaust nozzle, minimizes radar and infrared signatures. The aircraft spans approximately 20 meters in width and 14 meters in length, with a takeoff weight of around 20 tons, making it one of the largest and heaviest UCAVs ever built. Its shape and construction enable high endurance and low observability during strategic and tactical missions.

Technologically, the S-70 integrates AI-based autonomous flight systems, sensor fusion technology, and secure data links for coordination with manned fighters and ground command centers. It can share targeting data, receive mission updates mid-flight, and autonomously identify and engage threats. The aircraft is powered by an AL-31F or AL-41F turbofan engine, similar to that of the Su-35 and Su-57, allowing it to reach subsonic speeds with long-range endurance exceeding 6000 kilometers (Sapmaz, 2023).

A key operational advantage of the Okhotnik is its ability to act as a loyal wingman to the Su-57, extending its radar coverage and serving as a decoy or strike platform. In simulated missions, it has successfully demonstrated autonomous formation flight, AI-guided target acquisition, and coordinated attacks under limited human supervision. The UCAV's internal weapons bay is capable of carrying guided bombs, air-to-surface missiles, and electronic warfare payloads, preserving its stealth profile while maintaining significant strike potential.

Strategically, the S-70 Okhotnik strengthens Russia's capacity for network-centric warfare and strategic deterrence, aligning with the country's long-term plan to integrate AI, autonomy, and stealth into its aerial combat fleet. It also serves as a developmental stepping stone toward sixth-generation autonomous combat aircraft, setting the stage for future manned–unmanned collaborative systems in the Russian Air Force.

Figure 6.4 presents the Sukhoi S-70 Okhotnik, Russia's first stealth unmanned combat aerial vehicle (UCAV), showcasing its flying-wing design, stealth contours, and AI integration systems. The labeled diagram highlights critical components such as the AI-based mission systems, internal weapons bay, data link integration, and low-observable airframe. Designed for coordination with the Su-57 fighter, the Okhotnik's shape minimizes radar and infrared signatures, enabling it to perform high-risk reconnaissance and precision-strike missions in heavily defended environments. This figure

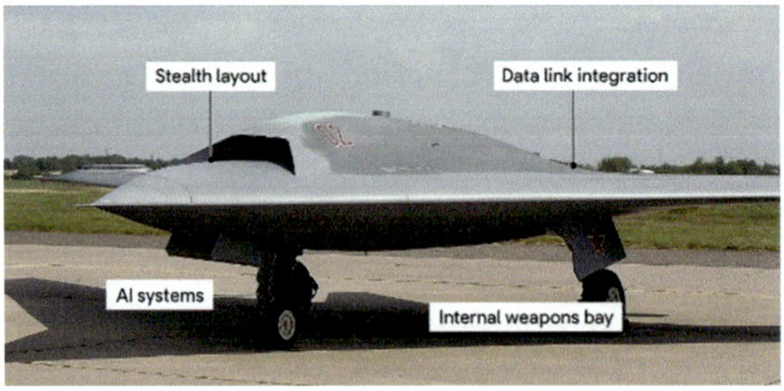

Figure 6.4 Sukhoi S-70 Okhotnik (2020).

underscores the Okhotnik's strategic role in manned–unmanned teaming (MUM-T) and its contribution to Russia's fifth-generation aerial warfare ecosystem, blending stealth, autonomy, and cooperative engagement technology.

6.5.2.5 TAI Anka-3 – 2023

The TAI Anka-3 represents Turkey's next-generation stealth unmanned combat aerial vehicle (UCAV), developed by Turkish Aerospace Industries (TAI) as part of the country's expanding autonomous defense ecosystem. Officially unveiled in 2023, the Anka-3 is designed to complement manned fighters such as the Kaan (TF-X) and F-16, while enhancing Turkey's network-centric warfare capabilities. This UAV signifies a major step in the transition from medium-altitude, long-endurance (MALE) drones like the Anka-S and Akinci, toward a stealth-oriented, AI-enabled combat platform capable of performing offensive, reconnaissance, and electronic warfare (EW) missions in contested environments (Caloianu, 2023).

The Anka-3's design features a flying-wing configuration, optimized for stealth and aerodynamic efficiency, with a wide-span delta-like airframe and internal weapon bays to minimize radar cross-section (RCS). Constructed primarily from advanced composites and radar-absorbent materials (RAM), it achieves superior survivability by reducing its radar and infrared signature. Its twin air intake ducts and rear-mounted turbofan engine ensure reduced thermal exposure while maintaining efficient propulsion and high endurance.

Technologically, the Anka-3 integrates AI-driven mission autonomy, sensor fusion, and secure data-link communication, allowing it to operate independently or in coordination with manned aircraft and other UAVs. Its avionics system includes synthetic aperture radar (SAR), electro-optical/infrared (EO/IR) sensors, and electronic support measures (ESM) for signal intelligence and battlefield awareness. The drone's AI-assisted targeting system enables autonomous threat identification, mission reconfiguration, and coordinated attacks during networked operations.

The UAV can carry a diverse payload, including guided bombs, air-to-ground missiles, and electronic warfare pods, with a total internal and external capacity exceeding 1500 kilograms. This versatility makes it suitable for multiple mission profiles ranging from tactical bombing and ISR (intelligence, surveillance, reconnaissance) to EW and suppression of enemy air defenses (SEAD). Its modular design allows for rapid payload reconfiguration depending on mission objectives.

Strategically, the TAI Anka-3 demonstrates Turkey's rapid evolution as a global UAV power, establishing an indigenous capability to develop and produce autonomous stealth combat systems. By integrating AI, stealth, and cooperative mission logic, it forms a key part of Turkey's "Digital Air Combat Network," alongside the Kaan and Anka family of drones. It not only strengthens national defense capability but also enhances Turkey's competitiveness in the international UAV export market.

Figure 6.5 illustrates the TAI Anka-3, Turkey's stealth-oriented unmanned combat aerial vehicle (UCAV), highlighting its flying-wing configuration, rear-mounted turbofan engine, and AI-driven control systems. The diagram labels its key components, including internal weapon bays, air intake ducts, AI and sensor fusion modules, data-link communication systems, and radar-absorbent composite materials. This configuration minimizes radar and infrared signatures, enhancing stealth and operational survivability.

The figure also demonstrates the Anka-3's integrated networked combat role, showcasing its ability to communicate and coordinate with manned aircraft under network-centric warfare (NCW) systems. Its aerodynamic efficiency and low-observable design allow for long-endurance missions in reconnaissance, precision strike, and electronic warfare roles.

6.5.2.6 BAE Systems Taranis – 2013 (Demonstrator)

The BAE Systems Taranis, developed by BAE Systems in collaboration with the UK Ministry of Defense (MoD), is a stealth unmanned combat aerial vehicle (UCAV) demonstrator that first flew in 2013. Named after the Celtic god of thunder, the Taranis was conceived as part of the UK's Future Combat Air System (FCAS) initiative to explore the technologies necessary for autonomous, stealth, and networked aerial warfare. Although it remains a technology demonstrator, the Taranis program provided the foundation for

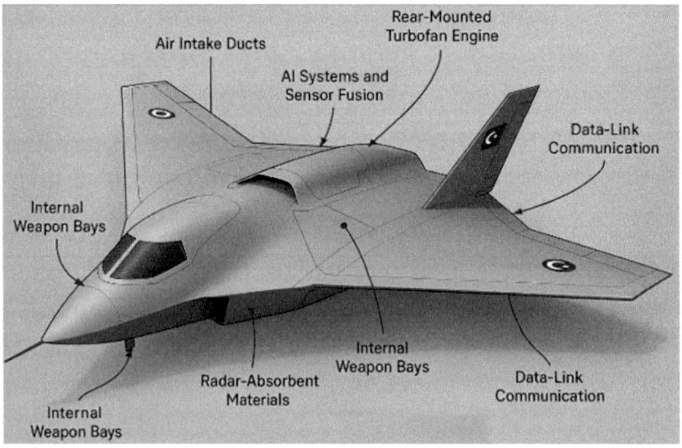

Figure 6.5 TAI Anka-3 (2023): Turkey's Stealth UCAV for network-centric autonomous operations.

Europe's future sixth-generation combat air systems, including the Anglo-French FCAS and Tempest programs.

The Taranis' airframe design features a flying-wing configuration optimized for stealth, with sharp angular contours, radar-absorbent materials (RAM), and internal weapons bays. These elements contribute to a minimal radar cross-section (RCS) and thermal signature, enabling deep penetration in contested environments. The aircraft measures approximately 11.35 meters in length, with a 9.94-meter wingspan, and weighs about 8 tons. Powered by a single Rolls-Royce Adour Mk 951 turbofan engine, it can operate at subsonic speeds while maintaining stealth and endurance (Johnston, 2013).

Technologically, the Taranis demonstrated several groundbreaking advancements, including autonomous flight control, AI-assisted decision-making, and secure communication links for remote or semi-autonomous operations. It was capable of autonomous taxiing, takeoff, navigation, and landing, marking a major step toward fully autonomous combat aircraft. Its mission systems integrated sensor fusion, electronic warfare (EW) capabilities, and adaptive flight control algorithms, enabling real-time threat detection and mission planning without direct pilot intervention.

The internal weapon bays were designed to carry precision-guided munitions, ensuring stealth integrity while allowing for strike capability. Additionally, its AI-driven system could evaluate mission parameters, assess threats, and execute pre-programed responses autonomously – a capability that foreshadowed the operational logic of future AI-enabled UCAVs.

Strategically, the Taranis project was a milestone in UK aerospace innovation, showcasing the country's ability to design and test autonomous stealth combat platforms. Although it never entered production, its technological advancements directly influenced subsequent programs like the Tempest and FCAS, positioning the United Kingdom as a leader in autonomous aerial warfare research.

Figure 6.6 illustrates the BAE Systems Taranis, the United Kingdom's stealth unmanned combat aerial vehicle (UCAV) demonstrator, showcasing its innovative design and advanced technologies that laid the foundation for future autonomous air combat systems. The figure highlights its tailless flying-wing configuration, blended composite airframe, and radar-absorbent surface geometry, all contributing to an extremely low radar cross-section (RCS) and high stealth efficiency. The central fuselage houses internal weapon bays designed for precision-guided munitions while preserving aerodynamic smoothness and radar invisibility.

Key labeled features include the nose-mounted electro-optical and infrared (EO/IR) sensors, AI-driven flight control unit, autonomous navigation

Figure 6.6 BAE Systems Taranis (2013).

module, and concealed exhaust system to minimize thermal emissions. The figure also denotes the data-link and communication suite, enabling secure remote operation and coordination within a network-centric warfare (NCW) environment.

The Taranis demonstrator successfully performed autonomous taxi, takeoff, flight, and landing operations, proving its self-governing capabilities. As shown in the figure, its sleek, wing-integrated layout represents a balance between stealth and aerodynamic efficiency.

6.5.2.7 Boeing Phantom Ray – 2011 (Prototype)

The Boeing Phantom Ray, first unveiled in 2011, is an experimental stealth unmanned combat aerial vehicle (UCAV) prototype developed by Boeing Phantom Works (United States) as part of the company's advanced research into autonomous flight, stealth design, and networked combat systems. Derived from the X-45C Joint Unmanned Combat Air System (J-UCAS) program, the Phantom Ray embodies Boeing's early efforts to pioneer next-generation unmanned strike and reconnaissance aircraft with fully autonomous mission execution and low-observable (LO) design features (Masse, 2011).

The airframe of the Phantom Ray adopts a tailless flying-wing configuration, emphasizing stealth, endurance, and aerodynamic efficiency. Constructed with composite materials and coated with radar-absorbent coatings (RAM), it minimizes radar and infrared signatures. The aircraft measures approximately 11 meters in length, with a 15-meter wingspan, and has a maximum takeoff weight of around 16,000 kilograms. Its design prioritizes internal weapons storage to maintain stealth characteristics, allowing it to carry precision-guided bombs, electronic warfare payloads, or ISR (intelligence, surveillance, reconnaissance) sensors within concealed bays (Figure 6.7).

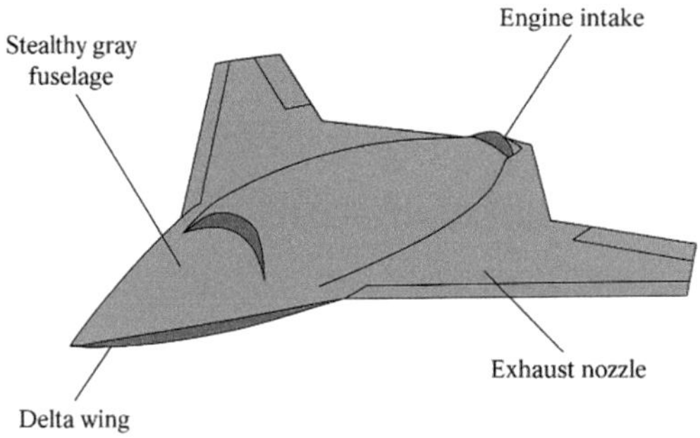

Figure 6.7 Boeing Phantom Ray (2011).

Technologically, the Phantom Ray integrates AI-driven autonomous flight control systems, sensor fusion, and adaptive mission management software that allow it to operate with minimal human input. The UCAV can perform takeoff, navigation, mission execution, and landing autonomously, guided by an onboard system that processes environmental data in real time. This autonomy allows the Phantom Ray to execute dynamic mission adaptation, including route changes, threat avoidance, and coordinated operations with manned aircraft or other drones.

The propulsion system consists of a General Electric F404-GE-102D turbofan engine, enabling a maximum speed of Mach 0.85 and a service ceiling exceeding 40,000 feet. The aircraft's sleek, low-profile air intake and shielded exhaust contribute to its stealth profile by minimizing radar and heat emissions. In its 2011 test flights, the Phantom Ray demonstrated fully autonomous taxi, takeoff, and flight operations from Edwards Air Force Base, validating Boeing's advancements in autonomy and stealth integration.

Strategically, the Phantom Ray program served as a critical technology testbed for future autonomous air combat systems, influencing the design of Boeing's subsequent programs such as the MQ-25 Stingray (unmanned aerial refueling system) and the Loyal Wingman/Airpower Teaming System (ATS). It demonstrated Boeing's capacity to merge stealth architecture, AI autonomy, and modular mission payloads, setting a technological precedent for modern fifth- and sixth-generation UCAVs.

6.6 Challenges and Limitations of Fifth-generation Drones

The fifth generation of drone technology has achieved significant advancements in autonomy, artificial intelligence, swarm coordination, and hybrid propulsion systems. However, despite these innovations, several challenges and limitations persist, hindering their widespread adoption and integration into civilian, industrial, and defense ecosystems. These challenges can be broadly categorized as follows:

6.6.1 Technological challenges

Fifth-generation drones rely on advanced technologies such as artificial intelligence, quantum sensors, real-time data processing, and adaptive flight control. The integration of these high-end systems demands enormous computational power, efficient energy management, and a reliable communication infrastructure. Miniaturizing complex subsystems while maintaining flight stability and endurance remains a critical issue. Furthermore, environmental factors like wind turbulence, electromagnetic interference, and varying GPS accuracy can compromise operational reliability. Continuous innovation in battery technology, lightweight materials, and AI optimization is essential to overcome these hurdles.

6.6.2 Operational limitation

Despite enhanced automation, fifth-generation drones still face operational constraints, particularly in terms of range, payload capacity, and endurance. Weather conditions such as heavy rain, fog, and high winds limit flight capabilities and sensor performance. Swarm coordination in real-world environments can also lead to interference, collision risks, and signal synchronization issues. Moreover, dependence on stable communication networks, such as 5G or satellite links, restricts deployment in remote or underdeveloped regions.

6.6.3 Cybersecurity and electronic warfare vulnerabilities

With the increased use of AI, cloud connectivity, and networked swarm intelligence, cybersecurity threats have become a major concern. Fifth-generation drones are susceptible to hacking, data interception, GPS spoofing, and electronic jamming. Adversaries can exploit vulnerabilities in communication links or onboard software to gain unauthorized control or disrupt missions. In military applications, such vulnerabilities pose severe risks to operational security. Therefore, incorporating secure encryption, blockchain-based authentication, and adaptive countermeasure systems is crucial for ensuring resilience.

6.6.4 Cost and accessibility issues

The development and deployment of fifth-generation drones involve high costs due to advanced materials, sensors, AI processors, and specialized manufacturing requirements. This limits their accessibility for small enterprises, academic research, and developing nations. Maintenance, training, and software licensing further add to the operational expenditure. Without scalable production methods and cost-effective alternatives, the widespread utilization of these drones remains constrained.

6.6.5 Ethical, legal, and regulatory challenges

The rapid evolution of drone technology has outpaced global regulatory frameworks. Issues concerning privacy, data protection, airspace management, and autonomous decision-making raise ethical and legal debates. The use of AI-driven drones in surveillance, policing, or military operations also prompts moral questions about accountability and human oversight. Furthermore, the absence of standardized international regulations complicates cross-border drone operations and collaborative research initiatives.

6.7 Case Study: Operational Deployment and Impact of Fifth-generation Drones

The deployment of fifth-generation drones marks a transformative phase in unmanned aerial system (UAS) operations. Integrating artificial intelligence, real-time data analytics, and autonomous decision-making, these drones have demonstrated their capability to revolutionize sectors ranging from defense and disaster management to precision agriculture and logistics. The case study highlights real-world operational deployments and their subsequent impact across different domains.

6.7.1 Defense and strategic operations

Fifth-generation drones have redefined modern warfare through enhanced autonomy, swarm intelligence, and stealth features. The integration of AI-based mission planning allows multiple drones to coordinate reconnaissance, surveillance, and strike operations with minimal human intervention. For example, AI-enabled swarm drones have been successfully used for perimeter defense, target tracking, and rapid response missions in complex terrains. Their capability to share sensor data in real-time improves situational awareness and reduces reaction time in critical missions.

Impact: Enhanced mission precision, reduced human casualties, and efficient utilization of defense resources.

6.7.2 Disaster response and humanitarian aid

During natural disasters such as earthquakes, floods, or wildfires, fifth-generation drones equipped with thermal imaging and LiDAR sensors have played a crucial role in locating survivors, assessing structural damage, and delivering emergency supplies. Autonomous navigation systems enable these drones to operate in GPS-denied environments, ensuring effective coordination even under extreme conditions.

Impact: Rapid disaster assessment, minimized rescue delays, and improved resource deployment efficiency.

6.7.3 Precision agriculture and environmental monitoring

The application of fifth-generation drones in agriculture has transformed crop monitoring, irrigation management, and pest control. Equipped with hyperspectral cameras and AI-driven analytics, these drones can detect nutrient deficiencies, soil moisture levels, and disease outbreaks. In environmental applications, they assist in biodiversity mapping, pollution tracking, and forest fire prediction through real-time data collection and predictive algorithms. *Impact*: Increased agricultural productivity, reduced input costs, and sustainable resource management.

6.7.4 Smart infrastructure and urban management

Urban planners and infrastructure agencies utilize fifth-generation drones for large-scale mapping, structural inspection, and traffic management. High-resolution imaging combined with AI-based defect detection helps identify wear and tear in bridges, pipelines, and high-rise structures without manual intervention. Additionally, swarm drones are being used for aerial monitoring of public events and traffic congestion analysis in smart cities. *Impact*: Improved urban safety, predictive maintenance, and data-driven governance.

6.7.5 Commercial logistics and industrial applications

Autonomous drones have revolutionized last-mile delivery, inventory monitoring, and industrial surveillance. Logistics companies are leveraging fifth-generation drones to transport lightweight goods across short distances, reducing delivery times and carbon emissions. In industries, drones integrated with IoT systems perform asset inspections, hazardous environment monitoring, and real-time inventory tracking.

Impact: Reduced operational costs, enhanced delivery efficiency, and safer industrial operations.

6.8 Summary

The chapter on fifth-generation drones presents an in-depth overview of the most advanced class of unmanned aerial systems, emphasizing their autonomous intelligence, adaptive functionality, and multi-domain integration. It defines their core characteristics, such as AI-driven decision-making, swarm coordination, and stealth capabilities, tracing their historical evolution from manually controlled UAVs to highly autonomous aerial platforms. This chapter outlines their principles and components, covering advanced hardware, AI-based software, and seamless system integration that enable independent and coordinated operations. Their applications span defense, environment, communication, agriculture, and commercial logistics, showcasing their transformative impact across sectors. Leading global models such as the AMCA, Kaan TF-X, KF-21 Boramae, Sukhoi S-70 Okhotnik, and BAE Taranis illustrate significant progress in stealth and intelligent design. However, challenges persist in areas of technology, cybersecurity, operational limitations, cost, and ethical governance. Real-world case studies in defense, disaster response, agriculture, and smart infrastructure highlight their efficiency and social relevance. Overall, fifth-generation drones embody the convergence of AI, aerodynamics, and digital communication, marking a pivotal step toward fully autonomous, intelligent aerial ecosystems that will define the future of unmanned flight.

Review Questions

1. Define fifth-generation drones in the context of UAV evolution.
2. What are the key characteristics that distinguish fifth-generation drones from earlier generations?

3. Explain the role of artificial intelligence in fifth-generation drone operations.
4. How does swarm intelligence enhance the capabilities of fifth-generation UAVs?
5. Describe the significance of autonomous flight control in modern drone systems.
6. What is meant by "multi-domain operability" in drone technology?
7. Discuss the importance of sensor fusion in fifth-generation UAVs.
8. Explain the role of stealth technology in advanced UAV design.
9. Trace the evolution of drone technology leading up to the fifth generation.
10. What technological breakthroughs enabled the development of fifth-generation drones?
11. How did advancements in communication networks (e.g., 5G) influence UAV evolution?
12. Identify key milestones in the historical timeline of autonomous drone systems.
13. Describe the basic principle of operation for fifth-generation drones.
14. What are the major hardware components of a fifth-generation UAV?
15. Explain the function of AI-based flight algorithms in autonomous drones.
16. Discuss the importance of real-time data analytics in UAV missions.
17. How does system integration improve the functionality of advanced drones?
18. Differentiate between onboard processing and cloud-linked analytics in UAVs.
19. What role do advanced propulsion systems play in fifth-generation UAVs?
20. Describe the interaction between navigation, control, and communication subsystems.

21. List the main military and defense applications of fifth-generation drones.
22. How are fifth-generation UAVs utilized in disaster response and humanitarian aid?
23. Discuss their role in precision agriculture and environmental monitoring.
24. Explain how drones contribute to smart infrastructure and urban management.
25. Describe the use of fifth-generation drones in commercial logistics.
26. How do UAVs support communication networks and data relaying?
27. What are the advantages of using drones for environmental conservation?
28. Provide examples of dual-use (civilian and military) applications of fifth-generation drones.
29. Mention any five notable fifth-generation drone models and their countries of origin.
30. Briefly describe the design and purpose of India's Advanced Medium Combat Aircraft (AMCA).
31. What are the key features of Turkey's Kaan (TF-X) drone system?
32. Describe the significance of the Sukhoi S-70 Okhotnik in Russia's UAV program.
33. What contributions has BAE Systems Taranis made to stealth drone technology?
34. How does the Boeing Phantom Ray serve as a prototype for future UAV innovations?
35. Identify the main technological challenges faced by fifth-generation drones.
36. What are the major cybersecurity and electronic warfare vulnerabilities of these drones?
37. Discuss the operational limitations associated with weather and environmental factors.

38. Explain how cost and accessibility issues affect the widespread use of advanced UAVs.
39. What ethical, legal, and regulatory challenges are linked with autonomous drone deployment?
40. Summarize the key outcomes of operational deployments of fifth-generation drones in defense, agriculture, and urban management.

7

Key Challenges in Drone Evolution

7.1 Introduction

The evolution of drone technology from its early experimental forms to the sophisticated fifth-generation unmanned aerial systems (UAS) has been driven as much by innovation as by the challenges encountered along the way. Each generation of drones has faced distinct technological, regulatory, and societal barriers that have collectively shaped the pace, direction, and purpose of development. Understanding these challenges provides critical insight into how drones have transformed from simple remotely piloted aircraft to intelligent, autonomous aerial systems capable of complex, multi-domain operations.

7.1.1 Technological constraints in drone advancement

From the beginning, drone evolution has been constrained by technological limitations, including propulsion efficiency, flight stability, endurance, and communication reliability. Early generations struggled with power sources, limited range, and mechanical precision. As automation advanced, AI integration, data processing speed, cybersecurity, and energy management became central challenges, particularly in fifth-generation drones.

Parallel to these, regulatory challenges have persisted across all eras. The absence of standardized frameworks for airspace management, privacy laws, and safety protocols has restricted large-scale civilian deployment. Governments have often been cautious, balancing innovation with concerns over misuse, surveillance, and national security.

Equally significant are societal challenges, arising from public perception and ethical debates. Issues related to privacy invasion, job displacement, and the military use of autonomous drones have fueled global discussions. The acceptance of drones within urban environments, agriculture, or public safety continues to depend on public trust and transparent regulatory governance (Agrawal, et al. 2024).

DOI: 10.1201/9788743812418-7

7.1.2 Technological milestones driven by limitation

Each generation of drones reflects a direct response to the limitations of its predecessors.

- *First-generation drones* (pre-1970s) were constrained by primitive control mechanisms and short flight times, which led to improvements in remote-control and propulsion systems.
- *Second-generation drones* (1970s–1990s) faced challenges of navigation and communication, prompting innovations like GPS integration and data-link technology.
- *Third-generation drones* (1990s–2010s) encountered issues in autonomy and sensor accuracy, leading to the development of onboard computing and AI-assisted stabilization.
- *Fourth-generation drones* introduced semi-autonomous capabilities but struggled with endurance and cybersecurity, driving research into advanced materials, hybrid power sources, and encrypted communication.
- *Fifth-generation drones*, though technologically superior, face the most complex challenges: cybersecurity threats, cost, ethical governance, and global policy harmonization, paving the way for the anticipated sixth generation, focused on human-machine collaboration and self-learning systems.

Table 7.1 illustrates how technological, regulatory, and operational challenges have continuously shaped the evolution of drone generations. Each generation's limitations have driven new research directions, leading to innovations in autonomy, energy systems, communication, and ethics, ultimately guiding the progression toward sixth-generation, self-evolving drones.

7.2 Technological Barriers

7.2.1 Power and energy limitations

One of the most persistent and defining challenges in drone evolution has been the limitation of power and energy sources, directly influencing flight duration, payload capacity, and operational efficiency. Despite significant advancements in propulsion systems and lightweight materials, the endurance of most drones remains restricted by battery energy density and power management constraints. Conventional lithium-polymer (Li-Po) and lithium-ion

Table 7.1 How challenges have shaped each generation of drones.

Drone generation	Period/ stage	Key challenges faced	Technological/strategic response	Resulting advancements and outcomes
1st generation (Early UAVs)	Pre-1970s	• Limited flight stability and control accuracy • Short range and endurance • Lack of sensors and automation	• Introduction of radio-controlled systems • Basic aerodynamic and propulsion improvements	• Established foundation for remotely piloted aircraft • Initial use in target practice and basic reconnaissance
2nd generation (remote-controlled systems)	1970s–1990s	• Unreliable communication links • Limited endurance and payload • Inadequate data transmission capability	• Incorporation of GPS and telemetry • Improved remote control and navigation systems	• Enhanced navigation accuracy and reliability • Transition from manually controlled to semi-autonomous UAVs
3rd generation (semi-autonomous drones)	1990s–2010s	• High dependence on human operators • Poor obstacle detection and sensor accuracy • Limited automation in mission tasks	• Integration of LiDAR, infrared, and high-resolution cameras • Introduction of onboard computing and flight stabilization algorithms	• Increased autonomy and environmental awareness • Expansion into mapping, surveillance, and research applications

(Continued)

Table 7.1 *Continued.*

Drone generation	Period/ stage	Key challenges faced	Technological/strategic response	Resulting advancements and outcomes
4th generation (intelligent networked UAVs)	2010s–2020s	• Cybersecurity vulnerabilities • Battery and power limitations • Integration with civilian airspace	• Application of AI for mission optimization • Swarm coordination algorithms • Development of hybrid power and secure communication systems	• Multi-drone coordination and real-time adaptability • Networked UAVs integrated with 4G/5G systems
5th generation (AI-driven and adaptive UAVs)	2020s–Present	• Energy efficiency and sustainability issues • Ethical and legal concerns • Data security and electronic warfare threats	• Deployment of deep learning and self-learning algorithms • Advanced stealth design and sensor fusion • Blockchain and quantum encryption for secure control	• Fully autonomous and intelligent UAV systems • Multi-domain operations (air, land, sea) • Ethical and regulated drone ecosystems
6th generation (projected – cognitive UAVs)	Future	• Human–drone collaboration and trust • Complex swarm governance • Integration into smart city air networks	• Cognitive AI with human-machine collaboration • Quantum computing and neural network processing • Adaptive swarm intelligence frameworks	• Self-evolving, ethically aware UAVs • Integration into urban air mobility systems • Sustainable, cooperative, and resilient aerial networks

batteries, though efficient and lightweight, still offer limited energy output relative to the power demands of high-performance UAVs. This restricts long-range missions, especially in surveillance, mapping, and delivery applications that require sustained flight over extended periods.

To overcome these limitations, researchers and developers have been exploring alternative energy systems that can enhance endurance while maintaining operational safety. Solar-powered drones have demonstrated promise in achieving prolonged flights by harnessing renewable energy, though efficiency depends heavily on weather and daylight conditions. Similarly, hydrogen fuel cells are emerging as viable solutions, offering higher energy density and cleaner operation compared to conventional batteries. Hybrid propulsion systems combining battery and combustion-based energy sources are also being tested to balance efficiency, power, and sustainability.

Another crucial area of innovation lies in wireless power transmission and energy harvesting technologies, which can potentially allow drones to recharge mid-air or while hovering over charging platforms. Moreover, advances in nanomaterials, solid-state batteries, and intelligent power management algorithms are improving energy storage and utilization efficiency.

7.2.2 Communication and data transmission

Efficient and secure communication and data transmission form the backbone of modern drone operations. From basic remote control in early UAVs to complex real-time data exchange in autonomous systems, communication networks have evolved to support increasingly sophisticated missions. However, as drones become more intelligent and interconnected, maintaining reliable, high-speed, and secure communication links has emerged as one of the most critical challenges in drone evolution.

A primary issue lies in ensuring real-time connectivity between drones, ground stations, and other networked systems, especially in complex or remote environments. Drones operating beyond visual line of sight (BVLOS) or in urban areas often experience signal interference, latency, and bandwidth limitations. These issues can compromise command responsiveness, data integrity, and mission safety. Additionally, terrain obstacles, weather conditions, and electromagnetic interference further degrade signal quality, posing serious challenges for continuous communication (Fang, et al. 2010).

With the increasing demand for high-definition imaging, telemetry, and sensor data streaming, the need for greater bandwidth and lower latency has become essential. The integration of 5G networks, satellite communication, and mesh networking architectures offers promising solutions by enabling seamless, low-latency connections among multiple drones and control systems. These technologies support swarm operations, real-time data analytics, and autonomous decision-making, allowing drones to communicate and coordinate effectively.

However, enhanced connectivity also introduces significant cybersecurity risks. Drones are vulnerable to hacking, jamming, and GPS spoofing, which can lead to unauthorized control or data breaches. To counter these threats, researchers are implementing end-to-end encryption, blockchain-based authentication, and quantum-secure communication protocols. Such systems aim to ensure data confidentiality and operational resilience against electronic warfare and cyberattacks.

7.3 Navigation and Autonomy Challenges

7.3.1 Precision in GPS-denied environments

One of the most significant navigation challenges in modern drone operations is maintaining positional accuracy and flight stability in GPS-denied environments. Global positioning system (GPS) signals are the cornerstone of drone navigation, enabling precise localization, route planning, and autonomous control. However, in many operational contexts – such as indoor facilities, dense urban areas, mountainous terrains, forests, or military zones – GPS signals become weak, obstructed, or deliberately jammed, leading to loss of positional awareness and mission failure.

To address this limitation, researchers have developed a range of alternative navigation technologies that compensate for GPS unavailability. Inertial navigation systems (INS), which rely on accelerometers and gyroscopes, provide short-term positional estimates based on motion tracking, though they tend to accumulate errors over time. Visual odometry (VO) and simultaneous localization and mapping (SLAM) techniques have gained prominence by using onboard cameras and LiDAR sensors to build real-time maps of the environment and estimate the drone's location relative to recognizable features. These methods allow drones to navigate autonomously through vision-based perception rather than relying solely on satellite signals.

Moreover, sensor fusion, the integration of multiple data sources such as barometric pressure sensors, magnetometers, infrared depth sensors, and radar, enhances accuracy and robustness in uncertain conditions. Recent advances in machine learning and AI-based navigation algorithms further improve real-time path estimation, enabling drones to recognize environmental patterns, adapt to obstacles, and predict movement trajectories without external guidance (Lu, et al. 2022).

In military and defense applications, where GPS denial may be intentional due to electronic warfare or signal jamming, adaptive strategies like terrain-relative navigation, radio-frequency mapping, and cooperative multi-drone localization are being implemented. These techniques allow a swarm of drones to share positional data and collectively maintain accuracy, even when external references are unavailable.

The ability to achieve precise navigation in GPS-denied environments remains a cornerstone of fully autonomous drone operations. The integration of AI-driven perception, multi-sensor fusion, and vision-based mapping techniques is steadily enhancing positional reliability under complex and constrained conditions. As research advances, these technologies will enable drones to operate intelligently and self-sufficiently, ensuring accurate localization, safe maneuvering, and mission continuity even in environments where traditional satellite-based guidance systems are unavailable.

7.3.2 Developing reliable obstacle detection and avoidance systems

For autonomous drones to operate safely and efficiently, especially in dynamic or cluttered environments, the development of reliable obstacle detection and avoidance systems is paramount. This capability enables drones to perceive their surroundings, identify potential hazards, and execute real-time maneuvers to prevent collisions, a critical requirement for operations in urban settings, industrial sites, forests, and confined indoor spaces.

Traditional obstacle avoidance relied primarily on ultrasonic and infrared sensors, which provided only short-range and limited-directional awareness. However, as drones began performing more complex missions, these basic systems proved inadequate. Modern designs now incorporate advanced sensor arrays including LiDAR (light detection and ranging), stereo vision cameras, millimeter-wave radar, and time-of-flight (ToF) sensors to create high-resolution, three-dimensional environmental maps. These sensors allow drones to detect objects of varying sizes and materials, even under poor lighting or weather conditions.

The integration of artificial intelligence (AI) and machine learning algorithms has significantly improved obstacle recognition and decision-making capabilities. Using techniques such as deep learning-based object detection and path prediction, drones can classify obstacles, anticipate movement, and plan optimal avoidance trajectories in real time. In swarm or coordinated operations, inter-drone communication systems further enhance safety by enabling drones to share environmental data and avoid collisions collectively.

Despite these advancements, challenges persist. Sensor performance can be affected by rain, fog, glare, or electromagnetic interference, and balancing computational demand with onboard power limitations remains a concern. Continuous improvements in edge computing, AI chip efficiency, and lightweight sensor integration are addressing these constraints, allowing faster data processing and more responsive flight control.

Ensuring robust and intelligent obstacle detection and avoidance is vital for achieving true autonomy in next-generation drones. With ongoing research in multi-sensor fusion, predictive modelling, and AI-enhanced perception systems, drones are steadily evolving toward a level of situational awareness and adaptive behavior comparable to that of human pilots, enabling safe navigation in both structured and unpredictable environments.

7.3.3 Real-time decision-making with minimal human intervention

The hallmark of fifth-generation and emerging sixth-generation drones lies in their ability to perform real-time decision-making with minimal or no human intervention. This capability transforms drones from remotely controlled machines into autonomous, intelligent agents capable of analyzing data, assessing risks, and executing actions independently. Such autonomy is vital for missions requiring rapid adaptation, such as surveillance, search and rescue, logistics delivery, and combat operations, where communication delays or human limitations can compromise outcomes (Okere, et al. 2025).

Real-time decision-making depends on a combination of advanced onboard computing, artificial intelligence (AI), and sensor fusion technologies. Drones equipped with high-speed processors and AI-driven algorithms can analyze incoming data from cameras, LiDAR, radar, and inertial sensors to understand their environment and make immediate navigational or operational decisions. For instance, an autonomous drone can detect an obstacle, re-plan its route, and continue its mission all within milliseconds without operator input.

A key enabler of this autonomy is machine learning (ML), particularly through reinforcement learning and deep neural networks, which allow drones to learn optimal responses from past experiences and adapt to new scenarios dynamically. Edge AI computing, where decision algorithms run directly on the drone rather than relying on cloud processing, ensures minimal latency and independence from network connectivity, which is crucial in remote or signal-restricted areas.

Despite significant progress, achieving consistent reliability in autonomous decision-making presents several challenges. The system must function effectively under uncertain conditions, such as unexpected obstacles, signal interference, or dynamic weather patterns. Ensuring that AI decisions align with safety protocols and ethical constraints is equally important, particularly in defense and civilian operations. To address these issues, developers are implementing fail-safe systems, redundancy architectures, and explainable AI frameworks that allow human operators to audit or override autonomous decisions when necessary.

Real-time, low-latency decision-making represents a critical step toward full autonomy in drone operations, where machines exhibit intelligent, context-aware, and ethical behavior. As AI, neuromorphic computing, and adaptive learning systems advance, future drones will be capable of self-governing missions, responding instantly to complex environments while maintaining human-level judgment, precision, and accountability.

7.4 Regulatory and Airspace Management Issues

As drone technology advances toward greater autonomy and widespread use, regulatory and airspace management issues have become some of the most complex challenges in the evolution of unmanned aerial systems (UAS). The rapid proliferation of drones across commercial, defense, industrial, and recreational sectors has outpaced the establishment of standardized laws and frameworks to ensure safe, ethical, and coordinated airspace integration. Effective regulation is essential to balance innovation, public safety, and privacy while enabling drones to coexist with traditional manned aircraft within national and international airspaces.

A major concern in current UAS regulation is the fragmentation of global legal frameworks. Different countries adopt varying rules for drone registration, permissible altitude, flight range, and operational zones, creating inconsistencies that hinder cross-border operations and global standardization. For instance, while some nations permit beyond visual line of sight (BVLOS) missions under regulated supervision, others restrict drone operations strictly to visual line-of-sight for safety reasons. The absence of harmonized international standards complicates commercial logistics, international research, and cooperative defense operations, where drones must transition across multiple jurisdictions.

Another critical issue lies in airspace integration and traffic management. As the number of drones increases, the risk of airspace congestion and mid-air collisions also rises, especially in urban areas and near airports. To mitigate this, regulatory agencies such as the Federal Aviation Administration (FAA), European Union Aviation Safety Agency (EASA), and the Directorate General of Civil Aviation (DGCA) in India are developing frameworks for unmanned traffic management (UTM) systems. These systems aim to manage drone identification, routing, altitude separation, and communication in real time, ensuring safe and organized coexistence between manned and unmanned aerial vehicles.

Regulatory concerns extend beyond safety to encompass privacy, security, and liability. The use of drones for surveillance or data collection has raised public concerns about unauthorized monitoring and data misuse. Governments and international organizations are now focusing on establishing clear data protection guidelines, geo-fencing protocols, and no-fly zones around sensitive areas to safeguard privacy and national security. Furthermore, determining legal accountability in cases of drone-related accidents or violations remains a pressing issue, particularly as autonomous systems increasingly make independent decisions.

Addressing these challenges requires international cooperation, adaptive policymaking, and technological innovation. Initiatives such as digital airspace mapping, remote identification systems, and AI-assisted traffic management are paving the way toward smarter regulatory ecosystems. As drones evolve into integral components of smart cities, logistics networks, and defense systems, the future of drone regulation will depend on establishing a unified, flexible, and forward-looking governance framework that ensures both innovation and responsibility in global airspace management.

7.4.1 Navigating global regulatory frameworks for UAVs

The exponential growth of unmanned aerial vehicles (UAVs) has created an urgent need for robust and harmonized global regulatory frameworks that govern their safe and ethical use. However, the rapid pace of drone innovation has outstripped the ability of international and national authorities to create uniform standards. As a result, UAV operations today are regulated under a complex mosaic of national policies, aviation laws, and international guidelines, leading to inconsistencies that hinder global interoperability, research collaboration, and commercial scalability.

Most nations have established civil aviation authorities such as the Federal Aviation Administration (FAA) in the United States, the European

Union Aviation Safety Agency (EASA) in Europe, and the Directorate General of Civil Aviation (DGCA) in India to oversee drone operations within their airspace. These agencies define rules for drone registration, pilot licensing, maximum flight altitudes, line-of-sight requirements, and operational restrictions near airports or sensitive zones. Despite these national efforts, there remains no universal legal framework that standardizes regulations across borders, resulting in operational barriers for international drone missions, logistics operations, and cross-border research programs.

International aviation organizations such as the International Civil Aviation Organization (ICAO) and Joint Authorities for Rulemaking on Unmanned Systems (JARUS) are actively working toward harmonizing global UAV regulations. Their goal is to establish a common set of operational, safety, and technical standards that ensure interoperability among countries and minimize the risk of airspace conflicts. ICAO's initiatives, including the development of UAS traffic management (UTM) and remote identification systems (Remote ID), represent major steps toward creating a globally recognized regulatory structure that can accommodate both manned and unmanned aircraft within shared airspace.

However, navigating these frameworks remains challenging due to varying national security concerns, technological readiness, and socio-economic priorities. Some countries impose strict surveillance and data-sharing policies for drones, while others adopt liberal approaches to encourage innovation and commercial use. This diversity creates difficulties for global manufacturers, service providers, and operators who must tailor their systems to meet multiple regulatory standards.

To ensure sustainable drone integration, there is a growing need for mutual recognition of certifications, cross-border data sharing protocols, and adaptive airspace policies. Emerging digital tools, such as AI-driven compliance monitoring and blockchain-based registration systems, are being explored to simplify regulatory adherence and enhance transparency.

Ultimately, effective navigation of global regulatory frameworks will depend on international collaboration and continuous policy evolution. Establishing unified safety, privacy, and operational standards will not only foster innovation but also ensure that UAVs operate responsibly within a secure, interoperable, and globally managed aerial ecosystem.

7.4.2 Airspace integration for both civilian and military drones

Integrating civilian and military drones into shared airspace presents one of the most critical challenges in modern aviation management. As UAVs become increasingly prevalent across commercial, industrial, and defense sectors, ensuring safe, coordinated, and interference-free airspace operations has become a top regulatory priority. Unlike traditional aircraft, drones vary widely in size, purpose, altitude range, and control mechanisms, making unified airspace management both technically and administratively complex.

Civilian drones are primarily used for logistics, mapping, surveillance, agriculture, and research, whereas military drones operate in strategic, reconnaissance, and combat roles, often under classified missions and restricted communication protocols. The coexistence of these two categories in shared or adjacent airspaces raises substantial concerns regarding flight safety, communication interference, and operational secrecy. Military drones, for instance, may employ encrypted or stealth communication systems that could disrupt civilian traffic monitoring networks or evade standard identification systems like Remote ID, complicating civil aviation oversight.

To address these issues, aviation authorities and defense organizations are working toward the development of integrated airspace management (IAM) systems, where both civilian and military UAVs can operate under coordinated frameworks without compromising security or mission objectives. This integration relies heavily on unmanned traffic management (UTM) architectures, which provide real-time tracking, altitude deconfliction, and automated route optimization. By employing advanced technologies such as AI-driven flight scheduling, satellite navigation, and digital airspace mapping, IAM systems aim to maintain safe separation between drones and manned aircraft while allowing simultaneous operations in dense airspaces.

Collaboration between civil aviation regulators and defense agencies is also essential to define segregated air corridors, priority zones, and dynamic access protocols. Some nations have adopted the concept of flexible use of airspace (FUA), wherein airspace is dynamically allocated to either civilian or military users based on real-time demand and operational necessity. This model enables more efficient use of limited airspace resources without compromising national security or public safety.

However, successful airspace integration faces challenges related to data sharing, interoperability, and jurisdictional boundaries. Defense agencies often restrict information about military operations, while civilian regulators prioritize transparency and safety. Bridging this gap requires secure data exchange protocols, standardized communication systems, and encrypted identification methods that allow mutual awareness without exposing sensitive mission data.

In the future, the convergence of AI-assisted UTM systems, digital twin simulations, and real-time airspace analytics will play a pivotal role in enabling smooth civilian-military airspace coexistence. A unified, adaptive, and technology-driven airspace management ecosystem will not only enhance operational safety and efficiency but also establish a foundation for next-generation aerial mobility, where drones of diverse functions, commercial, defense, and humanitarian can operate collaboratively in a shared sky.

7.4.3 Managing air traffic and ensuring safety in densely populated areas

As the deployment of drones expands into urban and suburban environments, managing air traffic and ensuring operational safety in densely populated areas has become a major regulatory and technological challenge. Unlike traditional aircraft that operate in controlled airspace at higher altitudes, drones frequently fly at low altitudes (below 400 feet) close to buildings, power lines, and public spaces, creating new complexities in urban air mobility (UAM) and unmanned traffic management (UTM) systems. The growing number of drones for delivery, surveillance, construction monitoring, and emergency response necessitates the creation of structured, intelligent air traffic systems to prevent collisions, protect privacy, and maintain public safety.

The primary difficulty lies in managing a large volume of simultaneous drone operations within confined airspace. Conventional air traffic control (ATC) systems are not designed for small, fast-moving, and autonomous UAVs that lack onboard pilots or transponders. To address this, regulators and technology developers are introducing UAS traffic management (UTM) frameworks, which utilize AI-based flight coordination, real-time tracking, and automated route optimization. These systems enable drones to communicate with each other and with central control nodes to maintain safe distances, avoid restricted zones, and respond dynamically to changing conditions such as weather or congestion.

Another essential aspect of airspace safety in populated regions is the establishment of geo-fencing and virtual boundaries, which prevent drones from entering prohibited zones like airports, government buildings, or residential areas. Additionally, remote identification (Remote ID) technologies are being implemented to track drones and their operators in real time, enhancing accountability and security. Redundant control systems and fail-safe mechanisms such as automatic return-to-home (RTH) and emergency landing protocols further minimize risks associated with signal loss or technical failure.

Ensuring safety in urban air environments also requires the integration of advanced sensing and communication technologies. Drones equipped with LiDAR, radar, and computer vision can detect obstacles such as skyscrapers, cables, and other drones, enabling adaptive flight path correction. Furthermore, the use of 5G and satellite-based communication allows for low-latency data transfer, ensuring that real-time situational awareness is maintained even in high-traffic zones.

However, despite these advancements, public safety and privacy concerns continue to challenge widespread drone integration in cities. The risks of mid-air collisions, unauthorized surveillance, or drone malfunctions in crowded areas demand stringent regulatory oversight, operator training, and enforcement mechanisms. Governments and aviation bodies are collaborating to create urban drone corridors, designated air lanes for drones complemented by digital air traffic maps and AI-driven deconfliction systems that enhance safety without stifling innovation.

The future of air traffic management in urban environments will rely on autonomous coordination between drones, manned aircraft, and ground infrastructure. Through the convergence of AI, IoT, and smart city technologies, drones will become part of a digitally connected aerial ecosystem, where safety, efficiency, and public trust are maintained through intelligent regulation, adaptive control, and transparent governance.

7.5 Ethical and Legal Dilemmas

The rapid advancement of drone technology, especially in autonomy, artificial intelligence (AI), and data-gathering capabilities, has brought forth a complex set of ethical and legal dilemmas that extend far beyond technical innovation. While drones have become indispensable tools in fields such as defense, logistics, environmental monitoring, and disaster management, their increasing independence and ubiquity raise critical questions about privacy, accountability, and the moral boundaries of technology use. These dilemmas highlight the tension between innovation and responsibility, where the benefits of drones must be balanced against their potential societal, legal, and humanitarian consequences.

One of the foremost ethical issues relates to privacy and surveillance. Drones equipped with high-resolution cameras, infrared sensors, and facial recognition systems can monitor individuals and environments without consent, challenging traditional notions of personal privacy and civil liberties. In both government and private operations, the unchecked use of drones for surveillance could lead to mass data collection and misuse, prompting urgent calls for stronger data protection laws and transparent operational guidelines.

Another major concern is legal accountability in the era of autonomous drones. As machines gain the ability to make decisions with minimal human input, determining who bears responsibility for accidents, damages, or unethical outcomes becomes increasingly complex. Legal frameworks developed for human-controlled systems often fail to account for AI-driven behavior, creating gaps in liability assignment between operators, manufacturers, software developers, and AI designers. Addressing this requires the evolution of adaptive legal standards, emphasizing shared accountability, rigorous testing, and ethical AI governance.

Perhaps the most controversial ethical challenge lies in the use of autonomous weapon systems. Drones capable of making lethal decisions without direct human command raise serious moral and humanitarian questions about the role of human judgment in warfare. Delegating life-and-death decisions to algorithms risks eroding moral accountability, increasing the potential for unintended harm and civilian casualties. Consequently, international bodies such as the United Nations (UN) and the International Committee of the Red Cross (ICRC) are advocating for global treaties that ensure "meaningful human control" over autonomous combat systems and adherence to international humanitarian law.

Furthermore, the widespread deployment of drones impacts societal ethics at multiple levels, ranging from surveillance culture and public trust to concerns about safety, bias in AI decision-making, and equitable access to technology. Without a clear ethical and legal framework, the same technology designed to protect and assist humanity could undermine privacy, security, and moral accountability.

7.5.1 Privacy concerns with widespread drone surveillance

The increasing use of drones for surveillance, monitoring, and data collection has sparked growing concern over privacy violations and the erosion of personal freedom. Modern drones, equipped with high-resolution cameras, night-vision systems, LiDAR sensors, and facial recognition technologies, can capture detailed images and data from considerable distances – often without the knowledge or consent of the individuals being observed. While these capabilities are invaluable in applications such as law enforcement, disaster management, and infrastructure inspection, they also raise profound ethical and legal questions about how, where, and by whom such surveillance should be conducted.

The most significant concern lies in the intrusive potential of continuous aerial monitoring. Unlike fixed surveillance cameras, drones are highly mobile, allowing them to access private spaces such as residential areas, backyards, and workplaces. This ability to collect real-time data across wide geographic areas can lead to unintentional or unauthorized breaches of privacy, especially when used by commercial operators or government agencies without transparent policies. In densely populated urban settings, the risk of mass surveillance where citizens' daily activities are recorded and analyzed poses a serious challenge to civil liberties and individual autonomy.

Legal frameworks in many countries have yet to fully address the privacy implications of drone surveillance. While some jurisdictions require flight permissions, geo-fencing, or data usage declarations, there remains a lack of comprehensive global standards regulating data ownership, consent, and retention. Moreover, advancements in AI-powered analytics now allow drones to identify and track individuals, vehicles, and patterns of movement, further blurring the line between legitimate security operations and invasive observation.

To mitigate these issues, policymakers and technologists are working toward establishing ethical and legal safeguards. These include implementing geo-fencing zones to restrict surveillance in sensitive or private areas, mandating data encryption and anonymization protocols, and requiring transparency in drone identification and operation through visible markings or public disclosure systems. Some nations are also exploring "privacy by design" principles, ensuring that privacy protection measures are built into drone hardware and software from the outset rather than added later as regulatory afterthoughts.

Ultimately, ensuring privacy in an era of pervasive drone surveillance demands a balance between security and individual rights. Clear policies, technological safeguards, and public awareness must work together to establish trust and accountability in drone operations. Without such measures, the unchecked use of aerial surveillance could lead to a culture of constant observation, undermining the very social and ethical foundations that technology aims to strengthen.

7.5.2 Legal responsibility and liability in autonomous drone operations

The growing autonomy of drones, driven by advancements in artificial intelligence, machine learning, and onboard decision-making, has introduced complex questions about legal responsibility and liability. Unlike conventional drones that require human pilots, autonomous drones can make real-time decisions with minimal human oversight, raising uncertainties about who should be held accountable when accidents, damages, or unlawful actions occur. This issue lies at the intersection of technology, law, and ethics, demanding the evolution of legal frameworks capable of addressing the unique nature of AI-driven systems.

Traditionally, responsibility for UAV-related incidents rested with the human operator or pilot, who was directly in control of the aircraft's actions. However, in the case of autonomous drones, the decision-making process is often delegated to algorithmic systems that interpret data, plan routes, and respond to environmental conditions independently. When a drone malfunctions, violates airspace regulations, or causes harm, it becomes difficult to determine whether liability lies with the operator, manufacturer, software developer, data provider, or AI algorithm itself. This phenomenon, often termed the "accountability gap," poses significant challenges to policymakers and insurers.

Legal systems around the world are beginning to address this ambiguity by exploring shared or distributed liability models. Under such frameworks, responsibility is proportionally assigned based on the source of failure:

- Manufacturers may be liable for hardware defects or design flaws.
- Software developers may bear responsibility for algorithmic errors or cybersecurity vulnerabilities.
- Operators may be accountable for misuse, negligence, or failure to maintain oversight.
- Data service providers may be implicated if inaccurate data leads to operational errors.

Several nations are also considering the establishment of AI accountability and certification standards, requiring developers to validate the safety, transparency, and explainability of autonomous systems before deployment. These initiatives aim to ensure that drones operate within predefined ethical and safety parameters, reducing the risk of harm while maintaining a clear chain of accountability. Additionally, the introduction of mandatory UAV insurance policies and flight data recorders ("black boxes") is being discussed as a means of documenting events and assigning responsibility post-incident.

A critical legal debate concerns whether autonomous drones driven by sophisticated AI should be recognized as independent legal entities capable of bearing limited responsibility, similar to corporate personhood in law. While this notion remains controversial, it reflects the ongoing need to adapt legal principles to technologies capable of autonomous decision-making.

To ensure accountability in the era of intelligent drones, future legal frameworks must promote transparency, traceability, and human oversight. Regulations should require all autonomous UAV systems to include mechanisms for human intervention and explainable AI decision logs. Ultimately, maintaining human responsibility within the operational chain is essential to uphold public safety, ethical governance, and trust in drone technologies that increasingly act and decide on their own.

7.5.3 Autonomous weapons systems and the ethics of drone warfare

The emergence of autonomous weapon systems (AWS) drones capable of identifying, targeting, and engaging adversaries without direct human control represents one of the most profound ethical challenges in modern warfare. These systems mark a paradigm shift from remotely piloted drones, where human operators retain decision-making authority, to machines that can make lethal decisions independently using artificial intelligence (AI), machine learning, and sensor fusion. While such technologies promise faster response times, reduced human casualties, and improved precision in combat, they also raise deep concerns about accountability, moral responsibility, and compliance with international humanitarian law (IHL).

A key ethical dilemma lies in the delegation of life-and-death decisions to algorithms. Human operators are traditionally bound by moral reasoning, empathy, and legal frameworks such as the Geneva Conventions, which emphasize proportionality, distinction, and necessity in the use of force. In contrast, autonomous systems operate based on programed logic and data patterns, devoid of emotional or ethical context. This raises the question: can a machine truly make morally justifiable decisions in war? The possibility of unintended civilian casualties due to misidentification, system malfunction, or data bias highlights the grave moral risks of removing human oversight from combat operations.

The issue of accountability further complicates the ethical landscape. When an autonomous drone commits an unlawful act, such as targeting non-combatants, determining liability becomes ambiguous. Should responsibility rest with the programer, manufacturer, military commander, or the state that deployed the system? This "responsibility gap" undermines established principles of international law, which rely on human accountability for wartime conduct. Without clear responsibility, the moral and legal fabric of warfare could erode, leading to unregulated and potentially indiscriminate violence.

Another major concern is the potential for an AI-driven arms race. Nations investing heavily in autonomous combat drones risk destabilizing global security, as the proliferation of such systems could lower the threshold for armed conflict. Autonomous weapons may make war seem more politically and economically acceptable, as they reduce human losses, but this convenience could lead to more frequent and less constrained military engagements. Additionally, the risk of AI system hacking, spoofing, or malfunction poses significant threats to international peace and stability.

To address these issues, international organizations and humanitarian groups are advocating for the establishment of global treaties and ethical regulations governing autonomous weapons. The United Nations (UN), International Committee of the Red Cross (ICRC), and several member states have called for maintaining "Meaningful Human Control (MHC)" over all lethal decision-making processes. This principle ensures that humans remain the ultimate decision-makers in the authorization of lethal force, preserving moral judgment and accountability. Moreover, transparency in algorithm design, adherence to IHL, and global collaboration on AI ethics are being emphasized as prerequisites for responsible military AI development.

In essence, the integration of autonomous drones into warfare challenges the very foundations of human ethics, legality, and accountability. While these systems may enhance efficiency and reduce human exposure to danger, they also risk dehumanizing conflict and eroding moral responsibility. The future of drone warfare must, therefore, be guided by strict ethical oversight, international cooperation, and the reaffirmation of human control, ensuring that technology serves to protect humanity rather than undermine its moral conscience.

7.6 Environmental Challenges

While drones are often promoted as efficient and environmentally friendly tools due to their small size, electric propulsion, and reduced emissions compared to conventional aircraft, their increasing proliferation introduces several environmental challenges that must be carefully addressed. These challenges encompass the impact of drones on wildlife and ecosystems, the issue of noise pollution, and the broader concerns of sustainable manufacturing and end-of-life disposal. As drone usage expands into agriculture, logistics, surveillance, and environmental monitoring, understanding and mitigating their ecological footprint has become essential for ensuring long-term environmental sustainability.

One major concern is the disruption of wildlife and natural habitats. Drones operating in forests, wetlands, and coastal regions can unintentionally disturb animals, especially birds and sensitive species. The presence of UAVs may cause stress responses, alter migration patterns, and disrupt nesting or breeding behavior. For example, studies have shown that the noise and motion of drones can startle wildlife, leading to energy expenditure or habitat avoidance. To minimize these effects, operators are advised to follow ethical flight protocols, such as maintaining safe altitudes, limiting flight durations, and avoiding operations during breeding seasons or near protected zones.

Noise pollution represents another growing issue, particularly in urban and suburban environments. Although smaller drones are quieter than traditional aircraft, their high-frequency buzzing can be intrusive and irritating, especially when operated in large numbers for delivery or monitoring purposes. Continuous drone operations in populated areas may contribute to acoustic pollution, affecting both human comfort and animal behavior. Manufacturers are therefore focusing on aerodynamic propeller design, noise-dampening materials, and optimized flight control algorithms to reduce acoustic emissions and ensure compliance with urban noise standards.

From a sustainability perspective, the manufacturing and disposal of drones pose additional environmental challenges. Most drones are constructed from composite materials, metals, and electronic components that are difficult to recycle. The widespread use of lithium-ion and lithium-polymer batteries, while energy-efficient, contributes to the growing problem of electronic waste and hazardous chemical disposal. Developing biodegradable materials, recyclable components, and eco-friendly battery alternatives such as solid-state or hydrogen-based systems could significantly reduce the environmental impact of future UAV production.

Furthermore, the increasing reliance on drones for various industries also raises concerns about their carbon footprint during production, operation, and maintenance. Although electric propulsion is relatively clean, the energy used to charge large drone fleets, especially from non-renewable sources, can offset some of their environmental benefits. Integrating renewable energy-based charging systems, solar-assisted operations, and sustainable logistics practices can help balance technological advancement with ecological responsibility.

7.6.1 Drones impact on wildlife and ecosystems

While drones have become valuable tools in environmental monitoring, wildlife conservation, and ecological research, their increasing presence in natural habitats has raised concerns about unintended ecological disturbances. The same attributes that make drones effective mobility, proximity, and silent observation, can also disrupt delicate ecosystems when not managed responsibly. Understanding and mitigating drones' impact on wildlife and ecosystems is therefore critical to ensuring that their use supports, rather than harms, environmental sustainability.

The most direct impact of drones on wildlife is behavioral disturbance. Many animal species, particularly birds, perceive drones as aerial predators due to their movement patterns and noise signatures. This can cause flight responses, nest abandonment, and increased stress levels, especially during breeding and nesting seasons. Research has shown that repeated drone exposure can alter feeding, migration, and mating behaviors in birds, mammals, and marine life. For instance, seabirds may abandon colonies, while marine animals such as seals and whales can exhibit avoidance behavior when drones operate at low altitudes.

Beyond behavioral stress, drones can also cause habitat disruption. Frequent low-altitude flights, particularly in protected areas, can disturb fragile ecosystems by affecting pollinators, dispersing dust or debris, and unintentionally damaging vegetation. In some cases, drones used for filming or tourism have entered sensitive zones, leading to long-term ecological imbalance. These disruptions, though often subtle, can accumulate over time and threaten local biodiversity.

To minimize ecological disturbance, researchers and regulatory bodies advocate for the implementation of wildlife-safe drone operation protocols. These include maintaining safe distances and altitudes, avoiding breeding or nesting seasons, using quiet propulsion systems, and obtaining proper environmental flight permissions. Some conservation programs have successfully adopted bio-acoustic and AI-based drones designed to operate quietly and non-invasively for wildlife monitoring.

7.6.2 Noise pollution

As drone operations expand across commercial, industrial, and public sectors, noise pollution has emerged as one of the most noticeable and growing environmental challenges. While drones are significantly quieter than traditional aircraft, the high-frequency noise generated by their propellers and motors can be intrusive, particularly in urban areas and natural habitats. This persistent background noise affects not only human populations but also wildlife, which relies heavily on acoustic cues for communication, navigation, and survival. Managing and mitigating drone-generated noise has therefore become a crucial objective in sustainable drone development.

The noise produced by drones primarily stems from rotor blade aerodynamics, motor vibration, and airflow turbulence. Multi-rotor drones, such as quadcopters, produce a continuous buzzing sound that intensifies with speed and proximity. In residential zones, this can cause public annoyance, psychological discomfort, and potential sleep disruption, particularly as the frequency of drone deliveries, aerial inspections, and surveillance operations increases. In wildlife environments, drone noise can disrupt animal communication patterns, trigger stress responses, and interfere with predator–prey interactions. Birds, for example, are especially sensitive to drone sounds, often altering flight or nesting behavior even at moderate altitudes.

To address these issues, researchers and manufacturers are developing multiple strategies for noise reduction and acoustic optimization. One key approach involves redesigning propellers to reduce aerodynamic turbulence. The use of serrated or curved blade edges, inspired by owl wing morphology, has shown promise in lowering noise levels without sacrificing lift efficiency. Similarly, variable-pitch propellers and low-RPM (rotations per minute) systems can balance thrust and noise output, making drones quieter during both takeoff and flight.

Another area of innovation is acoustic damping and motor insulation. By incorporating vibration-absorbing materials, noise-canceling housings, and optimized rotor-motor synchronization, engineers are able to significantly decrease mechanical noise transmission. Advances in electric propulsion systems, particularly the transition from brushed to brushless motors, have also contributed to quieter operation.

In addition to hardware improvements, operational and regulatory measures play a vital role in noise mitigation. Establishing minimum flight altitudes, noise-sensitive no-fly zones, and time-of-day restrictions can minimize community disturbance and ecological impact. Urban planners and aviation authorities are also exploring acoustic zoning policies for drones, similar to those used for traffic and aviation noise management.

Looking forward, the integration of AI-driven flight path optimization and adaptive noise prediction models will enable drones to adjust their altitude, speed, and trajectory dynamically to minimize acoustic impact in real time. Manufacturers are increasingly prioritizing "low-noise certification standards" as part of drone design, ensuring compliance with environmental and social sustainability goals.

In essence, tackling drone noise pollution requires a multifaceted approach combining engineering innovation, operational discipline, and regulatory foresight. Through these collective efforts, the next generation of drones aims to operate not only efficiently and safely but also quietly and harmoniously within both human and natural soundscapes.

7.6.3 Sustainable manufacturing and end-of-life disposal of drones

As drone technology advances and production scales up globally, the focus on sustainable manufacturing and responsible end-of-life disposal has become increasingly important. While drones are often associated with efficiency and environmental monitoring, their own manufacturing processes, material composition, and disposal methods present notable environmental challenges. The rising number of drones in civilian, industrial, and defense sectors means that without sustainable practices, the industry could contribute to resource depletion, electronic waste (e-waste), and pollution, undermining its environmental benefits.

Most commercial drones are built using lightweight composite materials, such as carbon fiber-reinforced polymers (CFRP), thermoplastics, and aluminum alloys. Although these materials provide strength and weight advantages, their production is energy-intensive, and recycling is complex and limited. Carbon fiber, for instance, is difficult to recover once bonded with resins, and improper disposal leads to long-term environmental persistence. Moreover, drones rely heavily on lithium-based batteries, which contain toxic elements like cobalt, nickel, and manganese. If not properly managed, these batteries can leak harmful chemicals into soil and water systems, posing significant ecological risks.

To mitigate these challenges, drone manufacturers and researchers are increasingly adopting sustainable design and production principles. This includes the use of eco-friendly, recyclable materials, such as bio-based composites, biodegradable polymers, and lightweight metals with high recyclability rates. Manufacturing processes are being optimized through additive manufacturing (3D printing), which minimizes material waste and energy consumption while allowing for precision engineering. Some companies are also experimenting with modular drone designs, where damaged or outdated components can be easily replaced or upgraded, extending the lifespan of the entire system and reducing waste.

The issue of end-of-life (EoL) disposal is equally critical. As drones reach obsolescence due to technological upgrades or mechanical failures, sustainable disposal methods must ensure material recovery, safe recycling, and minimal environmental footprint. Establishing drone recycling networks similar to electronic waste collection systems can facilitate proper dismantling, material separation, and reuse. Recycling programs that focus on battery recovery and metal reclamation are gaining traction, supported by new technologies that safely extract valuable components like lithium and rare earth metals for reuse in future production.

In parallel, the adoption of circular economy principles in drone manufacturing is reshaping sustainability practices. By emphasizing reuse, remanufacturing, and closed-loop supply chains, drone producers can significantly reduce raw material dependency and environmental impact. Governments and regulatory bodies are also playing a vital role by introducing eco-design directives, e-waste management regulations, and green certification standards for UAV manufacturers.

Ultimately, achieving sustainability in the drone industry requires a holistic approach that spans the entire product lifecycle from design and material selection to recycling and disposal. By integrating green materials, renewable energy sources, and efficient production technologies, drones can truly become tools that not only serve humanity but also align with global efforts to protect the planet. The next generation of UAVs must, therefore, embody the principles of environmental stewardship, circular economy, and lifecycle responsibility, ensuring that technological progress contributes to both innovation and ecological balance.

7.7 Public Perception and Acceptance

The widespread adoption of drone technology depends not only on regulatory frameworks and technical advancements but also on public perception and societal acceptance. Despite their immense potential across industries, ranging from agriculture and logistics to surveillance and disaster relief, drones often face skepticism rooted in privacy concerns, safety fears, and ethical apprehensions. The general public's acceptance of drones determines how smoothly these systems can be integrated into daily life, public spaces, and shared airspace. Therefore, building trust, transparency, and communication with society is as critical to drone development as improving technical performance.

A mixture of curiosity, utility, and apprehension shapes public attitudes toward drones. While people recognize drones' benefits in emergency response and innovation, many remain concerned about their possible misuse, such as unauthorized surveillance, noise disturbance, or interference with manned aviation. Addressing these fears through awareness, policy clarity, and responsible deployment is key to fostering confidence in drone technology.

7.7.1 Overcoming fears and concerns related to drone technology

Public resistance to drone usage often stems from perceived risks and uncertainties rather than direct negative experiences. Common concerns include privacy invasion, safety hazards, loss of jobs due to automation, and misuse for illegal activities. Accidents involving drones, such as mid-air collisions or crashes in public spaces, tend to receive significant media attention, amplifying these fears and influencing societal perception.

To overcome these challenges, stakeholders must adopt a transparent and educational approach. Public awareness campaigns should emphasize how drones operate, what safety features they employ (such as geofencing, obstacle avoidance, and fail-safe mechanisms), and how regulations ensure responsible usage. Highlighting positive applications, such as life-saving medical deliveries, precision agriculture, and disaster response, can also help shift perception from fear to appreciation.

In addition, visible compliance with laws and ethical guidelines, including privacy protection measures and data security protocols, can reassure citizens that drones are operated responsibly. Demonstrating a commitment to community safety, environmental sensitivity, and public engagement will be essential in dispelling myths and building informed acceptance.

7.7.2 Building trust in autonomous drones for commercial and personal use

As drones become increasingly autonomous, public trust in AI-driven systems has become a critical factor for commercial viability and social integration. People tend to be cautious about machines that make decisions independently, particularly in scenarios involving transport, deliveries, or surveillance. Questions of control, reliability, and accountability dominate public discourse about autonomous drone use.

To foster trust, developers must ensure that drones demonstrate consistent safety, predictability, and transparency in their behavior. Technologies such as redundant flight systems, real-time monitoring, explainable AI (XAI), and ethical programing frameworks can make autonomous operations more transparent and understandable to the public. Commercial operators should also communicate openly about operational boundaries such as flight zones, data handling practices, and human oversight protocols to demonstrate responsible usage.

Additionally, certification and third-party auditing of autonomous drone systems can help build credibility. When drones are visibly regulated, tested, and certified for safety and privacy compliance, users are more likely to accept them as reliable technologies. Over time, successful deployment in public-facing services, for example, parcel delivery, medical supply transport, or traffic monitoring, will normalize drones as trustworthy tools in everyday life.

7.7.3 Engaging with communities to address privacy and safety concerns

Community engagement is central to achieving long-term public acceptance of drones. By directly involving local populations in discussions about drone operations, flight planning, and data management, authorities and operators can cultivate mutual understanding and trust. Transparent communication about where and why drones are being used, as well as how collected data is stored and protected, helps reduce suspicion and resistance.

Organizing public consultations, demonstration events, and participatory policy workshops allows residents to voice their concerns and contribute to shaping drone-related regulations. This collaborative approach not only improves public confidence but also ensures that operational frameworks reflect societal values and local contexts. For example, involving communities in setting no-fly zones or determining acceptable surveillance practices creates shared ownership of drone policies.

Moreover, incorporating feedback mechanisms and grievance redressal systems enables the public to report misuse, technical disturbances, or privacy violations, reinforcing accountability. Continuous dialogue between industry, regulators, and citizens ensures that drone technology evolves in harmony with social expectations and ethical norms.

7.8 Security Threats and Vulnerabilities

As drones become more accessible, intelligent, and widespread, they also present increasing security threats and vulnerabilities. Originally developed for defense and industrial applications, drones are now used across various civilian sectors – ranging from logistics and infrastructure inspection to photography and law enforcement. However, this accessibility also enables malicious actors to exploit drone technology for unlawful or harmful purposes. Concerns include their potential misuse in criminal activities, cybersecurity breaches, and threats to national and public safety.

The challenge lies in balancing the advantages of drone innovation with effective risk management and defense mechanisms. With autonomous capabilities, long-range control, and the ability to carry payloads, drones can be misused in ways that are difficult to detect or prevent using traditional security frameworks. Addressing these risks requires a combination of technological countermeasures, legal enforcement, and international cooperation to safeguard both airspace integrity and public security.

7.8.1 Potential misuse of drones for illegal activities

The proliferation of affordable, high-performance drones has led to a growing risk of their misuse for criminal or terrorist purposes. Drones can be easily modified to perform illicit actions such as smuggling, espionage, unauthorized surveillance, and even weaponization. Their small size, maneuverability, and ability to operate remotely make them difficult to detect, especially in urban or restricted areas.

Common illegal activities involving drones include smuggling contraband into prisons or across borders, spying on sensitive installations, and unauthorized aerial photography or data collection. More concerningly, there have been reports of drones being used to carry explosives, deliver biological or chemical agents, and disrupt public events or air traffic. Such misuse not only poses immediate security risks but also undermines public trust in legitimate drone applications.

To combat these threats, governments and law enforcement agencies are implementing strict registration requirements, remote identification systems, and geofencing technologies that restrict drone operation in prohibited areas. Additionally, the use of AI-based surveillance and airspace monitoring systems enables real-time detection of suspicious drone behavior. International collaboration and intelligence sharing are also critical to prevent the weaponization and cross-border misuse of drones.

7.8.2 Hacking, spoofing, and hijacking drones: Addressing cybersecurity risks

With the increasing integration of AI, GPS, and wireless communication systems, drones are vulnerable to cyberattacks that target their software, navigation, or control networks. Hacking, GPS spoofing, and signal hijacking are among the most serious cybersecurity risks, allowing attackers to manipulate or take over drones mid-flight. Such intrusions can lead to data theft, mission disruption, or physical damage, particularly in defense, surveillance, and commercial delivery applications.

GPS spoofing involves transmitting false satellite signals to mislead a drone's navigation system, redirecting it off course or into restricted airspace. Similarly, hacking attacks may exploit weak encryption or unsecured data links to gain control of the drone's command system. Denial-of-service (DoS) attacks can overwhelm communication channels, causing drones to crash or lose connection with operators. These threats are particularly alarming for autonomous drones relying on cloud-based data or swarm coordination systems.

To mitigate such vulnerabilities, robust cybersecurity architectures are essential. Drones should employ end-to-end encryption, multi-factor authentication, and intrusion detection systems to safeguard communication links and control software. Blockchain-based command authentication and quantum encryption are emerging technologies offering enhanced data security and traceability. Additionally, adopting AI-driven threat detection algorithms can help identify anomalies or unauthorized access attempts in real time.

Regulatory agencies are also emphasizing cyber certification standards for UAVs, ensuring that both hardware and software components meet security benchmarks before deployment. Training operators in cyber hygiene and secure handling protocols further reduces exposure to cyber threats. Ultimately, strengthening drone cybersecurity is fundamental to preventing hostile exploitation and maintaining operational trust in autonomous aerial systems.

7.8.3 Counter-drone technologies and defense measures

The rise in malicious or unauthorized drone activities has driven the development of counter-drone technologies (C-UAS) designed to detect, track, and neutralize rogue drones. These systems are now vital for protecting critical infrastructure, airports, military bases, and public gatherings from potential drone threats. Counter-drone strategies encompass a combination of radar detection, radio frequency (RF) analysis, optical tracking, and active interdiction methods.

Detection and identification form the first line of defense. Advanced C-UAS systems use multi-sensor fusion combining radar, acoustic sensors, infrared imaging, and RF scanners to identify drones based on their flight pattern, sound signature, or communication frequency. Once detected, the next step involves neutralization or mitigation, which can be achieved through either soft-kill or hard-kill measures.

Soft-kill methods include jamming RF and GPS signals to disrupt communication between the drone and its operator or to force an emergency landing. Protocol manipulation can also take control of the drone and guide it safely to a designated area. Hard-kill solutions involve kinetic interception, such as using anti-drone projectiles, nets, or laser systems to physically disable or destroy the drone. Military-grade systems are increasingly incorporating directed-energy weapons (DEWs) and AI-assisted targeting to respond swiftly to swarming threats.

However, counter-drone measures must be carefully regulated to prevent interference with legitimate UAV operations or civilian communication networks. Governments and aviation authorities are now developing legal frameworks for C-UAS deployment, defining operational boundaries, response protocols, and data-sharing mechanisms among law enforcement, defense agencies, and civil aviation bodies.

Looking ahead, future defense measures will rely on AI-powered predictive threat modeling, digital twin simulations, and real-time swarm analysis to anticipate and neutralize drone-based risks. By combining technology, regulation, and international coordination, counter-drone systems will form a crucial component of global airspace security, ensuring that drones continue to serve as tools of innovation without becoming instruments of threat.

7.9 Addressing Weather and Environmental Factors

The performance and reliability of drones are heavily influenced by weather and environmental conditions, which pose significant challenges to safe and consistent operations. Unlike manned aircraft, drones, particularly lightweight and electrically powered models, are highly susceptible to variations in wind speed, precipitation, humidity, and temperature. These environmental factors can affect aerodynamics, sensor performance, battery life, and communication systems, limiting operational efficiency and mission success. Therefore, developing strategies and technologies to address these environmental constraints is crucial for enhancing drone endurance, resilience, and adaptability across diverse conditions.

Modern research and engineering efforts are increasingly focused on weatherproofing drone designs and integrating adaptive flight control technologies that allow UAVs to function safely in challenging conditions. By improving both hardware and software resilience, drones can expand their range of applications from military reconnaissance and disaster response to agriculture and logistics, regardless of environmental unpredictability.

7.9.1 Drone performance in extreme weather conditions

Weather extremes, such as strong winds, heavy rainfall, extreme temperatures, or dense fog, pose serious operational challenges for drones. High wind speeds can destabilize flight, strain motors, and reduce battery efficiency, often leading to crashes or mission failure. Even advanced stabilization algorithms may struggle to maintain steady positioning under gusty conditions. Similarly, rain and humidity can interfere with electronic components, degrade sensor accuracy, and increase the risk of short circuits. Icing and condensation on propellers or sensors during cold-weather operations can further compromise aerodynamics and visibility.

Temperature variations also significantly affect drone performance. Cold environments reduce battery capacity and chemical reactivity, shortening flight times and limiting power output. Conversely, high temperatures can cause battery overheating, material deformation, and reduced electronic efficiency. For drones operating in desert or tropical regions, managing thermal balance becomes essential to prevent system failure.

To mitigate these effects, drone manufacturers and researchers are adopting environmental hardening techniques such as sealed electronics, hydrophobic coatings, and reinforced airframes to protect components from moisture and debris. Additionally, AI-based flight management systems are being used to predict weather-induced instability and automatically adjust altitude, speed, or route to maintain optimal performance. The integration of real-time weather data into flight planning software allows operators to anticipate atmospheric conditions and modify missions accordingly.

Ultimately, ensuring consistent performance in extreme weather requires a holistic approach that combines durable hardware, intelligent control systems, and pre-flight environmental analysis. As drones evolve, their ability to adapt to variable climates will define their suitability for global operations, including those in remote or disaster-prone regions.

7.9.2 Advances in weatherproofing and adaptive flying technologies

To enhance reliability and operational flexibility, the drone industry is investing in weatherproofing and adaptive flight technologies that enable UAVs to withstand harsh environments. Weatherproofing involves designing drones with sealed fuselages, corrosion-resistant materials, and protective coatings that safeguard critical systems from moisture, dust, and temperature fluctuations. Modern waterproof drones can even function during light rainfall or over bodies of water, expanding their use in rescue, maritime, and environmental monitoring applications.

Adaptive flight technologies represent the next frontier in overcoming environmental challenges. These systems employ real-time environmental sensing, AI-driven flight control algorithms, and dynamic propulsion management to adjust flight parameters based on surrounding conditions. For instance, smart drones equipped with wind-compensation algorithms can modify rotor speeds asymmetrically to counter turbulence, while thermal regulation systems help maintain optimal internal temperatures for batteries and electronics.

Recent developments in bio-inspired design are also influencing drone adaptation. Mimicking the flight mechanics of birds and insects, engineers are creating flexible wings, morphing rotors, and self-healing materials that allow drones to recover stability after impact or adapt to shifting wind patterns. Moreover, the integration of machine learning models enables predictive responses, allowing drones to anticipate weather changes and optimize navigation paths for safety and efficiency.

In addition, collaborative drone networks and cloud-based weather intelligence platforms are improving environmental awareness across drone fleets. By sharing real-time meteorological data, multiple drones can collectively map local weather conditions and avoid hazardous zones.

These advancements in weatherproofing and adaptive flight are transforming drones into resilient, all-weather aerial systems capable of operating in environments once considered inaccessible. As future designs continue to evolve, drones will not only withstand the elements but also leverage environmental awareness as a strategic advantage, enabling safer, smarter, and more sustainable operations across all sectors.

7.10 Human–Drone Collaboration and Interface Issues

As drones become increasingly autonomous and integrated into civilian, industrial, and defense operations, the interaction between humans and drones, known as human–drone interaction (HDI), has emerged as a critical area of research and development. While automation has reduced the need for constant manual control, human oversight, decision-making, and supervision remain essential to ensure safety, reliability, and ethical compliance. Effective human–drone collaboration depends on intuitive interfaces, real-time communication, and mutual understanding between the operator and the autonomous system.

The goal of HDI design is to create systems where humans and drones can cooperate seamlessly, with drones interpreting operator intent accurately and humans understanding the drone's decisions, limitations, and feedback. However, challenges such as control complexity, interface overload, latency, and operator fatigue continue to limit optimal collaboration. Overcoming these challenges requires the development of user-centered interaction frameworks, adaptive control interfaces, and AI systems that can communicate with humans transparently and contextually.

7.10.1 Developing intuitive human–drone interaction systems

Intuitive human–drone interaction (HDI) systems are key to enabling effective and safe collaboration between operators and drones. These systems must bridge the gap between human cognition and machine autonomy, ensuring that communication is natural, responsive, and error-tolerant. Traditional drone control interfaces rely heavily on manual joysticks or tablet-based applications, which can be cognitively demanding in complex missions.

Recent advancements are introducing multimodal interaction methods, including gesture recognition, voice commands, eye-tracking, and augmented reality (AR) interfaces. These technologies allow operators to control drones intuitively without requiring constant manual input, making operations more efficient and immersive. For instance, AR headsets can display flight data, environmental maps, and drone perspectives in real time, allowing users to make informed decisions quickly.

AI-powered HDI systems are also evolving toward predictive and adaptive control, where drones can interpret operator intent and offer assistance proactively. Through machine learning, drones can learn user preferences and adjust their responses accordingly, improving overall mission efficiency and safety. Additionally, haptic feedback systems are being integrated into control devices to convey tactile information such as resistance, vibration, or force cues, enhancing situational awareness and precision.

Ultimately, developing intuitive HDI systems reduces operational workload, minimizes errors, and allows drones to act as collaborative partners rather than passive tools. This shift from control to cooperation is a defining feature of next-generation drone-human relationships.

7.10.2 Remote control challenges in autonomous and semi-autonomous drones

Despite rapid progress in autonomy, many drones still depend on varying levels of human supervision and remote control. Managing this balance between autonomy and oversight introduces several challenges related to communication latency, signal interference, situational awareness, and control reliability.

In semi-autonomous operations, the human operator must monitor the drone's behavior and intervene when necessary. However, as drones operate farther from visual range such as in beyond visual line of sight (BVLOS) missions, the operator's ability to perceive environmental hazards decreases. Latency in communication networks, especially when using satellite or long-distance radio links, can delay commands and sensor feedback, potentially leading to collisions or mission failure.

Furthermore, interface design complexity can hinder effective control during high-stress or time-sensitive operations. Overly detailed dashboards or poor visual displays may overwhelm operators, reducing reaction speed and situational awareness. As autonomy increases, maintaining human-in-the-loop (HITL) control becomes even more challenging, especially when multiple drones are managed simultaneously.

To mitigate these issues, researchers are developing adaptive remote control systems that employ AI-driven decision support tools to assist operators. These systems can automatically prioritize alerts, suggest corrective actions, or temporarily assume control during emergencies. Improvements in 5G connectivity, low-latency satellite communication, and redundant control channels are also enhancing reliability. Future control systems will likely emphasize shared autonomy, where humans and drones dynamically adjust their levels of control based on situational demands, ensuring both safety and efficiency.

7.10.3 Enhancing user experience in drone operation for various industries

A positive user experience (UX) is fundamental for widespread adoption of drones across industries such as logistics, agriculture, filmmaking, public safety, and infrastructure inspection. The effectiveness of drone systems depends not only on performance and autonomy but also on how intuitively and comfortably humans can operate them. Poorly designed interfaces or overly technical controls can limit the accessibility of drones, particularly for non-expert users.

Enhancing UX begins with simplifying operational workflows. Automated flight planning, one-touch takeoff and landing, and intelligent mission templates reduce user effort and the potential for human error. For industrial operators, customized control dashboards tailored to specific applications, such as crop monitoring or thermal inspection, can streamline operations and improve productivity.

Integration with AI-assisted analytics also enhances user satisfaction by translating raw sensor data into actionable insights. For example, in agriculture, drones equipped with machine learning can automatically assess crop health, while in construction, they can detect structural anomalies and generate inspection reports autonomously. These capabilities allow operators to focus on decision-making rather than manual control.

Additionally, attention to ergonomics, feedback systems, and visual clarity plays a vital role in reducing fatigue and enhancing control precision. The adoption of voice and gesture-based inputs, immersive AR interfaces, and real-time data visualization makes drone operation more interactive and accessible.

In the future, human-centric drone design will prioritize inclusivity, safety, and adaptability across diverse user groups. Whether for professionals managing fleets or individuals using drones recreationally, the key lies in creating systems that are intuitive, intelligent, and engaging, ensuring that drones seamlessly augment human capability while maintaining trust, control, and satisfaction.

7.11 Solutions and Future Trends in Overcoming Challenges

The evolution of drone technology has been shaped as much by innovation as by the persistent challenges faced across generations from energy limitations and navigation difficulties to cybersecurity, regulatory, and ethical complexities. As drones advance toward greater autonomy, connectivity, and intelligence, the focus is shifting toward developing sustainable, secure, and adaptive solutions that address these challenges while unlocking new possibilities. The future of drones will be defined by breakthroughs in energy systems, communication networks, artificial intelligence (AI), quantum computing, and global regulatory cooperation, ensuring that drones become safer, smarter, and more integrated into human society.

7.11.1 Emerging technologies addressing energy, navigation, and autonomy issues

Energy efficiency, precise navigation, and autonomous decision-making remain three of the most critical areas of improvement in drone technology.

New advances are paving the way for longer endurance, greater precision, and higher operational intelligence.

In the field of energy and propulsion, innovations such as hydrogen fuel cells, solar-powered systems, and hybrid-electric propulsion are extending flight duration while reducing carbon emissions. Solid-state batteries and graphene-based supercapacitors promise higher energy density and faster charging, addressing one of the biggest limitations of current UAVs: short flight times. Additionally, wireless charging pads and mid-air power transmission systems are under research to enable continuous operations without frequent battery replacements.

In navigation, progress is being made to overcome the limitations of GPS dependency. AI-driven visual odometry, LiDAR-based mapping, and simultaneous localization and mapping (SLAM) technologies now allow drones to operate accurately in GPS-denied environments, such as indoors or under dense canopies. The integration of quantum sensors and inertial navigation systems (INS) further enhances spatial awareness and precision.

Regarding autonomy, the introduction of edge AI computing enables drones to make real-time decisions without relying on external data processing. This allows for rapid adaptation to environmental changes, obstacle avoidance, and mission reconfiguration. Swarm intelligence is another transformative trend, allowing fleets of drones to collaborate autonomously using collective AI, thereby enhancing coordination, redundancy, and operational scalability. Together, these technologies are leading drones toward true self-reliance and multi-environment capability.

7.11.2 Innovations in regulatory frameworks and public engagement

As drone technologies advance faster than regulations, adaptive and harmonized policy frameworks are becoming essential to support safe, ethical, and socially acceptable operations. Governments and international bodies are now working toward global regulatory convergence to facilitate cross-border drone operations and streamline certification, licensing, and compliance processes.

Key innovations include the establishment of unmanned traffic management (UTM) systems for real-time airspace coordination and remote identification (Remote ID) requirements for accountability. Some nations are adopting risk-based regulatory models, where operational permissions are tailored to the mission's risk profile rather than fixed classifications. This flexibility enables innovation while maintaining safety.

Beyond regulation, public engagement and awareness are crucial for societal acceptance of drones. Governments and private operators are increasingly involving communities in consultations, policy feedback sessions, and awareness programs to address privacy, safety, and ethical concerns. Transparent communication regarding data collection, operational boundaries, and emergency procedures builds trust and reduces public apprehension.

Furthermore, digital regulatory platforms powered by AI and blockchain are simplifying drone registration, flight authorization, and compliance monitoring. This evolution toward smart governance ensures that drone innovation remains aligned with social values, human rights, and environmental stewardship.

7.11.3 The role of AI, IoT, and quantum computing in solving future drone challenges

The integration of artificial intelligence (AI), Internet of Things (IoT), and quantum computing will be pivotal in shaping the next generation of drones, enabling unprecedented levels of intelligence, connectivity, and computational power.

AI plays a central role in making drones more autonomous, efficient, and adaptive. Through machine learning, deep learning, and neural networks, drones can interpret sensor data, recognize patterns, and make real-time decisions during missions. Applications such as predictive maintenance, autonomous navigation, swarm coordination, and context-aware object detection demonstrate how AI is transforming drones from reactive machines into proactive, intelligent agents.

IoT technologies allow drones to become part of a connected ecosystem, where UAVs interact seamlessly with ground systems, satellites, and other devices. In smart cities, IoT-enabled drones are expected to play key roles in traffic management, infrastructure monitoring, precision agriculture, and emergency response, sharing data instantly across digital networks. The emergence of 6G connectivity will further enhance drone communication, enabling ultra-low-latency data exchange and global synchronization.

Quantum computing represents the next leap in computational capability, offering exponential improvements in data processing, encryption, and optimization. Quantum algorithms can help drones process complex flight dynamics, optimize swarm formations, and manage vast environmental datasets in real time. Quantum cryptography ensures secure communication resistant to hacking and spoofing, addressing one of the most significant cybersecurity threats in modern UAV systems.

The convergence of AI, IoT, and quantum technologies will transform drones into autonomous, intelligent, and secure aerial ecosystems capable of functioning harmoniously with human infrastructure. These advancements will not only overcome current challenges but also redefine the role of drones in shaping sustainable, connected, and resilient societies.

7.12 Summary

This chapter comprehensively explores the multifaceted challenges that have shaped the evolution of drone technology, from early experimental designs to modern autonomous systems. It begins with an overview of technological constraints and milestones, emphasizing how limitations in energy, communication, and design have historically driven innovation. The chapter then delves into critical technological barriers, such as limited battery capacity, inefficient data transmission, and the ongoing struggle for reliable operation in GPS-denied environments. Issues of autonomy, including obstacle detection, real-time decision-making, and AI-based flight control, highlight the transition from remote operation to intelligent self-governance. The discussion expands to regulatory and airspace management, focusing on global inconsistencies, civilian–military airspace integration, and the growing need for unified unmanned traffic management (UTM) systems. Ethical and legal dilemmas, particularly concerning privacy, liability, and the moral implications of autonomous warfare, underscore the societal responsibilities accompanying technological progress. Environmental challenges are analyzed through drones' impact on wildlife, noise generation, and sustainability concerns related to manufacturing and disposal. Public perception and acceptance are presented as key enablers of progress, requiring education, transparency, and community engagement. The chapter also examines security threats, including cyberattacks, illegal use, and counter-drone defense measures, as well as the influence of weather and environmental factors on drone reliability. Finally, it explores the importance of human–drone collaboration, intuitive interfaces, and user experience, before concluding with emerging solutions and future trends such as advancements in AI, IoT, quantum computing, and renewable energy systems that aim to overcome these barriers. Collectively, the chapter emphasizes that overcoming these interlinked technological, ethical, environmental, and societal challenges is essential to ensuring that drones evolve into sustainable, secure, and ethically governed aerial systems that harmoniously integrate with human and environmental ecosystems.

Review Questions

1. What are the major technological constraints that have historically influenced drone advancement?
2. How have technological limitations driven innovation in drone design and performance?
3. Explain how material, energy, and control challenges shaped the early generations of drones.
4. What role do computational and sensor technologies play in overcoming traditional drone limitations?
5. Discuss how power and energy limitations restrict drone endurance and payload capacity.
6. What are the current research trends in alternative drone power sources, such as hydrogen fuel cells and solar energy?
7. Explain the main challenges in maintaining real-time communication and data transmission for drones.
8. How do bandwidth, latency, and cybersecurity issues affect drone network reliability?
9. Why is precision navigation difficult in GPS-denied environments?
10. Describe how technologies like LiDAR and SLAM improve drone navigation accuracy.
11. What are the key design considerations in developing reliable obstacle detection and avoidance systems?
12. Explain the importance of AI and machine learning in real-time drone decision-making.
13. How does edge computing contribute to autonomous drone operations with minimal human intervention?
14. What are the key challenges in navigating global UAV regulatory frameworks?
15. Compare how different aviation authorities (FAA, EASA, and DGCA) regulate drone operations.
16. How does airspace integration differ between civilian and military drone applications?

17. Discuss the challenges of managing drone air traffic in densely populated urban areas.
18. What is the role of unmanned traffic management (UTM) systems in ensuring safe drone operation?
19. Identify the main privacy concerns associated with widespread drone surveillance.
20. How can governments balance national security with individual privacy rights in drone use?
21. Explain the challenges in assigning legal responsibility for autonomous drone operations.
22. What is the concept of the "accountability gap" in autonomous UAV systems?
23. Discuss the ethical implications of autonomous weapons systems in drone warfare.
24. What does "meaningful human control" mean in the context of military drone ethics?
25. How do drones impact wildlife behavior and natural ecosystems?
26. Describe the sources of noise pollution in drone operations and methods to reduce them.
27. Why is sustainable manufacturing important for drone production and disposal?
28. What materials and technologies are being explored to make drones more environmentally friendly?
29. What are the common public fears associated with drone technology?
30. How can public education and transparency improve acceptance of drone use?
31. Discuss how community engagement contributes to safer and more ethical drone operations.
32. Explain the importance of building trust in autonomous drones for commercial and personal applications.
33. Identify major ways drones can be misused for illegal or harmful activities.

34. What are hacking, spoofing, and hijacking in the context of drone cybersecurity?
35. Discuss the technologies and strategies used in counter-drone defense systems.
36. How do soft-kill and hard-kill countermeasures differ in counter-drone operations?
37. How do extreme weather conditions, such as wind, rain, and temperature, affect drone performance?
38. Explain how adaptive flight and weatherproofing technologies enhance drone reliability.
39. What are the major challenges in human–drone collaboration and interface design?
40. How can intuitive interaction systems and enhanced user experiences improve the safety and accessibility of drone operations across industries?

8

Evolutionary Impacts of UAVs on Modern Systems

8.1 Introduction

The evolution of *Unmanned Aerial Vehicles (UAVs)*, commonly known as drones, represents one of the most transformative technological advancements of the 21st century. From their origins as experimental military tools to their current role in commercial, industrial, and humanitarian applications, UAVs have profoundly influenced modern systems across multiple domains. Their evolution embodies the convergence of *aerodynamics, artificial intelligence, robotics, and data science*, redefining how humans interact with the sky, gather information, and execute complex operations with unprecedented precision and efficiency.

The *evolutionary impact of UAVs* extends far beyond aviation; it touches nearly every sector of modern life. In *defense and security*, drones have revolutionized surveillance, reconnaissance, and precision targeting, while in *civil and industrial applications*, they have become indispensable for infrastructure inspection, agriculture, environmental monitoring, logistics, and disaster management. The ability of UAVs to access remote or hazardous environments has dramatically enhanced operational safety, speed, and data accuracy. As automation and AI continue to advance, drones are evolving into *autonomous, networked systems* capable of performing multi-agent coordination and real-time decision-making without direct human control.

Equally significant are the *societal and economic dimensions* of UAV evolution. Drones have created new industries, enabled smart city initiatives, and reshaped fields such as cinematography, mapping, and emergency response. However, their rapid proliferation also brings challenges, including *airspace congestion, privacy concerns, ethical dilemmas, and cybersecurity threats*. Balancing innovation with safety, regulation, and environmental sustainability has become a central theme in UAV integration into modern systems.

DOI: 10.1201/9788743812418-8

This chapter explores how UAVs have evolved to influence *technological infrastructures, economic systems, defense operations, environmental management, and societal frameworks*. It examines their transformative roles, highlights emerging trends, and analyzes how drone technology continues to redefine global connectivity and intelligent automation. Ultimately, the chapter underscores that the *evolutionary impact of UAVs* is not merely technological; it is systemic and multidisciplinary, symbolizing the transition toward a *new era of aerial intelligence, autonomy, and sustainable innovation*.

8.1.1 Sectoral expansion and influence of UAV technologies

The evolution of unmanned aerial vehicles (UAVs) has catalyzed a profound transformation across multiple sectors, making them one of the most disruptive technologies of the modern era. Initially developed for military reconnaissance and surveillance, UAVs have rapidly expanded into *civilian, industrial, scientific, and humanitarian applications*, reshaping operational efficiency, data collection, and decision-making processes. Their versatility stems from a combination of *technological miniaturization, automation, and connectivity*, enabling their integration into diverse domains such as agriculture, healthcare, construction, logistics, disaster management, and environmental conservation.

In the *industrial and commercial sectors*, UAVs have become indispensable tools for precision operations. In agriculture, they are used for *crop health monitoring, irrigation assessment, and pesticide application*, optimizing resource use and boosting productivity. In logistics, drones have introduced *last-mile delivery solutions* that reduce cost and time while enhancing accessibility in remote or congested areas. The *construction and infrastructure industries* benefit from UAV-based inspection and 3D mapping, enabling real-time progress tracking and risk mitigation. Meanwhile, in the *energy sector*, drones are utilized to inspect power lines, pipelines, and wind turbines, minimizing downtime and ensuring worker safety.

The *public sector and scientific community* have also embraced UAV technologies for their adaptability and precision. Governments use drones for *traffic management, surveillance, and environmental monitoring*, while researchers employ them to study *climate change, wildlife habitats, and natural resource mapping*. In *healthcare and emergency services*, drones are deployed for *medical supply delivery, organ transport, and disaster relief*, bridging critical infrastructure gaps in underserved regions. The *media and film industries* have revolutionized content creation through aerial cinematography and live broadcasting, democratizing perspectives once limited to large-scale productions.

Overall, the expansion of UAV technologies reflects a *cross-sectoral evolution*, where drones have moved beyond specialized tools to become *integrated systems within broader digital ecosystems*. Supported by advances in *AI, IoT, cloud computing, and automation*, drones now serve as the aerial extensions of intelligent infrastructure, enhancing precision, reducing operational costs, and contributing to sustainability. As industries continue to digitize and automate, the influence of UAVs will deepen, reinforcing their role as a *core enabler of the Fourth and Fifth Industrial Revolutions*.

8.1.2 Industrial and societal transformation through drone innovation

Drone innovation has brought about a significant transformation in both industrial operations and societal practices, redefining how tasks are performed, services are delivered, and information is accessed. In the industrial domain, drones have introduced automation, precision, and real-time data intelligence, enabling organizations to operate more efficiently and safely. Sectors such as construction, mining, oil and gas, and transportation now rely on UAVs for surveying land, inspecting hazardous sites, optimizing workflows, and reducing human exposure to risk. These capabilities have not only improved operational reliability but have also reduced costs and accelerated project timelines. Similarly, in logistics and retail, drones are reshaping the last-mile delivery ecosystem, offering faster, eco-friendly deliveries that respond to the rising demands of e-commerce-driven economies.

Societally, drones are reshaping how communities connect with services and information. They have played a crucial role in emergency response, disaster management, and public safety, where real-time aerial insights help authorities assess situations and coordinate faster relief. In remote and underserved regions, drones are enabling equitable access to healthcare by delivering vaccines, blood samples, and medical supplies where road-based transport is slow or unavailable. Meanwhile, in environmental and conservation work, drones help monitor wildlife, track deforestation, and assess ecological changes with minimal disturbance. The cultural and creative sectors have also experienced transformation through aerial photography, immersive cinematography, and artistic drone light performances, expanding the boundaries of storytelling and public engagement.

As drones continue to evolve with advancements in AI, machine learning, computer vision, and edge computing, their role in industry and society is shifting from supportive tools to autonomous decision-making systems capable of collaborative and predictive functions. This marks a transition toward smart, interconnected ecosystems, where drones interact seamlessly with ground-based robotics, IoT devices, and digital infrastructure. However, this transformation also brings forward ethical, regulatory, and privacy considerations, requiring careful governance to ensure responsible and inclusive integration. Ultimately, drone innovation is not merely enhancing efficiency, it is reshaping the way modern societies work, interact, adapt, and progress in the digital age.

8.2 Economic Contributions of UAV Technology

The rise of unmanned aerial vehicles (UAVs) marks one of the most dynamic technological revolutions contributing to global economic growth in the 21st century. Drones have transitioned from experimental devices to *mainstream industrial assets*, generating economic value across multiple sectors through enhanced productivity, reduced operational costs, and the creation of new business ecosystems. The UAV market now represents a *multibillion-dollar global industry*, supported by rapid innovation in automation, artificial intelligence, and advanced materials. Beyond direct commercial impact, drones have also spurred the growth of *ancillary industries* such as software development, data analytics, component manufacturing, and pilot training. Collectively, these developments have redefined labor markets, stimulated entrepreneurship, and positioned drone technology as a critical enabler of economic modernization and digital transformation.

8.2.1 Growth trajectories and employment opportunities

The global UAV industry is experiencing exponential growth, driven by technological advancements, supportive government policies, and increasing demand for aerial intelligence across sectors. According to international market projections, the global drone market is expected to surpass USD 150 billion by the early 2030s, reflecting expanding applications in logistics, agriculture, construction, and defense. This growth is accompanied by a surge in employment opportunities from drone manufacturing and maintenance to data analysis, software engineering, and air traffic management.

New professional domains such as drone piloting, UAV systems integration, and drone-based surveying are emerging as specialized careers, requiring multidisciplinary expertise in aeronautics, robotics, and geospatial sciences. The establishment of drone training academies and certification programs has created pathways for technical skill development, particularly in developing economies where UAV adoption is accelerating. Moreover, the growth of drone startups, consultancy services, and operational platforms contributes to job creation across both urban and rural landscapes, supporting local economies and technological self-reliance.

Thus, UAV technology not only stimulates industrial innovation but also nurtures human capital and entrepreneurial ecosystems, making it a cornerstone of the global digital economy and Industry 5.0 transformation.

8.2.2 Sectoral economic gains from drone applications

Across industries, UAVs have proven to be cost-efficient multipliers of productivity and precision. In agriculture, drones equipped with multispectral sensors assist in crop health analysis, irrigation management, and pesticide application, helping farmers optimize yields and reduce input costs. In infrastructure and construction, drones perform aerial mapping, structural inspections, and progress monitoring, cutting survey times by up to 70% compared to traditional methods. Similarly, in logistics, UAVs streamline last-mile delivery operations, minimizing fuel consumption and reducing carbon footprints.

The energy and utilities sectors benefit significantly from drone-assisted inspections of power lines, pipelines, and offshore wind farms, improving maintenance accuracy and reducing downtime. In mining and resource exploration, UAVs offer precise terrain mapping and volumetric analysis, enabling better resource estimation and operational planning. Even in public administration and emergency management, drones save millions in disaster assessment, environmental monitoring, and law enforcement activities.

The integration of UAVs thus creates cumulative economic benefits across industries – lowering operational costs, improving data-driven decision-making, and opening new market opportunities. By transforming conventional processes into digitally optimized ecosystems, drones contribute directly to national productivity, economic resilience, and sustainable growth.

8.2.3 Market trends and future investment perspectives

The UAV market is evolving rapidly, with expanding investment from private corporations, government agencies, and venture capitalists seeking to capitalize on emerging drone applications. The market is characterized by continuous innovation in AI-driven autonomy, swarming technologies, 5G connectivity, and green propulsion systems, which are shaping the next generation of UAV solutions. Defense applications continue to dominate global spending, but the commercial drone segment, particularly in logistics, agriculture, energy, and infrastructure, has become the fastest-growing domain, attracting large-scale investments.

Countries like the United States, China, India, and members of the European Union are making substantial investments in developing UAV manufacturing clusters, test zones, and policy frameworks, fostering a competitive and innovation-driven environment. Strategic partnerships between drone manufacturers, software developers, and research institutions are accelerating product diversification and market penetration.

Looking ahead, the future of UAV investment will focus on sustainability, automation, and integration with emerging technologies such as quantum computing, blockchain-based logistics, and AI-enhanced decision systems. The next decade will likely witness the rise of drone-as-a-service (DaaS) business models, scalable UAV fleet management systems, and urban air mobility (UAM) platforms. Together, these trends underline the economic potential of UAVs as a core driver of industrial transformation, job creation, and digital globalization.

8.3 Drones and the New Era of Digital Warfare

The emergence of advanced unmanned aerial vehicles (UAVs) has ushered in a new phase of digitalized warfare where information dominance, autonomy, and persistence are as decisive as traditional firepower. Modern combat theaters increasingly rely on drones for intelligence, surveillance, reconnaissance (ISR), electronic warfare, and strike capabilities, often operating as distributed, networked platforms that collect and act on data in real time. This section examines how autonomous combat and surveillance systems are reshaping tactics and strategy, the tactical and cost advantages they bring, and the profound ethical and legal questions their use raises for states and militaries worldwide (Duffield, 2015).

Table 8.1 illustrates the progressive evolution of UAV roles in military contexts, moving from early remotely piloted reconnaissance platforms to highly autonomous and swarm-enabled systems. While initial UAVs served primarily as observational assets with limited endurance and intelligence capabilities, subsequent generations incorporated *networked communication, precision strike functionality, and AI-supported tactical decision-making*. The most recent shift toward distributed autonomous operations marks a transformative phase in digital warfare, where drones not only act as *force multipliers* but also participate in cooperative mission execution with minimal human oversight.

Table 8.1 Evolution of drone roles in modern warfare.

Generation/ stage	Primary role	Key capability	Technological enabler	Representative UAV systems
Early UAVs (pre-2000)	Basic reconnaissance and battlefield observation	Limited remote camera imaging and short-duration surveillance	Radio-controlled flight systems and basic telemetry	MQ-1 Predator, IAI Searcher
Networked UAVs (2000–2015)	Intelligence, surveillance, reconnaissance (ISR) with precision strike capability	Persistent surveillance and guided weapon delivery	Satellite communication, EO/IR imaging systems, laser designation	MQ-9 Reaper, IAI Heron, IAI Harpy
Semi-autonomous UAVs (2015–2022)	Integrated combat support and target coordination	Automated target tracking, patrol route optimization, and partial onboard AI	Embedded microprocessors, machine-vision sensors, advanced GNSS navigation	Bayraktar TB2, CH-4, Wing Loong II
Fully autonomous and swarm-enabled UAVs (2022–Future)	Cooperative combat operations, swarm-based offensive maneuvers	Self-organizing multi-drone decision systems and dynamic battlefield adaptation	Edge AI computing, swarm algorithms, secure datalink coordination	Loyal Wingman (Boeing), Sukhoi S-70 Okhotnik, ALTIUS, Shahed-136

8.3.1 The rise of autonomous combat and surveillance systems

The evolution of UAVs from manually piloted platforms to highly autonomous systems marks a pivotal shift in contemporary military strategy. Modern combat environments require rapid situational awareness, persistent surveillance, and precision engagement, all of which drones are uniquely equipped to provide. Advances in machine learning, pattern recognition, onboard processing systems, and sensor fusion have enabled drones to perform operations such as target identification, route optimization, surveillance patrolling, and battlefield reconnaissance with significantly reduced human intervention. Many state militaries now employ drones not only as tools for collecting intelligence but also as independent or coordinated strike units capable of engaging targets in contested environments.

The integration of UAVs into network-centric warfare has further transformed command and control hierarchies. Drones are now embedded into multi-layered battle networks where they act as data nodes collecting, transmitting, and interpreting real-time battlefield information. This has given rise to concepts such as "loyal wingman" systems, in which autonomous drones accompany crewed fighter jets to perform support or offensive roles, effectively multiplying air combat capabilities while minimizing risk to human pilots. Likewise, drone swarming frameworks allow dozens or hundreds of small UAVs to operate cooperatively, overwhelming air defense systems with distributed intelligence and coordinated movement (Ryan Jr, et al. 2019).

Additionally, the rise of autonomous surveillance drones has revolutionized information dominance, an increasingly decisive factor in modern warfare. High-altitude long-endurance (HALE) UAVs can patrol border regions, maritime stretches, and conflict zones continuously for days, providing granular intelligence that shapes strategic decisions. However, this autonomy also introduces new vulnerabilities, including dependence on satellite GPS links and exposure to electronic warfare attacks. Securing autonomy in contested communication environments remains a critical research direction for defense organizations worldwide.

Table 8.2 highlights the dual nature of UAV deployment in modern military operations, demonstrating how their tactical advantages must be carefully balanced against operational vulnerabilities. Drones offer significant benefits such as reduced personnel risk, lower operational costs, high-precision engagement, and continuous real-time surveillance, all of which enhance strategic flexibility and responsiveness on the battlefield. However, these advantages are countered by emerging threats, including cybersecurity risks, electronic warfare interference, target identification errors, and evolving counter-drone defense systems.

Effective mitigation requires integrating secure communication architectures, resilient navigation systems, human-supervised targeting procedures, and stealth or adaptive flight technologies. Thus, while UAVs function as key enablers of digital-era warfare, their deployment demands technical robustness, controlled autonomy, and careful compliance with operational protocols to ensure reliability and ethical application under combat conditions.

8.3.2 Tactical efficiency and operational cost benefits

Drones confer significant tactical advantages by reducing operational complexity, time, and resource expenditure. Compared to manned aircraft, UAVs have lower acquisition and maintenance costs, require fewer crew members, and avoid the logistical challenges associated with pilot life support

Table 8.2 Tactical advantages and operational risks of military UAV deployment.

Advantages of UAV use	Operational impact	Associated risks/ vulnerabilities	Mitigation and control measures
Reduced risk to human pilots	Allows execution of missions in high-threat environments without endangering personnel	Reliance on remote communication links, which may be disrupted	Encrypted communication channels and backup control systems
Lower operational and maintenance costs	Enables continuous surveillance and scalability of military operations	Exposure to electronic warfare, including jamming and signal interception	Hardened GPS systems, anti-jamming INS navigation, and spectrum monitoring
Precision strike capability	Minimizes collateral damage and improves target discrimination	Risk of misidentification due to algorithmic or sensor limitations	Human-in-the-loop targeting protocols and validation procedures
Persistent surveillance and reconnaissance	Provides real-time battlefield awareness and situational intelligence	Detection and counteraction by advanced anti-drone defense systems	Stealth airframe shaping, low-RCS materials, and dynamic flight path modulation

and training. This cost-effectiveness allows military forces to deploy large numbers of drones simultaneously, supporting continuous surveillance, rapid target acquisition, and highly flexible mission profiles. In large-scale operations, the ability to swiftly reposition drones across battlespaces provides superior adaptability compared to traditional assets.

On the battlefield, drones excel in precision targeting, which lowers collateral damage and enhances mission accuracy. Loitering munitions capable of surveilling an area and striking only after confirmation of a target have become particularly valuable in operations requiring high discrimination and proportionality. UAV-assisted artillery correction and real-time strike coordination have also improved the precision of indirect fire systems, enabling engagements at longer distances while minimizing risk to friendly forces. This contributes directly to tactical lethality while preserving compliance with engagement protocols.

Moreover, drones enhance force protection by allowing militaries to conduct reconnaissance and offensive operations without exposing personnel to hostile environments. Surveillance drones monitor improvised explosive device (IED) threats, patrol conflict zones, and provide overwatch for ground units during convoy and urban operations. Maritime drones support naval forces by detecting submarine movements, monitoring coastlines, and conducting anti-piracy patrols. However, these advantages require ongoing investments in cybersecurity, communication redundancy, maintenance infrastructure, and AI reliability, as operational efficiency depends heavily on system integrity.

8.3.3 Ethical and legal considerations in autonomous warfare

While UAVs offer clear operational benefits, they also raise deeply complex ethical and legal dilemmas, particularly when autonomy reduces human control in lethal decision-making. Traditional laws of armed conflict are grounded in human judgment, accountability, and moral agency. Autonomous drones challenge this framework by allowing algorithms to influence or execute engagement decisions. This gives rise to the concern that lethal force could be employed without meaningful human oversight, contradicting ethical norms and risking unintended harm due to algorithmic misinterpretation or data errors.

A central issue is the "accountability gap" determining who is responsible when an autonomous system makes a faulty or harmful decision. Is it the programer, the commanding officer, the political authority, or the machine's internal logic? Current legal systems are not fully equipped to address these distributed layers of responsibility. Additionally, the proliferation of low-cost autonomous weapons poses geopolitical risks, as non-state actors or unstable regimes may acquire systems capable of targeted violence without adequate oversight or regulation.

To mitigate these concerns, international organizations and research institutions emphasize the need to maintain meaningful human control (MHC) over all lethal operations. This requires that humans remain responsible for final engagement decisions, and that autonomous systems function as advisory or assistance tools rather than independent agents. There is also growing momentum toward international frameworks governing AI transparency, algorithmic accountability, kill-switch protocols, and autonomous weapons treaties. Establishing these norms will be essential to ensure that advancements in drone warfare align with humanitarian principles and global security stability.

Table 8.3 outlines the principal ethical and legal challenges associated with the increasing autonomy of drone warfare systems. As UAVs gain capabilities to select and engage targets with limited human oversight, fundamental concerns arise regarding *accountability, transparency, and adherence to international humanitarian norms*. The automation of lethal decision-making

Table 8.3 Ethical and legal challenges in autonomous drone warfare.

Core issue	Primary concern	Critical question	Guiding principle/ proposed safeguard
Autonomy in the use of lethal force	Reduced or absent human decision-making in engagement	Who is accountable for harm caused by an autonomous drone strike?	Ensure meaningful human control (MHC) over all lethal-decision processes
Algorithmic judgment and target identification	Risk of misclassification and unintended civilian harm	Can AI reliably distinguish lawful targets in dynamic environments?	Develop explainable AI and human-verified target authorization
Proliferation and accessibility of autonomous weapons	Increased risk of misuse by non-state actors or unstable regimes	How can access to lethal UAV systems be regulated effectively?	Strengthen international export controls, monitoring, and end-user verification
Compliance with international humanitarian law (IHL)	Maintaining proportionality and minimizing civilian casualties	Can autonomous systems apply humanitarian legal principles without human cognition?	Create global legal frameworks, incorporating IHL and ethical review in system design

challenges longstanding ethical expectations that human judgment, empathy, and responsibility must guide the use of force in conflict.

8.4 UAV Applications in Smart Agriculture

The integration of unmanned aerial vehicles (UAVs) into agricultural systems has significantly transformed the way farming operations are planned, monitored, and executed. Driven by the need to increase productivity while reducing resource consumption, UAV-based precision agriculture offers highly accurate, data-driven insights that enable farmers to make informed decisions at field, crop, and plant scales. These systems leverage aerial imaging, environmental sensing, and intelligent data analytics to enhance crop performance, ensure efficient use of inputs, and support sustainable agricultural practices. As global food demand continues to rise, UAVs play a crucial role in bridging the gap between traditional cultivation methods and digitally optimized farming systems.

Table 8.4 highlights the key technological components of UAV systems used in smart agriculture and explains how each contributes to improving agricultural productivity, decision-making, and resource efficiency. The table shows that UAVs are not merely aerial imaging platforms, but *integrated*

Table 8.4 UAV technologies and their functional benefits in smart agriculture.

UAV component/feature	Function in agriculture	Benefit to farmers
RGB cameras	Visual crop health inspection and documentation	Quick field assessment; early detection of abnormalities
Multispectral sensors (e.g., NDVI)	Plant health, vigor, and stress monitoring	Data-driven crop management; targeted interventions
Thermal cameras	Irrigation monitoring and moisture stress detection	Prevents water waste and improves drought resilience
AI/ML analytics platforms	Automated disease, pest, and nutrient diagnosis	Reduces expert dependency and time delay in analysis
Precision spraying drones	Targeted pesticide, fertilizer, and micronutrient application	Lower chemical usage, reduced cost, and improved environmental safety
Autonomous flight planning systems	Automated path planning and repeatable surveys	Consistent monitoring and reduced labor requirements

data-driven farm management tools. For instance, *RGB cameras* allow farmers to visually inspect crop conditions and detect abnormalities quickly, while *multispectral sensors* provide deeper insights into crop health by analyzing plant vigor, stress levels, and nutrient status information that is not visible to the naked eye. *Thermal imaging sensors* further expand monitoring capabilities by identifying water stress and guiding irrigation scheduling, thus supporting water conservation efforts.

8.4.1 Precision farming and yield enhancement

Precision farming focuses on optimizing agricultural efficiency by tailoring crop management to the specific conditions of each field section. UAVs equipped with high-resolution imaging systems, such as RGB, multispectral, hyperspectral, and thermal cameras, enable farmers to observe spatial variations in plant health that are not visible from the ground. This data helps detect nutrient deficiencies, pest infestations, disease spread, and growth anomalies at early stages, allowing timely corrective interventions.

In yield optimization, drones support variable-rate application strategies, where fertilizers, seeds, and irrigation are distributed only where needed, reducing waste and improving overall productivity. By continuously monitoring field conditions, UAVs help maintain uniform crop growth and minimize input costs. Furthermore, yield prediction models enhanced with drone-collected data provide reliable forecasts that support harvest planning, supply chain coordination, and market decision-making. Overall, precision farming enabled by drones enhances both productivity and environmental responsibility.

8.4.2 Remote sensing for soil, water, and crop management

Remote sensing with UAVs has emerged as a vital tool for comprehensive land and crop analysis. Soil characteristics such as texture, moisture content, and organic matter can be evaluated using drones equipped with multispectral and thermal sensors, enabling accurate mapping of soil variability. This assists in determining suitable planting zones, irrigation requirements, and soil treatment strategies.

In irrigation management, UAV thermal imaging identifies areas experiencing water stress, allowing the adjustment of irrigation schedules to prevent over- or under-watering. Similarly, drones enable precise monitoring of crop canopy temperature and root zone moisture conditions, offering real-time data to optimize irrigation efficiency. For pest and disease management, UAV-based vegetation index maps such as NDVI (normalized difference vegetation index) assist in early detection of stress patterns, enabling targeted biological or chemical interventions. Through these capabilities, UAV remote sensing enhances crop health assessment, input optimization, and resource conservation.

Table 8.5 presents a comparative assessment of traditional farming practices and UAV-assisted precision agriculture, emphasizing how drone technology enhances economic efficiency and sustainability in crop production.

Table 8.5 Economic and sustainability gains enabled by UAV agriculture.

Impact area	Traditional farming	UAV-assisted precision farming	Resulting benefit
Monitoring labor	Manual scouting required	Automated aerial scanning	Lower labor demand and faster coverage
Fertilizer and pesticide use	Uniform application	Targeted "apply where needed" precision spraying	Reduced chemical costs and soil preservation
Irrigation efficiency	Broad-area irrigation	Real-time moisture mapping and selective irrigation	Water savings and sustainability
Crop yield	Reactive crop management	Proactive and predictive crop health response	Higher yield and stable production output

The table shows that tasks such as field monitoring, irrigation management, and agrochemical application – once reliant on manual labor and generalized treatment are now optimized through data-driven, targeted interventions enabled by UAVs. This shift reduces labor requirements, minimizes fertilizer and pesticide usage through precise application, and improves irrigation efficiency by identifying moisture stress at specific field locations. As a result, farmers achieve higher crop yields, reduced input costs, and improved environmental stewardship, demonstrating that UAV integration supports both *economic viability* and *sustainable food production*, particularly in large-scale and resource-sensitive agricultural systems.

8.4.3 Economic efficiency and sustainable food production

UAV-based agriculture delivers substantial economic benefits by reducing input costs, minimizing labor-dependent operations, and improving crop quality and yield. Farmers save resources by applying fertilizer, pesticides, and water only where necessary, while UAV-enabled monitoring reduces the need for manual field scouting. This efficiency becomes especially significant in large-scale commercial farms and remote agricultural zones, where traditional monitoring methods are time-consuming and expensive.

Additionally, UAV-supported sustainable farming practices contribute to improved environmental stewardship. By reducing excessive agrochemical use, drones help protect soil fertility, irrigation water systems, and biodiversity. Their ability to support climate-smart agriculture through continuous monitoring of crop health under variable climate conditions positions UAVs as essential tools for resilient food production systems.

With global food requirements rising, UAV-integrated agriculture promotes sustainable intensification, ensuring higher output without expanding cultivated land. As drone technology continues to advance, supported by AI analytics, cloud-based farm platforms, and autonomous flight planning, UAV agriculture will play a defining role in securing long-term agricultural sustainability and global food security.

8.5 UAVs for Environmental Monitoring and Conservation

The use of unmanned aerial vehicles (UAVs) in environmental monitoring and conservation has expanded rapidly due to their ability to collect high-resolution data over large and often inaccessible landscapes. Drones offer a non-intrusive, cost-effective, and adaptable toolset for supporting environmental research, biodiversity protection, climate monitoring, and disaster

Table 8.6 Economic and sustainability outcomes of UAV-integrated agriculture.

Outcome area	UAV-enabled improvement	Economic benefit	Environmental/ sustainability benefit
Input resource use (fertilizers, pesticides)	Targeted spraying based on vegetation health data	Reduced chemical expenditure and operational cost	Lower soil contamination and prevention of water pollution
Water management	Real-time moisture and thermal mapping for selective irrigation	Reduced water consumption and energy use in pumping systems	Conservation of freshwater resources and improved drought resilience
Crop yield and quality	Early detection of stress, disease, and nutrient deficiencies	Higher and more consistent yields; improved market value	Healthier plant growth with reduced agrochemical dependency
Labor efficiency	Automated field scouting and monitoring flights	Lower labor requirements and reduced human field risk	Reduced manual workload and improved farm safety
Supply chain planning	Accurate yield forecasting and growth progress mapping	Optimized harvest scheduling and reduced post-harvest loss	More stable crop supply with lower waste footprint
Long-term farm sustainability	Continuous multi-season data collection for trend analysis	Data-backed investment and production planning	Builds resilient and climate-adaptive farming systems

management. Their capacity to operate in diverse terrains, from dense forests and coastal wetlands to arid deserts and mountainous regions, allows conservationists and ecologists to observe ecological processes in real time without disturbing habitats. As environmental threats intensify due to climate change, habitat fragmentation, and natural disasters, UAVs are becoming essential instruments in shaping sustainable ecosystem management and conservation strategies.

8.5.1 Aerial guardians

Drones have become indispensable in wildlife monitoring, enabling researchers to observe species behavior, track animal movements, and conduct population surveys with minimal ecological disturbance. Equipped with thermal imaging, high-zoom cameras, and acoustic sensors, UAVs provide real time insights into animal distribution patterns, mating behaviors, and migratory routes that would otherwise be difficult to document due to terrain challenges or safety risks. This is particularly valuable in monitoring endangered species in protected reserves, where ground observation can be invasive or impractical.

UAVs also assist in anti-poaching operations, offering rapid surveillance coverage and the ability to detect human intrusions before harm occurs. Many protected areas now utilize drones to patrol park boundaries, detect illegal logging, and monitor fires. By integrating drone surveillance data with geospatial software and AI-driven pattern recognition, authorities can predict poaching hotspots or detect suspicious movement patterns early.

In broader conservation contexts, drones support ecosystem health assessments, helping scientists evaluate land degradation, vegetation loss, and changes in water bodies. These capabilities make UAVs central to contemporary conservation science, empowering decision-makers to plan targeted interventions and monitor environmental recovery over time.

8.5.2 Eco-restoration from above

Reforestation initiatives increasingly employ UAVs capable of seed dispersal and micro-pellet planting, allowing rapid restoration of degraded landscapes. Drone-based reforestation systems can plant thousands of seeds in a single mission, accessing steep slopes, burned areas, or remote forests where manual planting is impractical. This approach significantly accelerates ecosystem recovery and reduces labor intensity.

In habitat mapping, UAVs equipped with multispectral and LiDAR sensors generate accurate 3D vegetation models and topographic maps, enabling researchers to detect subtle ecological shifts. Such mapping helps identify vulnerable ecosystems, migration corridors, and biodiversity hotspots requiring protection. Drones also enable precise biodiversity assessments, including tracking plant species composition and monitoring the spread of invasive species.

By combining UAV data with ecological modelling and machine learning, conservation planners can track ecosystem changes over time, evaluate restoration impact, and design conservation strategies that maximize both ecological function and resilience.

8.5.3 Environmental resilience

Drones play a critical role in disaster response by offering real-time aerial assessments that guide rescue and rehabilitation efforts. During floods, wildfires, earthquakes, and cyclones, UAVs map damage extent, locate survivors, assess blocked routes, and deliver emergency supplies to inaccessible zones. Their rapid deployment and ability to operate in hazardous environments significantly reduce response time and improve coordination between rescue agencies.

In pollution monitoring, drones equipped with gas sensors, particulate analyzers, and thermal cameras monitor air quality, industrial emissions, oil spills, and chemical leaks. They can conduct systematic environmental sampling at different altitudes, allowing scientists to detect pollution trends and identify contamination sources that are otherwise difficult to monitor from ground level.

By streamlining disaster assessment and pollution surveillance, UAVs support environmental resilience and sustainability, helping communities prepare for and respond to ecological challenges more effectively.

8.6 Drones in Healthcare and Emergency Response Systems

The integration of drones into healthcare and emergency response systems represents a transformative development in modern public service infrastructure. UAVs have demonstrated unparalleled advantages in reaching remote or inaccessible regions, reducing response times in critical medical scenarios, and supporting efficient disaster management. Their ability to bypass terrain limitations, traffic congestion, and damaged infrastructure makes them highly valuable in delivering medical supplies, providing situational awareness during emergencies, and supporting search-and-rescue missions. As healthcare systems move toward more resilient and technology-enabled models, drones play an increasingly central role in bridging accessibility gaps, accelerating intervention timelines, and enhancing the overall effectiveness of humanitarian and medical operations.

8.6.1 Drone-based medical logistics and organ transport

Drones have become highly effective tools for transporting essential medical supplies such as blood units, vaccines, diagnostic samples, and medications, particularly in areas where road-based transportation is slow or unreliable. UAVs are capable of reaching remote villages, mountainous terrain, and rural health centers that often experience logistical delays, thereby reducing the time-critical risks associated with emergency medical care. For example, drone-based blood and vaccine delivery programs in Rwanda and Ghana have significantly reduced delivery times from several hours to just minutes.

More recently, drones have been tested and deployed for organ transportation, where the viability of organs such as kidneys, hearts, and livers depends on strict time and temperature conditions. UAV-based organ transport bypasses road traffic constraints and reduces the risk of delays, increasing the likelihood of successful transplant outcomes. Specialized medical drones equipped with secure, insulated storage and real-time telemetry ensure environmental control, location tracking, and delivery verification.

By combining speed, reliability, and accessibility, drones support life-saving medical logistics, improving emergency responsiveness and enhancing healthcare equity in underserved regions.

Table 8.7 summarizes the diverse roles of UAVs in healthcare logistics, showing how drones serve as rapid, reliable connectors within medical supply networks. By overcoming distance, terrain, and transport delays, UAVs enhance the speed, accessibility, and responsiveness of essential healthcare services, particularly in remote or emergency-affected regions.

8.6.2 First responders in the air

In disaster response scenarios, drones significantly enhance situational awareness and operational coordination. Earthquakes, floods, landslides, wildfires, and storms often damage transport infrastructure, making ground-based

Table 8.7 Functional roles of drones in healthcare delivery.

Application area	Primary function	Key benefits	Examples of use
Medical supply transport	Delivery of blood, vaccines, medicines, and diagnostic samples	Reduces delivery time; bypasses traffic and terrain; supports emergency care	Rwanda and Ghana national drone delivery programs (Zipline)
Organ transport	Rapid transport of transplant organs under controlled conditions	Maintains organ viability; increases transplant success rate	Hospital-to-hospital organ relay flights in the United States and India
Mobile laboratory linkage	Transport of lab specimens between remote clinics and diagnostic centers	Faster diagnosis; improved disease surveillance	Drone-supported tuberculosis and malaria testing networks
Public health outreach	Distribution of health education materials and community alerts	Supports awareness campaigns in isolated regions	Drone-based health messaging in flood and cyclone-prone zones

assessment slow and unsafe. UAVs equipped with real-time video, thermal cameras, and LiDAR mapping systems allow emergency teams to rapidly assess affected areas, identify structural damage, monitor disaster progression, and locate trapped survivors.

Search and rescue drones can detect heat signatures, broadcast audio instructions, and deliver first-aid kits or communication devices to individuals in need. In wildfire management, UAVs provide continuous aerial surveillance to track fire spread patterns, supporting more effective containment strategies. In flood and cyclone zones, drones help emergency agencies map safe evacuation routes and evaluate water levels without risking human lives.

By increasing the speed and precision of data gathering, drones improve the efficiency, coordination, and safety of disaster relief efforts, ultimately reducing loss of life and recovery times.

Table 8.8 outlines how drones significantly enhance disaster response efficiency by providing rapid situational awareness, mapping hazardous environments, and supporting coordinated rescue and relief efforts. Their ability to operate in dangerous or inaccessible areas helps protect rescue personnel, prioritize response actions, and accelerate recovery operations.

Table 8.8 Drone contributions to emergency response and disaster management.

Emergency operation area	Drone capabilities	Operational impact	Outcome/ advantage
Search and rescue missions	Thermal imaging, audio broadcasting, aerial mapping	Locating survivors quickly; guiding rescue teams	Faster rescue operations and reduced risk to rescuers
Disaster impact assessment	High-resolution imaging, LiDAR terrain mapping	Real-time damage estimation and hazard identification	Informed decision-making for relief planning
Wildfire monitoring	Heat signature detection, firefront perimeter tracking	Supports containment strategy; monitors fire spread	Reduced wildfire damage and improved firefighter safety
Flood and cyclone response	Water level mapping, route clearance analysis	Identifies accessible routes for relief supply movement	Efficient evacuation and supply chain coordination
Pollution and environmental hazard monitoring	Gas sensors, plume tracking, and spill detection	Identifies contamination hotspots and affected zones	Targeted cleanup and reduced ecological harm

8.6.3 Expanding medical access in remote and rural areas

Many communities worldwide face challenges in accessing consistent and timely healthcare due to geographic isolation, inadequate infrastructure, or conflict conditions. Drone-based healthcare initiatives help close this access gap by connecting hospitals, clinics, and medical supply hubs with rural and hard-to-reach populations. UAVs serve as aerial healthcare extensions, supporting telemedicine programs where diagnostic samples can be remotely analyzed, and medical advice can be delivered through hybrid drone-and-digital networks.

Additionally, drones play a crucial role in public health monitoring. They assist in vector control programs, mapping mosquito breeding zones, and aiding disease surveillance in malaria, dengue, and cholera-prone areas. Drones can also distribute educational materials and public health advisories during outbreaks.

As global health challenges and climate-related disruptions continue to strain traditional healthcare systems, drones offer a scalable, flexible, and resilient solution for ensuring healthcare continuity, particularly for vulnerable and marginalized populations.

8.7 Drones in Urban Infrastructure and Smart City Development

The increasing integration of UAVs into urban environments reflects a broader shift toward digitally networked, sensor-driven smart city ecosystems. Drones offer a unique advantage in observing, analyzing, and interacting with complex urban systems from the air, providing real-time data that can support infrastructure maintenance, transportation planning, public safety, and sustainable city management. Their ability to operate above-ground-level constraints makes them valuable assets in densely populated urban areas where traditional monitoring or inspection methods are costly, slow, or hazardous. As urban populations expand and cities pursue smarter, more efficient service models, drones are emerging as essential tools for creating resilient, responsive, and technologically advanced urban environments.

8.7.1 Aerial Inspectors

Urban infrastructure, such as bridges, flyovers, railway tracks, pipelines, cellular towers, and power lines, requires regular inspection to ensure structural safety and operational continuity. Traditionally, these inspections rely on scaffolding, cranes, manual climbing, or temporary shutdown methods that are labor-intensive, time-consuming, and sometimes dangerous. Drones equipped with high-resolution imaging, thermal sensors, and LiDAR mapping systems allow engineers to conduct detailed structural assessments quickly and without service interruption.

These aerial inspections detect micro-cracks, corrosion, misalignments, insulation faults, and heat leakage patterns, enabling proactive maintenance before failures occur. Moreover, repeated drone surveys create digital time-series records that support predictive maintenance models and digital twin infrastructure simulations. By minimizing inspection time, reducing worker exposure to hazards, and lowering maintenance costs, UAV-based monitoring enhances the reliability and sustainability of critical urban infrastructure networks.

8.7.2 Integrating UAVs into smart cities

Drones play an important role in improving urban traffic monitoring and mobility planning. Real-time aerial imagery enables traffic control centers to track vehicle flow patterns, identify congestion hotspots, monitor accidents, and evaluate the effectiveness of traffic signal timings. This dynamic visibility supports faster decision-making and adaptive traffic regulation, improving urban mobility efficiency.

In urban planning, UAV-based 3D mapping provides accurate spatial data for land-use analysis, construction planning, zoning, and environmental impact assessment. Drones help planners visualize terrain conditions, assess building density, identify heat islands, and monitor green space changes. This contributes to sustainable city design and supports data-driven policy-making.

By linking drones to smart city IoT networks, city authorities can integrate UAV-collected data into public transportation planning, emergency services coordination, and environmental monitoring platforms, making cities more responsive and interconnected.

8.7.3 Urban air mobility

Urban air mobility (UAM) represents the next frontier in drone-enabled smart cities, where passenger-carrying autonomous aerial vehicles will transport people across urban regions, bypassing congested road networks. Companies and governments are developing electric vertical take-off and landing (eVTOL) air taxis that can operate from small skyports integrated into city infrastructure. These systems promise to significantly reduce travel times for inter-city commutes and emergency medical evacuations.

Pilot programs in cities such as Dubai, Singapore, Paris, and Los Angeles demonstrate growing interest in UAM. However, large-scale adoption requires advancements in airspace integration, noise mitigation, battery performance, public safety regulations, and societal acceptance. Air traffic management systems must evolve to support low-altitude aerial corridors shared by drones and eVTOL vehicles.

Despite these challenges, UAM holds the potential to redefine the future of transportation. It aligns with sustainability goals by supporting electric propulsion mobility and contributes to a more efficient, multi-layered transportation framework in smart cities.

8.8 UAV Integration in Logistics and Distribution Systems

The integration of unmanned aerial vehicles (UAVs) in logistics and distribution systems represents one of the most significant advancements in modern supply chain transformation. Drones offer the ability to transport goods swiftly over short distances, bypassing ground-based limitations such as road congestion, difficult terrain, and limited transport infrastructure. Their use in last-mile delivery, warehouse coordination, real-time inventory monitoring, and emergency logistics has positioned UAVs as a strategic tool for enhancing the efficiency, responsiveness, and sustainability of supply chain operations. As global commerce increasingly prioritizes speed, reliability, and reduced environmental impact, UAVs are poised to play a central role in shaping future logistics ecosystems.

8.8.1 The evolution of last-mile delivery through drones

Last-mile delivery, the final leg in the distribution chain, has traditionally been the most time-consuming and costly part of logistics. Drones have introduced a transformative model where small-to-medium payloads can be transported directly from distribution centers to end-users or service nodes. E-commerce companies, healthcare logistics providers, and food delivery platforms have begun deploying UAV delivery fleets to shorten delivery times, particularly in congested urban settings or remote rural regions where traditional transportation is inefficient.

Early implementations focused on experimental and pilot projects, such as Amazon Prime Air and Google Wing trials, which validated UAV feasibility in delivering parcels weighing a few kilograms. Subsequent operational systems, such as Zipline in Africa, demonstrated the scalability of drone delivery for essential goods like medical supplies. Autonomous navigation, GPS precision landing, and route optimization algorithms have further improved the reliability and safety of these operations. As the demand for rapid fulfillment grows, drones are emerging as a core enabler of hyperlocal and same-day delivery services, addressing both logistical efficiency and customer expectations.

8.8.2 Real-time supply chain visibility and efficiency with UAVs

Beyond point-to-point delivery, UAVs contribute to end-to-end supply chain visibility through their ability to gather real-time data across distribution and storage environments. In warehouses and industrial yards, drones equipped with RFID scanners, barcode readers, and computer vision systems perform inventory audits, stock localization, and yard monitoring with minimal human intervention. These autonomous inspections reduce the need for manual stock-taking, prevent inventory errors, and support proactive supply chain planning.

In large-scale logistics networks, drones monitor vehicle movement, fleet parking zones, shipping dock conditions, and storage distribution layouts, providing continuous situational awareness. Integration with cloud-based supply chain platforms allows decision-makers to track inventory movement and detect disruptions instantly. This real-time connectivity strengthens predictive maintenance, demand forecasting, and route scheduling, improving operational throughput and reducing overall logistics costs. By functioning as aerial data collectors and delivery agents, UAVs enhance both the physical flow and information flow within the supply chain.

8.8.3 Large-scale implementation challenges in drone logistics

Despite their advantages, scaling UAV logistics systems to widespread commercial use presents several challenges. Regulatory constraints related to airspace access, flight beyond visual line of sight (BVLOS), and operations over populated areas remain significant barriers in many countries. Ensuring public safety, privacy protection, and controlled airspace integration requires coordinated development of unmanned traffic management (UTM) systems and adaptive regulatory frameworks.

Technical limitations also influence scalability. Current UAV models face payload capacity restrictions, limited battery endurance, weather sensitivity, and requirements for robust communication systems. Establishing reliable charging or battery-swapping infrastructure and secure landing/drop-off sites is essential for network-level deployment. Additionally, concerns related to cybersecurity, fleet coordination algorithms, noise pollution, and public acceptance require strategic solutions.

While large-scale implementation is complex, the rapid advancement of eVTOL cargo drones, 5G communication networks, automated ground handling systems, and AI-based airspace coordination indicates that UAV logistics will continue to expand. With coordinated regulatory evolution and infrastructure investment, drone logistics can become a core component of global supply chain transformation.

8.9 UAV Applications in Media and Visual Communication

UAVs have dramatically reshaped the media, entertainment, and communications industries by enabling new forms of visual production, live broadcasting, and storytelling. Traditionally, aerial imaging required cranes, helicopters, or large film rigs, techniques that were often expensive, logistically constrained, and limited in maneuverability. The introduction of compact, stable, GPS-enabled drones equipped with high-resolution cameras and real-time transmission systems has democratized aerial cinematography, making dynamic overhead perspectives widely accessible to filmmakers, journalists, content creators, and advertisers. Beyond creative applications, UAVs now function as rapid-response information tools, capturing events in real time and providing live situational visuals during news reporting, disaster coverage, and public communication.

8.9.1 Aerial imaging and real-time broadcast technologies

Drones are now standard tools in cinematic and broadcast production due to their ability to capture smooth, fluid aerial footage from angles that were historically difficult or impossible to obtain. Equipped with 4K/8K stabilized gimbal cameras, drones enable filmmakers to perform dynamic tracking shots, low-altitude sweeps, bird's-eye panoramas, and complex environmental fly-throughs with precision and repeatability. This capability has influenced film language, with drone shots becoming a hallmark of modern visual storytelling across documentaries, feature films, commercials, and live sports coverage.

Real-time broadcast technologies further expand UAV utility. With integrated low-latency video transmitters, drones can stream aerial footage directly to broadcasting centers, allowing real-time coverage of public events, ceremonies, sporting tournaments, and emergency situations. This provides instant access to a wider visual context, improving narrative clarity and enabling broadcasters to respond quickly to unfolding developments. Drones, therefore, function not only as imaging devices but also as information delivery systems, linking aerial visuals to public communication networks.

8.9.2 Drones in news collection and field journalism

In journalism, drones have become critical for gathering imagery and situational information in areas where human access is limited, dangerous, or restricted. UAVs allow field reporters to document natural disasters, conflict zones, protests, remote settlements, and hazardous industrial sites without endangering personnel. Their ability to fly low and access narrow or obstructed areas provides unique vantage points that complement ground-based reporting and satellite imagery.

Drones also support investigative and data-driven journalism, enabling long-duration surveillance for evidence collection in environmental violation cases, illegal mining operations, deforestation monitoring, and pollution mapping. The integration of UAV footage into news broadcasts adds depth and credibility to reporting by providing visually verifiable spatial context. However, their use in journalism also requires strong adherence to ethical and legal standards, particularly regarding privacy rights, flight authorization, public safety, and responsible narrative framing.

8.9.3 Creative and commercial uses in art and advertising

Beyond news and entertainment, drones have introduced innovative artistic and commercial possibilities in visual design and experiential communication. In advertising, UAVs enable attention-grabbing aerial product showcases, large-scale landscape branding shots, and immersive storytelling experiences. Drone light shows, where fleets of UAVs equipped with LED modules perform coordinated formations, have begun to replace fireworks at major celebrations, representing a sustainable, programmable, and visually synchronized art form.

In contemporary art practice, drones are used as interactive performance instruments, mobile projection platforms, and choreographed aerial sculpture components. Artists have explored drone soundscapes, drone-based painting, and aerial motion graphics, expanding the aesthetic language of kinetic and spatial art. These forms highlight the drone's unique position at the intersection of technology, perception, and creative expression, enabling entirely new modes of visual communication.

8.10 Case Study

8.10.1 Case study: Singapore – drone-enabled traffic monitoring and urban planning

Singapore has emerged as a global leader in integrating UAVs into *smart city governance*, particularly in the areas of traffic management and urban spatial planning. The *Land Transport Authority (LTA)* deploys drones to conduct real-time monitoring of major road networks, flyover corridors, and expressways. These drones feed *live aerial video and traffic density maps* to an integrated traffic operations control center, enabling authorities to identify congestion hotspots and respond promptly to accidents or bottlenecks. This airborne surveillance capability supports *adaptive traffic signal control*, dynamic route redirection for public transit fleets, and faster incident clearance, helping to maintain smooth traffic flows in one of the world's most densely populated urban environments (Bhattarai, 2024).

Beyond traffic control, Singapore's *Urban Redevelopment Authority (URA)* utilizes UAV-based 3D mapping and multispectral imaging to support *urban planning and land-use optimization.* Drone-generated digital elevation models and heat island mapping help planners evaluate rooftop solar potential, monitor green cover distribution, and design ventilation corridors that improve city-wide airflow. These insights are especially important in Singapore, where land scarcity demands *high-precision urban space analysis.*

The UAV data ecosystem is integrated with the nation's *Smart Nation Sensor Platform*, allowing seamless coordination between transportation, environmental, and emergency agencies. This approach exemplifies how drones serve as *aerial data infrastructure,* not isolated tools, but active components of a connected urban decision-making network. Singapore's case demonstrates that when strategically governed and digitally synchronized, UAVs can significantly enhance *mobility efficiency, environmental planning, and sustainable city management.*

8.10.2 Case study: Dubai – urban air mobility (UAM) trials with passenger drone taxis

Dubai is recognized as one of the first cities in the world to actively test and demonstrate urban air mobility (UAM) through the use of autonomous passenger-carrying drones. The Dubai Roads and Transport Authority (RTA), in collaboration with international aerospace firms such as Volocopter and EHang, began flight trials of passenger drone taxis as early as 2017. These trials demonstrated the feasibility of short-range, point-to-point aerial mobility within dense urban environments, showcasing Dubai's commitment to developing futuristic and technology-driven transportation systems (Madi, et al. 2024).

The drones used in Dubai's UAM trials are typically electric vertical take-off and landing (eVTOL) aircraft equipped with redundant motors, autonomous navigation systems, and emergency backup protocols. These systems are designed to transport one or two passengers over short intra-city distances without the need for conventional runways. During the test flights, the aircraft operated from specially designed skyport stations, highlighting new possibilities for compact aerial mobility infrastructure integrated into urban space. These skyports are envisioned to link business districts, airports, tourism hubs, and healthcare facilities, significantly reducing travel time compared to road transport.

To support safe operations, Dubai has invested heavily in developing low-altitude air traffic management frameworks, drone certification requirements, emergency response protocols, and digital tracking systems. These regulatory measures are being coordinated alongside the development of unmanned traffic management (UTM) systems to ensure secure flight corridors for both autonomous drones and future eVTOL air taxis. Despite remaining challenges, including battery endurance, weather sensitivity, noise control, and large-scale public acceptance, Dubai's strategic approach places it at the forefront of global UAM development.

This case illustrates how policy vision, regulatory planning, and public-private innovation can accelerate the adoption of next-generation aerial transportation. Dubai's UAM trials mark an important milestone toward transforming urban mobility from ground-constrained transport to a multi-layered air-ground network, representing a significant step toward the smart cities of the future.

8.10.3 Case study: France (Paris) – drone mobility for 2024 Olympic transport support

In preparation for the 2024 Summer Olympic Games, Paris has positioned itself as a major testing ground for drone-based urban mobility and aerial logistics systems. The French Civil Aviation Authority (DGAC), in collaboration with aerospace manufacturers and transport agencies, has initiated controlled flight corridors designed specifically for drone mobility operations. These corridors are intended to facilitate medical supply transport, rapid emergency response, VIP mobility support, and live event monitoring during the large-scale international event. The initiative reflects France's commitment to integrating UAV systems within regulated urban airspace to enhance operational efficiency in high-density environments.

A key component of the Paris Olympic drone mobility plan is the deployment of electric vertical takeoff and landing (eVTOL) aircraft from companies such as Volocopter, which are being tested for short-distance passenger transit across selected city zones. Specially designated vertiports compact landing and take-off platforms integrated into strategic locations – are being developed to support these flights. These vertiports are linked with Paris's broader transportation networks, demonstrating how aerial mobility systems may eventually integrate seamlessly with metro, bus, and rail services to create a multi-layered, hybrid transportation ecosystem.

Beyond passenger movement, UAVs will also support logistical and surveillance functions during the games. High-resolution imaging drones will provide real-time aerial monitoring of crowd movement, traffic flow, and venue security conditions. Medical logistics drones may be used to rapidly move defibrillators, blood units, and emergency equipment between event zones and medical stations, particularly when ground pathways are congested. These applications underscore the flexibility of UAVs in managing large events where timely and coordinated response systems are critical (KA AL-Dosari, et al. 2025).

While the Paris Olympic initiative demonstrates strong innovation potential, it also reveals several challenges requiring careful management. These include airspace safety coordination, noise considerations, public trust, regulatory harmonization with European Union aviation laws, and cybersecurity protection for autonomous vehicles. Nevertheless, Paris's proactive deployment showcases how international events can serve as catalysts for accelerated UAM testing and adoption, laying the foundation for long-term implementation of drone mobility in major world cities.

8.10.4 Case Study: United States – UAV-based infrastructure inspection in New York and California

In the United States, UAVs have become essential tools for monitoring and maintaining critical urban and industrial infrastructure, particularly in states such as New York and California, where aging networks and complex terrain present challenges for conventional inspection methods. Public agencies and utility companies now deploy drones to inspect bridges, power transmission lines, pipelines, rail infrastructure, and high-rise building façades, replacing traditional inspection processes that often require scaffolding, helicopter flyovers, or manual climbing. These drone-based inspection systems significantly reduce operational risk, cost, and inspection time, while producing high-resolution, geo-referenced imagery and structural condition datasets that support predictive maintenance.

In New York, the New York State Department of Transportation (NYSDOT) uses UAVs equipped with LiDAR scanners and optical zoom cameras to evaluate bridge joints, steel truss integrity, and corrosion hotspots in complex structural segments that are difficult or dangerous to examine manually. This approach eliminates the need to close roadways or erect temporary platforms, allowing inspections to be completed safely and with minimal interruption to urban mobility. UAV-generated digital models are incorporated into building information modeling (BIM) systems and digital twin platforms, enabling long-term structural monitoring and lifecycle planning (Duque, et al. 2018).

In California, the energy utility company Pacific Gas & Electric (PG&E) utilizes UAVs to inspect thousands of kilometers of overhead power lines that traverse mountainous, forested, and wildfire-prone regions. Equipped with thermal cameras and vegetation encroachment detection algorithms, these drones help identify frayed lines, overheated transformers, and hazardous tree overgrowth conditions. By detecting risks early, UAV-based monitoring contributes directly to wildfire prevention and grid reliability, which are major public safety concerns in the region.

These programs illustrate how UAVs support a shift toward data-driven infrastructure resilience. By improving inspection frequency, accuracy, and safety, UAVs enable municipalities and utility providers to transition from reactive repairs to predictive maintenance strategies. Despite challenges such as regulatory airspace limitations and the need for skilled data analysts, the success of UAV-based inspection initiatives in New York and California demonstrates the long-term value of integrating drones into urban asset management and public safety infrastructure.

8.10.5 Case study: India – Smart City Mission UAV integration

India's Smart City Mission has encouraged the adoption of UAV technologies to support urban governance, planning, and public service delivery in rapidly growing metropolitan areas. Cities such as Hyderabad, Varanasi, Surat, Bhopal, and Bengaluru have integrated drones into civic workflows to enhance monitoring, mapping, and decision-making capabilities. One of the major applications is high-resolution urban mapping, where UAVs generate 3D city models and geographic information system (GIS) layers to aid in land-use planning, infrastructure expansion, and redevelopment projects.

These maps allow municipal authorities to identify encroachments, assess informal settlements, plan road widening projects, and optimize public space usage with greater accuracy than traditional survey methods (Raman, et al. 2021).

UAVs have also become valuable tools in solid waste management and sanitation monitoring, particularly in cities such as Surat and Indore. Drones are used to track waste collection vehicle routes, identify illegal dumping hotspots, and monitor landfill capacity. By overlaying drone data with civic performance dashboards, city administrations have improved route scheduling, increased accountability among waste contractors, and strengthened environmental hygiene interventions under the national "Clean India" campaign.

Another impactful application is disaster risk assessment and flood resilience planning. For example, in Hyderabad and Chennai, UAVs are used to monitor lake boundaries, detect water encroachments, and model drainage flows to prevent urban flooding. During emergency situations such as the Kerala floods (2018), drones were deployed for search and rescue coordination, damage assessment, and relief distribution, helping authorities respond more rapidly and effectively. Similarly, in Varanasi, drone-based Ganga riverfront mapping supports both heritage conservation and water quality monitoring efforts.

The integration of UAVs under India's Smart City Mission highlights a shift toward data-driven urban governance, where real-time spatial intelligence guides policymaking. While challenges remain including airspace regulatory constraints, operator training requirements, and cybersecurity safeguards the expanding role of drones demonstrates India's commitment to leveraging affordable, scalable, and technologically advanced solutions to build more resilient, efficient, and citizen-responsive urban systems.

8.10.6 Case study: Zipline – medical logistics in Rwanda and Ghana

Zipline represents one of the most successful large-scale implementations of UAV technology in medical logistics. Launched in Rwanda in 2016 and later expanded to Ghana in 2019, Zipline's drone delivery network was designed to solve the problem of delayed access to life-saving medical supplies in regions where geographic barriers, road conditions, and limited transportation infrastructure often slow emergency healthcare response. Operating from strategically located distribution canters, Zipline uses fully autonomous fixed-wing drones capable of flying long distances to deliver essential medical products such as blood units, vaccines, antivenom, plasma, and emergency medications to remote clinics and district hospitals (Atiga, et al. 2024).

The drones follow GPS-guided flight paths and operate on a hub-and-spoke distribution model. Supplies are packed into specialized temperature-controlled cargo bays to maintain medical integrity during flight. Upon reaching the destination, the drone does not land; instead, it performs a mid-air drop using biodegradable parachutes, ensuring rapid delivery even where landing infrastructure is absent. Once delivery is complete, the drone autonomously returns to the distribution canter for recovery, battery replacement, and reloading. This system drastically reduces travel time what traditionally required three to six hours by road can now be completed in 15 to 30 minutes, even in mountainous or flood-affected regions.

Zipline's operations demonstrate measurable impact on public health outcomes. Rwanda's national blood distribution efficiency improved significantly, with documented reductions in maternal mortality during childbirth, emergency trauma fatalities, and stock-out incidents in rural health centers. In Ghana, the Zipline network supports over 2000 health facilities, enabling continuous vaccine availability during the national immunization drive and COVID-19 pandemic. The program is coordinated through national health ministries, making it a flagship example of how UAV logistics can be institutionally integrated into public healthcare systems, rather than functioning as isolated pilot projects.

The success of Zipline highlights the critical role of policy support, coordinated infrastructure planning, and local healthcare workforce collaboration in enabling UAV logistics at national scale. It also illustrates how drone-enabled medical supply chains can improve equity in healthcare access, strengthen national resilience during emergencies, and offer a model for other countries seeking to modernize essential service delivery.

8.10.7 Case study: Amazon Prime Air – commercial parcel delivery in the United States

Amazon Prime Air represents one of the most notable efforts to incorporate UAVs into commercial last-mile delivery networks. Announced in 2013 and refined through multiple phases of testing, Prime Air aims to deliver packages weighing up to approximately 2.5 kilograms directly to customers within 30 minutes of ordering. This initiative is driven by Amazon's broader logistics strategy of achieving ultra-fast, on-demand delivery while reducing reliance on traditional road-based courier networks, which are increasingly affected by traffic congestion, fuel costs, and environmental sustainability concerns.

Prime Air utilizes autonomous multirotor delivery drones equipped with computer vision, onboard collision avoidance systems, object-recognition sensors, and encrypted navigation software. These drones operate from local fulfilment nodes and fly directly to the delivery destination, where they safely lower packages into designated landing zones near the recipient's home. The drones maintain strict geofenced flight corridors and are programed to detect obstacles such as trees, wires, buildings, and moving objects. Several trial deployments have been launched in California, Texas, and select regions in the United Kingdom, under Federal Aviation Administration (FAA) and Civil Aviation Authority (CAA) regulatory oversight (Welch, 2015).

Despite technological readiness, large-scale implementation has faced challenges tied to regulatory approval, public airspace integration, noise control, varying suburban landscapes, and community acceptance. The FAA requires stringent safety certification for beyond-visual-line-of-sight (BVLOS) operations and flight over populated areas. Additionally, addressing privacy concerns and building trust among residential communities remain critical for urban-scale expansion. Amazon continues to work with aviation authorities to develop operational corridor frameworks, risk mitigation protocols, and drone identification tracking procedures.

Nonetheless, Prime Air has demonstrated the technical feasibility and service potential of drone-based parcel delivery. The program signifies a shift toward digitally coordinated, automated, and decentralized delivery networks that may eventually complement or replace ground-based logistics under certain scenarios. If regulatory and social acceptance hurdles are resolved, Amazon Prime Air could help establish UAV delivery as a mainstream component of commercial logistics, accelerating faster, cleaner, and more efficient distribution networks for consumer goods.

8.10.8 Case study: Google Wing – hyperlocal delivery in Australia and the United States

Google's Wing project is one of the most advanced examples of hyperlocal drone delivery, focusing on small, fast deliveries within residential and suburban communities. Unlike long-range medical or commercial parcel networks, Wing's system is designed to support short-distance, high-frequency deliveries such as groceries, medicines, food items, and household essentials directly to consumers' homes. Wing began community-scale operations in Canberra, Australia, where favorable regulatory conditions and suburban neighborhood layouts provided an ideal testbed. The program later expanded to additional regions in Australia and select cities in the United States, including areas in Texas and Virginia.

Wing drones are highly automated, lightweight, and equipped with precision flight controls and noise-reduction rotors, designed to minimize disturbance in densely populated neighborhoods. Drones are dispatched from small distribution hubs called "nests", where they autonomously take off, travel to delivery locations, and lower packages on tethers to avoid landing on the ground. This contactless delivery system reduces risks of theft, landing mishaps, or property interference. Orders are placed through a mobile application, allowing users to track drone flight progress in real time (Zenz, et al. 2024).

One of the most significant achievements of Wing is its community integration approach. Before deployment, Wing conducted public outreach programs to explain drone operations, safety systems, privacy measures, and noise control strategies. Feedback mechanisms were incorporated to adjust flight routes and reduce perceived neighborhood impact. This community-informed operational model helped Wing gain local acceptance and demonstrated that drone logistics can coexist harmoniously with residential lifestyles.

Wing's hyperlocal delivery network has had measurable positive outcomes, including reduced road congestion, lower carbon emissions compared to vehicle-based delivery, faster delivery turnaround times, and improved convenience for households. Its model provides a scalable blueprint for future neighborhood-level logistics, especially in cities seeking low-emission, efficient, and digitally integrated delivery ecosystems.

8.10.9 Case study: DHL Parcelcopter – mountain and island logistics in Germany

The DHL Parcelcopter initiative stands as one of the earliest and most consistent real-world demonstrations of drone-based logistics in remote and hard-to-access geographic regions. Launched by Deutsche Post DHL Group, the project was developed to address delivery challenges in mountainous Alpine villages and isolated North Sea islands, where conventional road or ferry transport is slow, weather-dependent, and operationally demanding. These regions presented an ideal testing ground for UAV logistics due to their low population density, irregular transport schedules, and harsh climatic conditions.

DHL designed a hybrid VTOL (vertical take-off and landing) UAV capable of transitioning between multirotor flight for take-off/landing and fixed-wing flight for extended cruising. This allowed the Parcelcopter to achieve both high maneuverability in confined landing zones and longer flight range needed for cross-water or high-altitude missions. Delivery operations were fully automated: parcels were loaded into the drone's secure cargo bay, dispatched from dedicated "SkyHub" stations, and flown autonomously to their destination. The system was supported by a drone-specific weather monitoring and flight authorization network, ensuring safe operation even under challenging environmental conditions (Baginova, et al. 2017).

During winter months, the Parcelcopter played an essential role in transporting medicine, emergency supplies, repair parts, and essential consumer goods to residents in regions where roads were snow-blocked or ferry routes disrupted. The project demonstrated that drone delivery significantly reduced transit time and weather-related failures, while also lowering the reliance on fuel-intensive transport vehicles. Moreover, local residents received deliveries more reliably and sustainably, supporting both daily living and emergency preparedness.

While the Parcelcopter program has not yet transitioned to nationwide deployment due to regulatory limitations and operational cost considerations, it remains a landmark example of how UAV logistics can strengthen supply resilience in remote areas. The DHL experience provides valuable insights into infrastructure design (SkyHubs), hybrid drone engineering, community acceptance strategies, and integration with national airspace systems. The lessons learned from Parcelcopter continue to influence emerging UAV cargo networks across Europe and beyond.

8.10.10 Case Study: CNN – drone-based news reporting and crisis coverage

CNN has been a global pioneer in the professional integration of UAVs into broadcast journalism. In 2016, the network formally launched CNN Air, a dedicated drone operations unit designed to enhance aerial reporting capabilities across news, documentary, and investigative programing. The goal was to improve situational awareness, visual accuracy, and narrative clarity during fast-evolving news events. Drones quickly became essential in covering natural disasters, humanitarian emergencies, protest movements, and conflict zones, where conventional ground-based or helicopter-based reporting was limited by access, cost, or safety conditions.

During hurricanes along the US coastline, drones enabled CNN reporters to capture real-time high-resolution imagery of damaged neighborhoods, flooded regions, and affected infrastructure, offering a wider context that helped emergency agencies and the public assess the severity of the situation. Similarly, during wildfire seasons in California, CNN's drone footage revealed fire spread patterns, evacuation challenges, and structural damage, supporting both crisis communication and relief coordination. These cases demonstrated the value of drones in delivering verified visual evidence during rapidly developing crises.

In addition to real-time live broadcasting, CNN has used drones extensively in investigative storytelling. UAVs have been deployed for environmental impact investigations, including pollution mapping, illegal mining monitoring, and coastal erosion documentation. In such cases, drone footage provided spatially and visually grounded evidence, strengthening the credibility of investigative findings and public accountability reporting (Niarchos, et al. 2022).

However, CNN's drone operations also underscore the importance of ethical and regulatory adherence in journalism. The network maintains strict compliance with FAA flight permissions, privacy protections, and newsroom editorial review protocols to ensure that aerial reporting does not intrude upon individuals or interfere with emergency operations. This illustrates how drones can be powerful journalistic tools when integrated responsibly, guided by clear professional standards and legal frameworks

8.10.11 Case study: National Geographic – wildlife documentaries and exploration filmmaking

National Geographic has been one of the most influential adopters of UAV technology for wildlife documentation, remote exploration, and environmental storytelling. Drone-based cinematography has enabled filmmakers and researchers to capture natural environments and animal behaviors that would otherwise be inaccessible due to geographical challenges, safety risks, or the need to avoid disrupting ecosystems. UAVs equipped with high-resolution, thermal, and multispectral cameras allow for silent, non-intrusive observation of wildlife habitats, enabling footage that combines scientific precision with visual artistry.

In regions such as the African savannas, drones have allowed National Geographic teams to film large migratory herds, predator–prey interactions, and nocturnal hunting behavior from aerial angles that minimize human presence and disturbance. Similarly, in the Arctic and sub-Arctic zones, UAVs have been used to track the movements of polar bears, seals, and whale pods across drifting sea ice, providing valuable material for both documentary narratives and ecological research. In tropical environments, such as the Amazon rainforest, drones have enabled canopy-level surveys, mapping deforestation patterns, and documenting indigenous cultural landscapes without requiring disruptive ground expeditions (Bouse, 1998).

National Geographic has also made significant use of drones in extreme and high-risk exploration filmmaking. UAVs have been deployed to survey active volcanoes, deep canyon systems, glacial crevasses, and high-altitude mountain terrains areas that are either too dangerous or physically unreachable for traditional film crews. These missions support scientific research teams by generating 3D terrain models, thermal activity maps, and time-lapse environmental change data, demonstrating the dual role of drones in research and visual communication.

The National Geographic case highlights the unique capacity of UAVs to bridge scientific documentation and public education. By providing visually compelling, accurate, and immersive footage, drone cinematography strengthens public awareness of biodiversity conservation, climate change impacts, and the fragility of global ecosystems. This integration of technology, environmental science, and storytelling has helped shape a new era of ethical, observational, and data-driven nature filmmaking.

8.10.12 Case study: Drone Light Shows – art, public performance, and sustainable entertainment

Drone light shows have emerged as a modern alternative to traditional fireworks, transforming public celebrations, cultural festivals, and artistic performances through coordinated aerial choreography. Companies such as Intel Drone Light Shows, SkyMagic, and Celestial design fleets of hundreds to thousands of lightweight drones equipped with programmable LED illumination systems. These drones perform synchronized movements controlled by centralized software platforms that calculate flight paths, formation patterns, and color transitions in real time. The result is a dynamic, three-dimensional visual spectacle that can create images, animations, symbols, and narrative scenes in the night sky.

One of the most notable advantages of drone light shows is their environmental sustainability. Unlike fireworks, which produce smoke, chemical residue, noise pollution, and wildlife disruption, drone shows are low-noise, emission-free, reusable, and programmable. Cities such as Shanghai, Singapore, Seoul, Dubai, and London have replaced major holiday fireworks with drone-based performances in order to reduce ecological impact and improve public safety. The Tokyo 2020 Olympic Games Opening Ceremony featured a drone formation of Earth floating above the stadium, symbolizing global unity through technological innovation an image widely recognized as an icon of modern aerial art (Waibel, et al. 2017).

Drone-based performances have also redefined the commercial and cultural marketing landscape. Corporations use drone light shows to launch products, celebrate brand milestones, and create immersive skyline-scale experiences. Tourism bureaus employ them to highlight city identity, heritage, and public messaging during festivals or national celebrations. Artists and creative directors increasingly collaborate with drone technologists to produce aerial ballet compositions, integrating soundscapes, storytelling, and synchronized movement. These performances expand the boundaries of public art by merging aerodynamic engineering, lighting design, digital choreography, and visual narrative composition.

However, drone light shows require careful planning, regulatory permissions, reliable GPS environments, and weather-stable conditions. Airspace coordination, battery safety, and crowd proximity rules are essential operational considerations. Despite these challenges, drone light shows represent a new paradigm in sustainable entertainment, demonstrating how UAVs can function not only as technical tools but also as mediums of expressive, cultural, and communicative art.

8.10.13 Case study: Bollywood and Hollywood – aerial cinematography in film production

The introduction of UAVs into professional film production has significantly transformed visual storytelling in both Bollywood and Hollywood, enabling filmmakers to achieve dynamic aerial perspectives that were previously limited by the high cost and logistical constraints of crane rigs, cable-cams, and helicopters. Drones equipped with stabilized gimbal-mounted 4K/6K/8K cameras provide filmmakers with unmatched flexibility in designing sweeping landscape shots, rapid chase sequences, top-down architectural reveals, and immersive environment flyovers. Their compact size allows for fluid camera motion in tight or complex locations such as narrow city streets, dense forests, and rugged terrains while maintaining smooth, cinematic image quality.

In Hollywood, films such as *Skyfall*, *Fast & Furious 7*, *Jurassic World*, and *The Revenant* employed drone cinematography to capture large-scale landscapes and high-action sequences without resorting to risky helicopter maneuvers. Drones enabled operators to film close-proximity shots at high altitudes and variable speeds while reducing both production time and operational hazards. This shift has contributed to cost efficiency, improved safety protocols, and creative freedom in modern filmmaking workflows, allowing directors to visualize scenes from new aerial vantage points.

In Bollywood, drone cinematography has become a powerful narrative and aesthetic tool, especially in epic-scale films and musical sequences. Films such as *Baahubali*, *Padmaavat*, *RRR*, and (*Uri: The Surgical Strike*) used drones to capture dramatic battlefields, royal cityscapes, mass choreography, and nature panoramas, enhancing visual grandeur and storytelling intensity. Bollywood cinematographers have also utilized drones in song sequences, wedding scenes, and location montages, where aerial movement enhances emotional and spatial context while reducing dependence on cranes and static wide shots.

The adoption of drones in cinematic production has also encouraged the development of hybrid production workflows, where aerial footage is integrated with CGI, motion capture, and digital compositing to create seamless visual environments. Regulatory frameworks in India and the United States now allow drone filming under licensed operators, flight plans, and safety compliance measures, ensuring responsible use in public and studio environments.

8.11 Summary

This chapter explored the transformative impact of unmanned aerial vehicles (UAVs) across modern systems, highlighting their rapid expansion from specialized tools to essential assets in multiple industrial, social, and technological sectors. UAVs have driven significant economic growth, generating new employment domains, improving sectoral productivity, and attracting sustained global investment. Their role in digital warfare has introduced autonomous surveillance and combat capabilities, alongside critical debates on ethics and accountability. In agriculture, UAVs enable precision farming, remote sensing, and sustainable food production, while in environmental conservation they serve as aerial guardians, supporting habitat monitoring, eco-restoration, and disaster resilience. UAVs have further strengthened healthcare systems through rapid medical logistics, organ transport, and emergency response operations. In urban environments, drones enhance infrastructure inspection, traffic intelligence, and smart city planning, and the rise of urban air mobility signals a future of aerial public transport. Logistics networks increasingly depend on UAVs for last-mile delivery and real-time supply chain visibility, despite regulatory and scalability challenges. In media and creative industries, drones have redefined aerial cinematography, news reporting, and large-scale artistic performances. The global case studies from Singapore's smart city monitoring and Dubai's air taxi trials to Zipline's medical networks in Africa, Amazon and Wing's delivery ecosystems, DHL's remote region logistics, CNN's crisis reporting, National Geographic's wildlife cinematography, and drone-based artistic light shows demonstrate how UAVs are reshaping operational landscapes worldwide, marking their role as a central force driving the next era of technological and societal evolution.

Review Questions

1. Define UAVs and explain their growing role in modern technological ecosystems.
2. How have UAV technologies expanded across different sectors of society?
3. Discuss how UAV innovation has influenced industrial transformation.
4. What societal challenges have UAVs helped to address in recent decades?
5. Explain the relationship between UAV adoption and digital modernization.
6. Describe the global growth trajectory of the UAV market.
7. What new employment opportunities have emerged due to UAV adoption?
8. Identify three sectors that have economically benefited from UAV integration.
9. How have UAVs contributed to reducing operational costs in industry?
10. Discuss investment trends shaping the future of UAV development.
11. What are autonomous combat drones, and how do they differ from remotely piloted systems?
12. Explain the tactical advantages drones provide in modern military operations.
13. How do drones reduce operational risks for armed forces?
14. Discuss an ethical concern associated with autonomous drone warfare.
15. What legal challenges arise when delegating lethal decisions to AI-driven UAVs?

16. Define precision farming and describe the role of UAVs in it.
17. How do drones assist in soil and irrigation monitoring?
18. Explain how UAV use supports sustainable food production.
19. Describe two benefits of using multispectral sensors in agriculture.
20. How do UAVs improve yield consistency and farm decision-making?
21. Explain how UAVs support wildlife monitoring and anti-poaching efforts.
22. Describe drone-based reforestation and seed-dispersion technologies.
23. How do drones contribute to disaster impact assessment and environmental resilience?
24. Discuss an example of drones used for habitat mapping.
25. What environmental advantage do drones offer compared to ground-based survey methods?
26. How do drones enhance medical supply delivery in remote regions?
27. Explain the importance of drone-based organ transport.
28. Describe the role of UAVs in search-and-rescue operations.
29. How do drones expand healthcare access during natural disasters?
30. What communication technologies support drone operations in emergency medicine?
31. How do drones improve infrastructure inspection efficiency and safety?
32. Describe how UAVs contribute to traffic monitoring within smart cities.
33. What is urban air mobility (UAM), and why is it significant for future transportation?
34. Identify challenges associated with integrating UAVs into urban airspace.
35. Explain how UAV-generated 3D mapping supports urban planning.
36. How do drones reduce last-mile delivery time in logistics networks?
37. Describe the use of UAVs in warehouse inventory monitoring.

38. What scalability challenges exist for widespread drone logistics adoption?
39. Explain how drones improve real-time supply chain visibility.
40. Discuss how UAV delivery supports sustainability goals.

9

Future of Drones

9.1 Introduction

The rapid evolution of unmanned aerial vehicles (UAVs) marks one of the most significant technological transformations of the 21st century, reshaping how societies operate, communicate, travel, and maintain critical systems. While drones initially emerged as specialized tools for military reconnaissance, they have since expanded into nearly every sector agriculture, logistics, healthcare, infrastructure, media, conservation, and urban planning. As advancements in artificial intelligence, autonomous navigation, swarm coordination, and energy-efficient propulsion continue to accelerate, drones are transitioning from remotely controlled machines into fully networked, intelligent aerial systems integrated seamlessly into daily life. The next era of drone development will not simply involve improved aerial platforms, but the creation of cooperative sky-based ecosystems, where UAVs communicate with ground networks, smart cities, and each other in real time.

The future of drones will be shaped by their ability to operate more autonomously, safely, and sustainably within shared airspace. Emerging concepts such as urban air mobility (UAM) and autonomous cargo corridors envision drones transporting passengers, medical supplies, and commercial goods across structured aerial highways. Meanwhile, bio-inspired flight mechanics, hybrid-electric propulsion, ultra-light materials, and quantum sensor-based navigation signal a shift toward more efficient, powerful, and climate-responsive UAV designs. At the same time, challenges involving air traffic regulation, privacy, cybersecurity, public trust, and ethical deployment must be addressed to ensure responsible integration.

As drones evolve from useful tools to infrastructure-level technologies, they will redefine mobility, emergency response, environmental management, and global communication systems. The future of UAVs lies not only in technological innovation, but in the emergence of aerial societies where the sky becomes an active, structured, and intelligently managed domain supporting human well-being, economic growth, and sustainable planetary stewardship.

DOI: 10.1201/9788743812418-9

9.1.1 Evolution and future pathways of drones

The evolution of drones reflects a steady progression from manually piloted remote aircraft to highly autonomous, intelligent aerial systems integrated into complex human environments. Early UAVs were primarily developed for military reconnaissance, where their value lay in extending human vision and reach without risking pilot safety. Over time, advances in GPS navigation, onboard computing, sensor miniaturization, and communication networks enabled drones to transition into civilian applications such as mapping, agriculture, logistics, infrastructure inspection, and public safety. Today, UAVs operate as data-driven platforms, capable of real-time sensing, analysis, and coordinated flight behavior.

Looking ahead, drone development is expected to move toward greater autonomy, seamless integration into shared airspace, and interconnectivity with smart city systems and digital infrastructure. Future UAVs will likely operate as fleets, communicating with one another through swarm intelligence, supporting tasks ranging from environmental restoration to synchronized cargo transport. Technologies such as electric vertical take-off and landing (eVTOL) vehicles, AI-based decision systems, advanced battery and hydrogen fuel cells, and quantum navigation sensors will push drones toward longer endurance, higher payload efficiency, and safer autonomous operation. As drones become more embedded in daily life from urban air mobility networks to emergency healthcare delivery their evolution will shape new societal norms, regulatory frameworks, and human technology relationships, making the sky an increasingly active and structured domain of modern civilization.

9.1.2 Future growth factors for drone technology

The continued growth of drone technology will be shaped by the interaction of technological innovation, policy development, economic demand, and societal acceptance. A major driver of future advancement is the rapid improvement in artificial intelligence and autonomous navigation systems, which will enable drones to independently plan routes, avoid obstacles, coordinate in swarms, and make real-time operational decisions. Progress in battery technology, hydrogen fuel systems, lightweight structural materials, and energy-efficient propulsion will increase flight endurance and payload capacity, expanding the range of viable commercial and industrial applications. At the same time, the expansion of high-speed communication networks, including 5G and satellite-based connectivity, will allow drones to operate safely and reliably over long distances beyond visual line of sight.

Meanwhile, the rise of urban air mobility (UAM) and smart city development will integrate drones into everyday public infrastructure, allowing aerial delivery services, emergency medical transport, and urban inspection operations to become routine. Increased adoption in sectors such as agriculture, disaster relief, power grid maintenance, environmental conservation, logistics, and entertainment demonstrates strong multi-industry demand for UAV solutions. However, the pace of drone growth will also depend on the development of supportive regulatory frameworks, including airspace management systems, operational certification standards, cybersecurity protection, and clear guidelines for privacy and safety. With coordinated efforts across technology, government, and industry, drones are positioned to evolve from specialized tools into permanent, interconnected elements of global economic and social systems, shaping the future of mobility, infrastructure, and digital service delivery.

9.2 AI and autonomous drones

The integration of artificial intelligence (AI) has been the most transformative step in the evolution of unmanned aerial vehicles, shifting their role from remotely operated tools to independently functioning aerial systems. While early drones required constant human guidance, modern UAVs are increasingly capable of perceiving, analyzing, and responding to their environments in real time. This transition is driven by rapid advancements in computer vision, sensor fusion, machine learning, neural network-based decision systems, and onboard processing power, enabling drones to navigate complex surroundings, recognize objects, avoid obstacles, and execute missions with minimal human input. As a result, drones are now able to operate in environments where GPS may be unreliable, communication links may be intermittent, or conditions may change unpredictably.

AI-driven autonomy is reshaping major application domains, including military operations, disaster response, logistics, agriculture, environmental conservation, media production, and smart city management. In military and security contexts, AI-powered drones support swarm coordination, reconnaissance, and collaborative combat operations, often functioning alongside manned aircraft. In civilian sectors, autonomy enables drones to perform precision agriculture, infrastructure inspection, and real-time traffic monitoring, while consumer-level drones use automated flight tracking to capture complex imagery without skilled piloting.

However, the rise of autonomous drones also brings forward important ethical, legal, and regulatory considerations related to safety, accountability, privacy, and airspace management. Ensuring responsible and secure deployment requires well-defined protocols, intelligent traffic control systems, and transparent human oversight frameworks. As AI continues to advance, drones will increasingly function as self-adapting aerial agents within interconnected digital ecosystems, marking a decisive step toward the development of intelligent, collaborative, and fully integrated aerial infrastructures of the future.

9.2.1 Full autonomy: Drones making real-time decisions without human intervention

Full autonomy in drones refers to the ability of UAVs to conduct entire missions independently, without the need for direct or continuous human control. This capability is enabled through the integration of advanced onboard sensors, real-time data processing, and AI-driven decision-making algorithms, which together allow the drone to perceive its surroundings, analyze situational variables, and adapt its actions accordingly. Fully autonomous drones can take off, navigate, avoid obstacles, carry out assigned tasks, and land safely even in unfamiliar or dynamic environments.

Such autonomy is especially valuable in high-risk or inaccessible areas, including disaster zones, remote terrain, deep forests, conflict regions, and offshore industrial sites, where human piloting may be dangerous or impractical. The drone continuously interprets input data from visual, thermal, LiDAR, and inertial sensors to generate a dynamic map of its environment and make instantaneous decisions such as rerouting during adverse weather or selecting alternative flight paths when obstacles appear.

As autonomous systems continue to develop, drones are gradually becoming self-reliant aerial agents capable of coordinating in groups, collaborating with ground systems, and fulfilling mission objectives with minimal oversight. This advancement represents a significant step toward intelligent, scalable, and adaptive UAV networks, forming the foundation for future applications in logistics, urban air mobility, defense, healthcare, and smart cities.

9.2.2 AI advancements in drone navigation, object detection, and mission planning

Artificial Intelligence has significantly enhanced the core operational capabilities of drones by improving their ability to navigate environments, interpret sensory information, and plan missions intelligently. Modern UAVs employ AI-driven navigation systems such as simultaneous localization and mapping (SLAM), which allow drones to build and update maps of unfamiliar environments while determining their position without depending solely on GPS. This is especially critical in indoor settings, dense urban areas, forests, and GPS-denied environments, where traditional navigation methods are unreliable.

AI-powered object detection and recognition systems further enable drones to identify terrain features, vehicles, structures, animals, and human activity with high precision. Through computer vision and deep learning models, drones can differentiate between obstacles and safe flight paths, detect structural damage, locate disaster survivors, assess crop health, or track moving targets all in real time. These perception capabilities have transformed drones into autonomous sensing platforms with applications in surveillance, agriculture, infrastructure inspection, and environmental monitoring.

In addition to perception and navigation, AI has improved mission planning and task execution. Algorithms can now optimize flight routes to conserve energy, avoid hazards, and complete objectives efficiently. Multi-drone coordination systems enable fleets of UAVs to divide complex tasks, share information, and cooperate as functional units, whether mapping large geographic areas or conducting search-and-rescue operations. As AI continues to advance, drones are expected to gain greater operational intelligence, adaptability, and collaborative capability, expanding their role in both civilian and strategic mission environments.

9.2.3 The role of machine learning and neural networks in drone optimization

Machine learning and neural networks play a central role in enhancing the performance, adaptability, and efficiency of modern drones. Unlike traditional programed systems that follow fixed instructions, machine learning enables drones to learn from experience, improve with continued use, and respond intelligently to changing environments. By analyzing large datasets such as flight histories, environmental patterns, sensor recordings, and visual imagery drones develop predictive capabilities that allow them to anticipate obstacles, adjust flight behavior, and refine operational strategies over time.

Neural networks, particularly convolutional neural networks (CNNs) and recurrent neural networks (RNNs), support advanced processing tasks such as real-time object detection, image classification, motion prediction, and route optimization. CNNs help drones recognize shapes, detect anomalies, and interpret visual scenes with high accuracy, which is crucial for activities such as surveillance, agricultural crop analysis, and infrastructure inspection. Meanwhile, reinforcement learning models enable drones to self-optimize navigation paths, improving stability, reducing energy consumption, and enhancing precision during complex maneuvers.

These learning-driven capabilities are foundational for emerging UAV applications such as autonomous swarms, collaborative robotics, precision delivery, and adaptive environmental monitoring. As machine learning algorithms become more efficient and computational hardware becomes lighter and faster, drones will increasingly operate as intelligent, self-adjusting aerial systems capable of making informed decisions with minimal human oversight. This shift marks a significant step toward highly resilient, adaptive, and scalable drone networks that can function reliably in diverse real-world conditions.

9.3 Swarm Drones and Collective Intelligence

Swarm drones represent a transformative advancement in UAV systems, where multiple drones operate together as a coordinated, intelligent collective rather than as individually controlled units. Inspired by natural swarm behaviors seen in flocks of birds, schools of fish, and insect colonies, swarm UAVs function through distributed decision-making, allowing each drone to respond to local environmental cues and communicate with others to achieve shared objectives. Instead of relying on a central controller, swarm intelligence enables drones to self-organize, reallocate tasks, maintain formation, and adapt dynamically if individual units are lost or conditions change. This significantly increases system resilience, mission efficiency, and operational scalability, making drone swarms suitable for tasks that are too complex, large, or dangerous for single UAVs or human-operated systems to manage alone.

The emergence of advanced artificial intelligence, real-time communication networks, cooperative path-planning algorithms, and lightweight autopilot systems has accelerated the practical deployment of swarm drones for both military and civilian applications. Swarm-based UAVs can perform simultaneous multi-point monitoring, area coverage, search operations, precision mapping, coordinated reconnaissance, and synchronized task execution far more efficiently than traditional aerial systems. As research evolves, swarms are expected to become integral components of smart battlefield networks, automated disaster relief frameworks, large-scale agricultural systems, and urban sensing infrastructures.

However, the increasing autonomy of swarm systems also brings forward significant considerations, including reliable communication protocols, secure swarm coordination, airspace deconfliction, energy management, and ethical responsibility in high-stakes applications such as defense. Ensuring that swarm drones behave predictably, safely, and transparently remains essential as they move from experimental demonstration to real-world deployment. Ultimately, swarm drones represent a shift toward collective aerial intelligence, where the sky becomes not merely a domain for flight, but a cooperative, adaptive, and shared operational environment.

9.3.1 Future applications of drone swarms in complex missions

Future applications of drone swarms are expected to expand significantly across domains where large-scale coordination, situational adaptability, and rapid coverage are essential. In environmental monitoring, swarms can simultaneously survey vast forest regions, assess wildfire spread, measure atmospheric pollution levels, and monitor coastal ecosystems in real time. Their ability to collect synchronized multi-point data enables high-resolution environmental modeling far beyond the capacity of single UAVs. In disaster response and humanitarian relief, swarms can rapidly deploy to locate survivors, map damage zones, identify safe evacuation paths, and deliver essential supplies to multiple locations at once, reducing response time and improving operational efficiency during floods, earthquakes, and large-scale accidents.

In smart agriculture, coordinated swarms can perform multi-field crop analysis, precision spraying, pest monitoring, and irrigation assessment in parallel, increasing agricultural productivity while reducing resource waste. Similarly, infrastructure and industrial operations may rely on swarms for pipeline inspection, power grid monitoring, construction site mapping, and mine surveying, where synchronized drone routes increase reliability and speed. Looking forward, swarm systems are anticipated to contribute to urban air mobility networks, where autonomous aerial fleets may coordinate traffic regulation, logistical routing, and real-time city sensing.

As swarm algorithms advance, drones will transition from support systems to autonomous mission participants, capable of developing shared strategies and adapting to uncertainties on the field. These future applications highlight the potential of drone swarms to function as distributed, resilient, and intelligent aerial systems, suited for complex and evolving mission environments that require real-time collaboration at scale.

9.3.2 Collaborative AI in swarm operations: Communication, coordination, and adaptability

The effectiveness of drone swarms depends on collaborative artificial intelligence, which allows multiple UAVs to communicate, coordinate, and adapt their actions as a unified system. Rather than relying on a single controlling node, swarm drones operate through distributed decision-making, where each drone exchanges real-time data regarding its position, velocity, environment, and mission status with nearby units. This constant information flow enables the swarm to maintain stable formations, avoid collisions, adjust flight patterns, and reassign tasks dynamically as mission conditions evolve. The communication framework is typically supported by mesh networking and low-latency inter-drone protocols, ensuring that the swarm functions cohesively even in changing or unpredictable settings.

Coordination within the swarm is driven by multi-agent AI models, where each drone follows simple behavioral rules that collectively produce complex group behavior. Algorithms drawn from swarm intelligence theory, reinforcement learning, and decentralized control systems allow the drones to distribute workload efficiently, for example, dividing survey areas, tracking multiple moving targets, or surrounding an object of interest. When one drone encounters a malfunction or threat, the swarm can automatically reconfigure its formation, demonstrating system-level resilience that is not achievable with single-UAV operations.

Adaptability is a defining strength of collaborative AI in swarm systems. By learning from real-time data and shared experiences, drone swarms can adjust to weather changes, avoid new obstacles, react to adversarial interference, and modify mission priorities without external intervention. This collective adaptability makes swarms particularly valuable for dynamic operations such as search-and-rescue, battlefield reconnaissance, wildfire monitoring, and synchronized infrastructure inspection. As collaborative AI continues to advance, swarm drones are expected to evolve into highly autonomous, self-organizing aerial networks capable of complex problem-solving in diverse mission environments.

9.3.3 Military and civilian uses for swarm drone systems

Swarm drone systems hold significant potential in both military and civilian sectors, with applications that benefit from wide-area coverage, coordinated task execution, and resilient distributed intelligence. In military operations, drone swarms are emerging as a key component of next-generation combat strategy, offering capabilities such as surveillance across contested terrain, electronic warfare support, precision strikes, and decoy maneuvers designed to overwhelm or confuse enemy defense systems. Their ability to function as a collective network, rather than as isolated aircraft, allows swarms to maintain mission continuity even when individual drones are disabled. Swarms can also coordinate with crewed aircraft in manned – unmanned teaming, enhancing the reach, safety, and tactical flexibility of combat forces.

In civilian contexts, swarm drones provide valuable support for disaster response, environmental monitoring, scientific research, and large-scale industrial operations. During earthquakes, wildfires, or floods, swarms can rapidly map damage, locate survivors, and deliver supplies to multiple areas simultaneously. In agriculture, coordinated swarms can perform field scanning, crop health assessment, and targeted pesticide or fertilizer application more efficiently than single-drone missions. For infrastructure management, swarms can inspect long linear assets such as pipelines, highways, and power grids in synchronized formations, accelerating inspection timelines and improving safety.

Swarm drones are also increasingly utilized in creative and cultural applications, such as synchronized aerial light shows and public art performances, where large numbers of drones form dynamic visual patterns in the sky. These displays highlight the potential of swarms not only as operational tools but also as expressive, programmable mediums of communication and cultural expression.

Collectively, the military and civilian uses of swarm drone systems demonstrate how collaborative autonomous aerial networks can perform complex, large-scale activities that exceed the capacity of individual UAVs. As swarm technologies continue to mature, they are expected to play a central role in shaping future security structures, emergency response frameworks, sustainable resource management systems, and public cultural experiences.

9.4 Urban Air Mobility (UAM) and Drone Taxis

Urban air mobility (UAM) represents a new paradigm in transportation where autonomous and semi-autonomous aerial vehicles operate within city environments to transport passengers and goods. Enabled by advances in electric vertical take-off and landing (eVTOL) systems, lightweight composite airframes, high-density battery technologies, and AI-based flight control, UAM seeks to reduce traffic congestion, shorten travel times, and introduce sustainable alternatives to traditional road transport. Unlike conventional aircraft,

UAM vehicles require minimal take-off space and can operate from compact hubs known as vertiports, making them suitable for dense metropolitan regions. Governments, industries, and smart city planners worldwide are preparing for UAM integration as part of next-generation multimodal mobility networks, where air and ground transport systems operate in synergy.

UAM development is shaped by three essential dimensions. First, the rise of passenger drones highlights the feasibility of short-range intra-city flights for airport transfers, emergency medical transport, and business district mobility. Second, cities must address infrastructure challenges, including vertiport placement, safe aerial corridor design, air traffic management, noise zoning, and charging logistics. Third, successful deployment requires robust regulatory frameworks governing airworthiness certification, autonomous operation permissions, flight safety protocols, and cybersecurity standards, alongside securing public trust and community acceptance.

Together, these developments signal a transformative shift in how cities may function in the near future where the sky evolves from passive space into a dynamic, structured, and intelligently managed mobility layer, complementing ground transportation and reshaping urban movement patterns.

9.4.1 The rise of passenger drones for urban transportation

The emergence of passenger drones, commonly referred to as eVTOL (electric vertical take-off and landing) air taxis, marks a significant step toward redefining how people move within and between urban environments. These aircraft are engineered to take off and land vertically, eliminating the need for runways and enabling operations from compact vertiports located on rooftops, transit hubs, commercial complexes, or dedicated platforms. Driven by advancements in electric propulsion, lightweight composite materials, autonomous flight systems, and smart navigation sensors, passenger drones aim to provide fast, direct aerial travel over congested road networks, thereby reducing commute time and easing pressure on urban transportation systems.

Several companies such as Volocopter, Joby Aviation, EHang, Lilium, and Airbus have developed and tested full-scale passenger drone prototypes capable of transporting one or more passengers safely and efficiently over short-to-medium intra-city distances. Pilot trials in cities like Dubai, Singapore, Paris, and select US regions demonstrate growing readiness for operational deployment. These passenger drones employ redundant safety systems, distributed electric propulsion units, autonomous stabilization controls, and low-noise rotor configurations, making them suitable for dense residential and commercial environments.

As passenger drones progress toward commercialization, they are increasingly viewed as a central component of urban air mobility (UAM) networks. Their integration promises to introduce rapid, low-emission, point-to-point air travel, particularly for airport transfers, medical emergency transport, and inter-business district mobility. While widespread adoption will require advancements in vehicle certification, vertiport infrastructure, airspace integration, and public trust, the rise of passenger drones signals the beginning of a multi-layered urban mobility ecosystem, where aerial transportation becomes as accessible and routine as ground-based travel.

9.4.2 Infrastructure challenges for integrating drones into cities

The successful integration of passenger drones into urban environments requires the development of specialized infrastructure capable of supporting safe, efficient, and coordinated aerial operations. Central to this is the establishment of vertiports designated take-off and landing facilities strategically located on rooftops, transit stations, commercial centers, and peripheral transport hubs. These vertiports must provide not only boarding and disembarkation platforms, but also charging stations, maintenance bays, passenger handling areas, and digital communication interfaces that connect aircraft to traffic management networks. Designing these facilities in dense cities presents spatial, architectural, and zoning challenges, particularly where building height regulations and land-use constraints are tightly controlled.

In addition to physical infrastructure, aerial traffic management is a critical requirement for large-scale urban drone operations. Cities will need unmanned traffic management (UTM) systems capable of monitoring flight paths, regulating airspace corridors, preventing collisions, and coordinating interactions between drones, traditional aircraft, and ground-based aviation services. This requires real-time data exchange supported by high-speed communication networks, including 5G, satellite connectivity, and low-latency ground relay systems. Furthermore, operating in urban landscapes introduces additional risks such as signal interference from tall buildings, sudden wind turbulence between structures, and noise footprint concerns that must be addressed through robust engineering and operational guidelines.

Energy infrastructure is another determining factor, as eVTOL vehicles require rapid charging or modular battery-swapping systems to support frequent, high-turnover operations. Cities must also consider emergency protocols, flight diversion routes, cybersecurity protections, and privacy safeguards, ensuring public confidence in the system's reliability and safety. These combined challenges highlight that the integration of drone taxis is not only a technological initiative but a complex urban planning and policy challenge, requiring coordinated progress across engineering, transportation design, regulatory frameworks, and community engagement.

9.4.3 Regulatory frameworks for UAM systems and public acceptance

The transition from experimental drone taxi demonstrations to widespread urban air mobility (UAM) services depends heavily on the development of clear, robust, and adaptive regulatory frameworks. Aviation authorities such as the Federal Aviation Administration (FAA) in the United States, the European Union Aviation Safety Agency (EASA) in Europe, and the Directorate General of Civil Aviation (DGCA) in India are currently designing standards for eVTOL airworthiness certification, pilot or operator licensing requirements, autonomous flight allowances, airspace integration procedures, and emergency safety protocols. These frameworks must address the unique characteristics of low-altitude urban flight, where drones operate in close proximity to buildings, pedestrians, and private spaces. Central to regulation is the establishment of unmanned traffic management (UTM) systems that coordinate drone taxis with conventional aircraft and ground-based transportation networks, ensuring safe and predictable aerial flows within complex city airspace.

Beyond legal oversight, public acceptance plays a decisive role in determining how quickly UAM systems are adopted. Concerns regarding noise levels, privacy, flight safety, crash risk, system reliability, and visibility in residential areas must be addressed through transparent communication and carefully managed pilot deployments. Community-engagement strategies such as demonstration flights, noise footprint trials, neighborhood consultation, and public education help citizens understand both the benefits and limitations of aerial mobility. Additionally, early UAM operations are likely to prioritize high-value use cases such as medical emergency transport and airport–city transfers, helping to build trust by demonstrating clear social and practical advantages.

Ultimately, the long-term success of UAM depends on the alignment of technical maturity, regulatory readiness, and societal confidence. As standards evolve, and as cities incorporate UAM into broader smart mobility planning, drone taxis have the potential to become a trusted, efficient, and widely utilized mode of urban transportation, reshaping how people move within future metropolitan landscapes.

9.5 Quantum Computing and Advanced Algorithms in Drones

The next major leap in drone capability lies in the integration of quantum computing and advanced algorithmic intelligence, enabling UAVs to handle complex decision-making tasks that exceed the processing limits of traditional onboard systems. Current drones rely primarily on classical processors to manage navigation, image interpretation, and mission execution; however, these systems face computational bottlenecks when operating in unpredictable environments, under heavy data loads, or in real-time multi-agent coordination. Quantum computing, with its ability to process information in parallel and evaluate multiple solution pathways simultaneously, has the potential to revolutionize drone autonomy, allowing UAVs to perform high-level reasoning, rapid optimization, and secure communication at unprecedented speeds.

By complementing quantum computational capabilities with machine learning, neural networks, and swarm coordination algorithms, drones can evolve into intelligent aerial systems capable of perception-driven decision-making, scaling from single-unit operations to large cooperative fleets. The convergence of these technologies paves the way for drones to function not merely as remote-controlled tools, but as adaptive, self-optimizing agents embedded in civilian, industrial, and defense infrastructures worldwide.

9.5.1 How quantum computing could revolutionize drone processing capabilities

Quantum computing introduces the possibility of extremely high-speed data analysis and parallel computation, enabling drones to evaluate complex sensory environments in real time. Quantum processors operate using qubits, which can hold multiple states simultaneously, allowing them to calculate and compare countless variables instantaneously. For UAVs, this means stronger capabilities in object recognition, environmental mapping, autonomous navigation, dynamic routing, and predictive analytics. Additionally, quantum-based encryption can secure drone communication systems against cyberattacks, ensuring data integrity and mission reliability in both civilian and military contexts.

9.5.2 Solving complex problems in real-time with quantum-enhanced drones

Quantum-enhanced drones excel at solving optimization challenges that involve many interdependent variables. For example, during disaster response, fleets of UAVs may need to determine the fastest rescue routes, identify safe landing zones, assess infrastructure damage, and coordinate supply delivery all simultaneously. Quantum algorithms allow drones to compute these scenarios in real time, enabling rapid adaptive decision-making under dynamic conditions. In surveillance and reconnaissance, quantum processing allows UAVs to analyze large-scale visual data streams, detect anomalies, and classify objects with high precision even in cluttered or changing environments. This capacity supports greater autonomy, efficiency, and operational safety.

9.5.3 Future applications in logistics, surveillance, and disaster response

The integration of quantum computing into drone systems holds promise across a wide range of future applications. In logistics, quantum-enabled UAVs could optimize cargo routing networks, reduce delivery times, and coordinate large autonomous fleets. In surveillance and environmental monitoring, drones could detect subtle atmospheric changes, analyze migration patterns, or monitor border and maritime zones with unprecedented analytical depth. In disaster response, quantum-capable drones could rapidly evaluate terrain conditions, locate survivors using multispectral analysis, and coordinate rescue teams through real-time swarm intelligence. These future scenarios underscore how quantum computing will enable drones to serve not only as tools of observation or delivery, but as strategic decision-support platforms across global systems.

Table 9.1 provides a comparative overview of three generations of UAV system development. Classical drones rely heavily on manual control and are limited in autonomous capability. AI-enhanced drones introduce learning-based perception, adaptive navigation, and partial autonomy, enabling more sophisticated civilian and industrial applications. Quantum-enabled drones represent the next evolutionary stage, where advanced computation, ultra-secure communication, and real-time optimization allow UAVs to operate collaboratively and intelligently within dynamic and complex environments. This progression highlights the shift from drones as remotely managed devices to self-adaptive, computing-driven aerial systems capable of supporting high-impact missions across future smart infrastructures.

9.6 5G and Beyond: Communication Networks for Drones

The evolution of drone capabilities is increasingly shaped by advancements in wireless communication networks, which determine how drones connect,

Table 9.1 Comparison of classical, AI-enhanced, and quantum-enabled drone systems.

Feature	Classical drones	AI-enhanced autonomous drones	Quantum-enabled next-generation drones
Core processing system	Standard microprocessor-based computation	Onboard AI chips/ neural processors	Hybrid quantum-classical processors
Decision-making	Human-controlled or semi-automated	Real-time autonomous decision-making	Parallel computation of multiple scenarios simultaneously
Navigation capability	GPS-dependent; limited obstacle avoidance	SLAM-based navigation and adaptive path planning	Hyper-optimized routing and navigation in complex + dynamic environments
Perception/ object recognition	Basic camera feed; manual interpretation	Computer vision + deep learning for object detection	Quantum-enhanced pattern recognition for high-complexity visual analysis
Communication security	Vulnerable to spoofing and signal interference	Encrypted digital communication with authentication	Quantum key distribution (QKD) and post-quantum cryptography
Fleet/swarm coordination	Difficult; requires centralized control	Multi-agent coordination with distributed AI	Highly scalable swarm intelligence with real-time collective optimization
Operational environment	Stable, predictable conditions	Works in partially dynamic environments	Capable of operating in highly dynamic, uncertain, or contested environments
Primary applications	Photography, hobby use, basic mapping	Agriculture, inspection, logistics, surveillance	Advanced logistics, defense autonomy, disaster intelligence, UAM control networks

share data, and coordinate operations in real time. Traditional radio and satellite-based control systems limit drone performance due to signal range constraints, limited bandwidth, and high communication latency, especially in complex environments. The arrival of 5G networks, and the development of future communication infrastructures such as 6G, low-earth-orbit (LEO) satellite networks, and edge computing platforms, have introduced the possibility of continuous, high-speed, and low-latency connectivity for UAV systems. These networks enable drones to perform data-intensive, long-range, and cooperative missions, making reliable communication a foundational requirement for the next phase of drone integration in civilian, industrial, and defense domains.

5G connectivity supports the transmission of high-resolution video, telemetry, environmental sensor outputs, and mission planning data in real time, allowing drones to respond instantly to changing conditions. This is especially important for beyond visual line of sight (BVLOS) operations, where drones must fly autonomously over long distances while maintaining stable command and control links. Enhanced network capacity also allows multiple drones to operate collaboratively, enabling swarm coordination, synchronized inspection, and shared mapping operations. As communication systems move toward hybrid architectures that combine terrestrial 5G networks with satellite relays and edge-cloud computing nodes, drones will gain the ability to operate seamlessly in both dense urban environments and remote terrain.

Future high-speed communication infrastructures will not only improve drone performance but also shape urban air mobility systems, where passenger drones and delivery networks rely on precise airspace coordination and continuous control. The shift to low-latency, high-capacity networks positions communication technology as the central nervous system of intelligent aerial mobility ecosystems. Thus, the advancement of drones is now inseparable from the development of next-generation communication networks, which will determine how drones integrate safely, efficiently, and broadly into the built environment.

9.6.1 How 5G and future networks will enable real-time drone connectivity

The introduction of 5G communication networks significantly enhances the real-time operational capabilities of drones by providing high bandwidth, low latency, and stable connectivity across large and complex environments. Unlike previous networks, where drones frequently faced signal delays or loss of control over long distances, 5G enables near-instantaneous data exchange between UAVs, ground control stations, cloud servers, and other drones in the vicinity. This allows drones to transmit high-resolution video, multispectral sensor data, geolocation information, and flight telemetry continuously and without interruption, which is essential for tasks such as live surveillance, precision agriculture, infrastructure inspection, and emergency response.

Future communication systems particularly 6G, satellite-integrated IoT networks, and edge-cloud computation frameworks are expected to extend this connectivity even further. These networks will allow drones to operate seamlessly in rural wilderness, maritime zones, mountainous regions, and dense urban environments where conventional radio communication struggles. With reliable real-time connectivity, drones can perform beyond visual line of sight (BVLOS) missions safely, coordinate with air traffic control systems, and interact dynamically with other aerial and ground units. Thus, 5G and future networking technologies form the backbone of high-autonomy, intelligent, and scalable drone ecosystems, enabling drones to become integrated, responsive actors within smart city and global mobility infrastructures.

9.6.2 Enhanced control and data transmission for long-range and complex drone missions

Advanced communication networks, such as 5G, play a crucial role in enabling drones to undertake long-range, multi-phase, and high-complexity missions that require uninterrupted command links and continuous data exchange. Traditional radio and Wi-Fi systems provide only limited range and often suffer from latency, signal interference, and reduced reliability when drones operate beyond visual line of sight. In contrast, 5G networks offer high-capacity data channels and real-time responsiveness, allowing UAVs to receive flight instructions, environmental updates, and mission adjustments instantaneously during operation. This ensures that drones can navigate accurately through changing conditions, avoid obstacles, and respond to mission-critical updates without manual intervention delays.

High-speed data transmission is particularly essential when drones are equipped with high-resolution cameras, LiDAR scanners, thermal sensors, or multispectral instruments, as these systems generate large volumes of data that must be processed and transmitted quickly for real-time decision-making. Through 5G and future hybrid networks that combine cellular, satellite, and edge computing, drones can stream data directly into cloud-based analytical platforms, enabling ground teams or automated AI systems to interpret results and update mission parameters immediately. This capability is vital in infrastructure inspection, precision agriculture, maritime surveillance, cross-border logistics, and emergency relief operations, where situational awareness and fast information feedback determine mission success.

By supporting both command-and-control stability and high-throughput data exchange, advanced communication networks transform drones into intelligent, fully networked aerial assets capable of performing coordinated and data-driven missions over extensive operational ranges. As network coverage and reliability continue to expand, long-range drone missions will become increasingly central to industrial monitoring, environmental management, national security, and global supply chain logistics.

9.6.3 The impact of high-speed, low-latency communication on drone operation

High-speed, low-latency communication networks fundamentally enhance the safety, responsiveness, and autonomy of drone operations. When drones can exchange data with control centers and other aircraft within milliseconds, they gain the ability to detect obstacles, adjust flight paths, and execute coordinated maneuvers in real time. This level of responsiveness is especially critical in dynamic or congested environments, such as urban air corridors, industrial sites, and disaster zones, where conditions can shift rapidly. Low latency reduces the risk of navigation errors, collision events, and system delays, enabling drones to function with far greater reliability and precision.

In autonomous and swarm-based operations, low-latency networking enables synchronized decision-making, allowing multiple UAVs to share sensor data and flight information instantly. This coordination supports complex missions such as multi-drone inspections, search-and-rescue grids, and intelligent traffic monitoring, where timing accuracy is essential. Similarly, in urban air mobility (UAM) systems, where passenger drones and cargo eVTOL vehicles must maintain strict spacing and follow structured air routes, high-speed communication ensures safe separation and integration with air traffic systems.

Furthermore, low-latency networks allow drones to rely on real-time cloud and edge computing, reducing the need for heavy onboard processors and enabling more powerful AI-based navigation, perception, and analytics to occur remotely. As a result, drones become lighter, more energy-efficient, and capable of more advanced decision-making. Ultimately, high-speed communication serves as the digital nervous system of future drone ecosystems, enabling UAVs to operate not as isolated aircraft, but as coordinated, intelligent, and seamlessly networked components of broader mobility and data infrastructure.

Table 9.2 compares major communication technologies relevant to drone connectivity and coordinated aerial operations. While 4G networks support only basic remote control and short-range missions, 5G introduces high-speed and low-latency connectivity, enabling real-time video streaming, BVLOS operations, and coordinated drone swarms. 6G networks, currently under research, are expected to support ultra-intelligent drone ecosystems

Table 9.2 Comparison of 5G, 6G, and satellite communication capabilities for drone operations.

Parameter	4G (legacy)	5G (current integration phase)	6G (future development)	Satellite/LEO networks
Latency	High (50–100 ms)	Very low (1–10 ms)	Ultra-low (<1 ms expected)	Moderate (20–50 ms depending on orbit)
Bandwidth/data rate	Limited data capacity	High data capacity, suitable for HD video streams	Extremely high data capacity, supports multi-sensor + AI inference	Variable; depends on satellite bandwidth and link stability
Network coverage	Urban and semi-urban areas	National networks; strong urban coverage	Global coverage envisioned with 6G mesh integration	Global, including remote and ocean regions
Drone swarm coordination ability	Low – delayed synchronization	High – supports real-time multi-UAV coordination	Very high – autonomous swarms with AI-distributed control	Moderate – dependent on line-of-sight and relay nodes
BVLOS flight support	Unreliable without relay systems	Supported with stable command-and-control links	Optimized for seamless BVLOS and autonomous routing	Strong in remote regions; weaker in dense urban areas
Integration with UAM/eVTOL systems	Not feasible	Feasible with supportive UTM systems	Fully integrated, core infrastructure layer	Supplemental support for long-range and inter-city missions
Primary use stage	Legacy operations	Current implementation and scaling	Research, testing, emerging pilot networks	Operational for remote sensing, maritime, defense, and expeditions

capable of autonomous decision-making and citywide aerial mobility integration. Meanwhile, satellite and LEO (Low Earth Orbit) networks expand operational range to remote, maritime, and mountainous environments. Together, these communication systems form the backbone of future adaptive, scalable, and globally connected UAV infrastructures.

9.7 Space Exploration and Drones

Drones are poised to play a critical role in the future of space exploration, offering mobility, adaptability, and autonomy in environments where traditional rovers, landers, and human astronauts face severe physical and operational constraints. Planetary surfaces such as those on Mars, the Moon, Titan, and asteroids present rugged terrain, thin or non-existent atmospheres, extreme temperature variations, and communication delays that make ground-based navigation slow and risky. Aerial and hopping drones, however, can navigate three-dimensional space, bypass obstacles, and survey wide regions quickly, making them ideal for scientific exploration, environmental mapping, and mission support operations.

Advances in lightweight composite materials, radiation-hardened avionics, autonomous navigation, multispectral imaging systems, and compact power sources are enabling drones to function effectively beyond Earth. Missions such as NASA's Ingenuity on Mars and the forthcoming Dragonfly mission to Titan demonstrate how drones can expand the scope of planetary science, discovering new features, identifying suitable landing sites, and collecting data that rovers and human teams cannot easily access. Over time, drones are expected to support resource prospecting, construction of extraterrestrial infrastructure, and logistical transport in support of sustained human presence in space.

As exploration shifts toward multi-planetary habitation and industrial utilization, drones will serve as autonomous partners, capable of performing hazardous, repetitive, or high-precision tasks. Their ability to operate in extreme environments marks a new phase in off-Earth robotics one where the sky, even on other worlds, becomes an active domain of scientific discovery and human expansion.

9.7.1 The potential role of drones in extraterrestrial exploration and planetary research

Drones offer significant advantages for extraterrestrial exploration because they can move freely across planetary surfaces and through atmospheric environments, where conventional rovers and land vehicles face severe limitations. Planetary terrains such as crater rims, canyon walls, volcanic fields, ice cliffs, and cavern openings are often inaccessible or hazardous for wheeled systems, greatly restricting the range and speed of surface exploration. In contrast, aerial and hopping drones can quickly traverse difficult landscapes, collect data from multiple locations in a short period, and provide high-resolution mapping across areas that would take rovers weeks or months to cover.

Equipped with multispectral cameras, atmospheric sensors, ground-penetrating imaging systems, and mineral detection instruments, drones can play a critical role in studying planetary geology, climate evolution, surface composition, and Astro biological potential. They can scout landing sites for future missions, locate promising regions for sample collection, and identify natural shelters such as lava tubes that could protect future astronauts from radiation. Because communication delays in deep space limit real-time piloting, extraterrestrial drones must rely on autonomous navigation and onboard decision algorithms, allowing them to adapt to terrain changes, weather variations, and unforeseen hazards independently.

By extending the reach of exploration missions, drones can greatly increase scientific return, enabling researchers to develop a more complete understanding of planetary histories, potential resources, and suitability for future human habitation. Thus, drones are not merely supporting tools, but critical scientific platforms in the expansion of humanity's presence beyond Earth.

9.7.2 Developing drones for missions on Mars

Designing drones for extraterrestrial missions requires overcoming environmental, mechanical, and operational challenges that differ significantly from conditions on Earth. Each planetary body presents unique constraints such as atmospheric density, gravity levels, surface composition, temperature extremes, and radiation exposure that directly influence drone structure, propulsion, and control systems. For example, Mars has a thin atmosphere, roughly 1% the density of Earth's, which reduces aerodynamic lift and requires drones to use large, lightweight rotor blades, high rotational speeds, and highly efficient power systems to maintain flight. NASA's Ingenuity Mars Helicopter demonstrated that even under such conditions, controlled flight is possible when drones are ultra-lightweight and supported by strong autonomous navigation algorithms.

On the Moon, the absence of an atmosphere eliminates the possibility of aerodynamic lift entirely, prompting the development of propulsion-based "hopper" drones that use small rocket thrusters or compressed gas bursts to make controlled jumps across the surface. These systems must be engineered to withstand extreme temperature fluctuations, abrasive lunar dust, and low-gravity dynamics, where even minimal thrust can create large motion changes. Similarly, future drone missions planned for Titan, Saturn's largest moon, will leverage its dense atmosphere and low gravity to enable highly energy-efficient flight. NASA's upcoming Dragonfly mission exemplifies this approach, using a dual-rotor octocopter to explore Titan's organic-rich landscapes and study prebiotic chemical processes.

Beyond planetary surfaces, drones designed for asteroids and orbital infrastructure will rely on micro-propulsion systems, cold-gas thrusters, ion engines, and robotic manipulators to maneuver in microgravity environments. They may be deployed to survey mineral-rich asteroids, support in-space construction, or assist in assembling solar power or fuel processing stations. These missions require drones to be radiation-hardened, autonomously intelligent, and capable of operating with intermittent communication links, since human command inputs are delayed over interplanetary distances.

The development of drones for space exploration thus represents a multidisciplinary engineering frontier, integrating aerospace design, robotics, artificial intelligence, astrobiology, and planetary science. As these technologies mature, drones will become essential partners in expanding human presence beyond Earth, enabling exploration, science, and early-stage infrastructure for future settlements on Mars, the Moon, and deep-space environments.

9.7.3 Drones in space mining and infrastructure development

As space agencies and private companies move toward establishing a sustained human and industrial presence beyond Earth, drones are expected to play a central role in space mining, resource assessment, and infrastructure development on the Moon, Mars, and near-Earth asteroids. In these environments, drones offer the ability to survey wide areas rapidly, identify resource deposits such as water ice, metallic ores, and volatile compounds, and map terrain to support site planning for mining operations or construction activities. Their aerial or micro-propulsion mobility allows them to reach locations inaccessible to rovers or human teams, including crater interiors, steep cliffs, lava tube entrances, and asteroid surfaces with irregular gravitational fields.

Drones will act as essential tools for early-stage resource evaluation, guiding where to deploy robotic miners, excavation devices, and processing units. In low-gravity and vacuum environments such as asteroids, drones equipped with ion thrusters or cold-gas microjets can provide precise maneuvering for material sampling, anchoring, and surface stabilization tasks. On the Moon and Mars, autonomous drones may support the construction of habitats, radiation shelters, solar power arrays, and communication relay towers, using regolith-based 3D printing and coordinated robotic assembly. They can also assist with transport of small payloads and spare components, extending logistical reach across scattered work sites.

In addition to physical operations, drones play a crucial role in real-time monitoring and maintenance of extraterrestrial infrastructure. They can inspect mining equipment for mechanical wear, monitor dust accumulation on solar panels, survey structural stability of habitats, and detect hazardous environmental changes such as radiation spikes or regolith shifting. Over time, networks of cooperative drones may form autonomous industrial ecosystems, coordinating with ground robots and orbital platforms to maintain continuous productivity without constant human oversight.

By enabling efficient resource extraction, adaptive construction, and operational resilience, drones will be instrumental in laying the foundation for long-duration scientific bases, commercial mining sites, and eventual human settlements in space. Their capabilities transform them from exploratory tools into core enablers of interplanetary industry and infrastructure development.

9.8 Case Study

9.8.1 Military swarm drones – cooperative autonomy in combat environments

Military swarm drones exemplify the application of distributed autonomy to achieve force-multiplying effects on the battlefield. Unlike single-platform UAVs, swarm systems consist of many relatively inexpensive, networked micro- or small-UAVs that cooperate to accomplish reconnaissance, electronic warfare, decoying, target acquisition, and precision strike missions. Each unit operates with local sensing and decision rules while sharing state information with neighbors; collectively the swarm self-organizes, reassigns tasks, and adapts to loss or interference without a single point of failure.

Operational employment of swarms emphasizes rapid area coverage and resilience. In reconnaissance roles, hundreds of small drones can distribute across a theater to generate high-resolution, multi-perspective sensor coverage, enabling commanders to detect movement, map terrain, and update situational awareness continuously. For offensive operations, swarms can perform saturation or distributed-attack tactics presenting many simultaneous threats to overwhelm air defenses and complicate targeting. Swarms are also used as electronic warfare nodes to create coordinated jamming patterns, and as decoys to draw enemy fire or reveal air-defense positions, preserving higher-value manned assets (Grimal, et al. 2018).

Technically, effective military swarms integrate four elements: robust local autonomy (collision avoidance, formation keeping), reliable inter-agent communication (mesh or relay architectures resilient to jamming), distributed tasking algorithms (multi-agent planning and dynamic reallocation), and scalable logistics (rapid rearming, deployment, and recovery). Real-world demonstrations and development programs have highlighted these requirements: air-launched micro-swarms to penetrate contested airspace, vehicle- or sea-based launchers for tactical rapid-deploy, and mixed-size swarms that combine long-endurance "motherships" with expendable sensor/strike nodes.

However, operationalizing swarm warfare raises important challenges and constraints. Communications in contested electromagnetic environments must resist jamming, spoofing, and interception; autonomy must be predictable, verifiable, and safe to prevent unintended escalation or fratricide; and mission assurance requires fault-tolerant designs so that degraded units do not compromise the whole. Logistics and sustainment mass-producing, transporting, recharging, and reloading large numbers of drones are nontrivial, especially in austere or expeditionary settings. Ethically and legally, swarms that carry lethal payloads intensify debates about accountability, target discrimination, and the delegation of lethal decisions to machines, prompting calls for clear doctrine, human-in-the-loop safeguards, and international norms.

Future trajectories point toward increasingly capable, heterogeneous swarms that combine sensing, communications, and effectors; tighter human–machine teaming where commanders set intent and swarms autonomously translate goals into tactics; and improved resilience through learning algorithms that adapt to adversary countermeasures. When carefully governed and technically hardened, military swarm drones can provide superior situational awareness, operational agility, and risk mitigation for personnel yet their deployment must be accompanied by rigorous oversight, robust cyber and electronic-warfare protections, and international dialogue on acceptable use.

Figure 9.1 illustrates a heterogeneous military drone swarm operating under a coordinated autonomous mission framework. A large, long-endurance mothership UAV functions as the central airborne command platform, distributing tasks and mission objectives to multiple classes of smaller unmanned

Figure 9.1 Heterogeneous military drone swarm architecture.

aerial vehicles. Surrounding the mothership are sensor drones, equipped with cameras and reconnaissance payloads used for terrain mapping, target identification, and situational awareness. Also present are electronic warfare drones, designed to conduct jamming, deception, and communication disruption against opposing forces, and loitering munition drones, which serve as precision strike units capable of engaging high-value targets. Communication between all drones occurs through a distributed mesh network, allowing real-time coordination, dynamic role reassignment, and resilience even if individual drones are disabled. The figure demonstrates how swarm intelligence enables drones to operate collectively as a flexible, adaptive, and mission-co-operative aerial combat system, rather than as isolated platforms.

9.8.2 Loyal Wingman UAV – Human–machine teaming in air combat

The Loyal Wingman concept embodies a shift from isolated unmanned systems toward tightly coupled manned–unmanned teaming (MUM-T) in air combat, where autonomous or semi-autonomous UAVs operate alongside crewed aircraft as cooperative partners rather than as remote tools. A loyal wingman acts on high-level intent issued by a human pilot or mission commander and translates that intent into tactical actions: conducting forward reconnaissance, carrying sensors or effectors, serving as electronic-warfare nodes, absorbing enemy fire as expendable assets, or extending the sensor/weapon reach of a manned platform. This model retains human responsibility for strategic decision-making while delegating hazardous, routine, or highly distributed tasks to the unmanned partner.

Technically, loyal wingman platforms combine robust autonomy, secure low-latency communications, redundant flight control, and modular payload architectures. They must be able to maintain formation with manned fighters, adapt to changing tactical contexts, execute coordinated maneuvers, and accept dynamic tasking with minimal supervisory bandwidth. Key capabilities include autonomous formation keeping, automated collision avoidance, cooperative target handoff, and autonomous execution of delegated missions under pre-approved rules of engagement. Communications and data links are designed for graceful degradation (e.g., operating with intermittent connectivity) and for resilience against electronic attack; embedded AI provides local decision logic when link to the human controller is constrained (Stensrud, et al. 2024).

Operational roles for loyal wingmen are diverse and mission-dependent. In reconnaissance and intelligence roles they can advance ahead of crewed aircraft to map threats and provide standoff sensing. As weapons carriers or decoys they can carry precision munitions or perform suppression of enemy air defense (SEAD) tasks that would otherwise risk human pilots. In contested environments they can act as sensor nodes to extend detection ranges, perform electronic-attack missions to blind adversary sensors, or loiter to provide continuous overwatch while crewed aircraft undertake higher-value tasks. Importantly, loyal wingmen enable task distribution and tempo amplification a single manned aircraft can control a small package of coordinated UAVs to achieve broader situational effect than the manned asset alone.

Despite the operational promise, numerous challenges must be addressed for practical deployment. Ensuring predictable, verifiable autonomy under combat stress is essential to avoid unintended escalation or fratricide; this requires rigorous validation, transparent decision logic, and rehearsed human override mechanisms. Communications must be highly secure and jam-resistant, with contingency modes for degraded connectivity. Logistics mass production, maintenance, rapid rearming, and forward basing pose practical constraints. Finally, ethical, legal, and doctrinal issues arise when autonomous systems are employed in lethal roles: clear chains of accountability, human-in-the-loop or on-the-loop safeguards, and international norms will be required to govern acceptable uses.

In sum, loyal wingman UAVs represent a pragmatic evolution toward collaborative air combat architectures: they preserve human strategic control while multiplying force effectiveness through autonomous teammates. When integrated with robust command-and-control, hardened communications, and mature autonomy safeguards, loyal wingmen can increase mission resilience, reduce pilot risk, and expand tactical options but widespread adoption will depend equally on technological maturation and careful operational, legal, and ethical governance.

Figure 9.2 illustrates the Loyal Wingman concept in which a manned fighter aircraft serves as the command platform, issuing mission intent and tactical direction to a group of autonomous unmanned aerial vehicles (UAVs). The Loyal Wingman UAVs perform complementary mission roles such as reconnaissance, electronic warfare, and precision strike while maintaining secure two-way communication links with the pilot. This cooperative architecture enhances situational awareness, extends combat reach, and reduces operational risk to human aircrew.

9.8.3 DJI ActiveTrack – civilian autonomy and real-time object tracking

DJI's ActiveTrack technology represents one of the most widely adopted real-time autonomous tracking systems in the civilian drone sector, demonstrating how AI-driven perception and flight control can make complex aerial operations accessible to non-expert users. ActiveTrack enables a drone to automatically identify, lock onto, and follow a moving subject such as a person, vehicle, animal, or sporting activity while maintaining stable framing and obstacle avoidance. Using computer vision, motion prediction algorithms, and multi-directional sensing, the drone continuously analyses the subject's movement and environment, adjusting its flight path autonomously to keep the target in view (Hansen, et al. 2024).

ActiveTrack operates in multiple tracking modes, including Trace (following behind or alongside a subject), Parallel (flying beside the subject at constant distance), and Spotlight (keeping the camera locked on the subject while the drone moves independently). The system uses deep-learning-based image classification to distinguish the tracked target from background objects

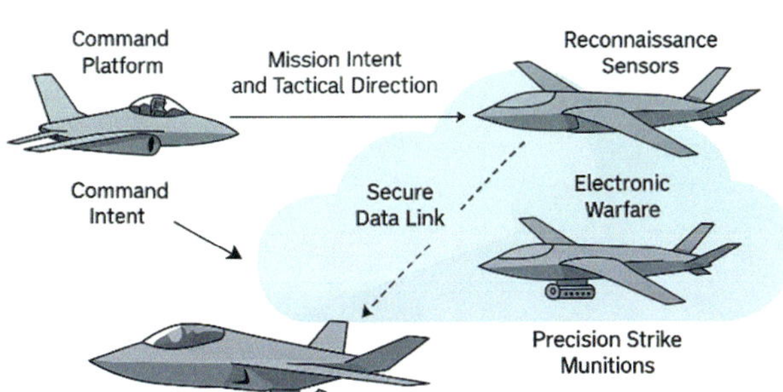

Figure 9.2 Loyal Wingman UAV – human–machine teaming in air combat.

and employs onboard sensing such as stereo cameras, infrared sensors, and downward-positioning modules to avoid collisions with trees, buildings, and terrain. This allows the drone to intelligently navigate complex environments where manual piloting would require considerable skill.

The impact of ActiveTrack is seen most prominently in film production, action sports recording, tourism, journalism, wildlife documentation, inspection, and personal hobby use. By removing the need for continuous joystick control, ActiveTrack democratizes aerial cinematography, enabling users to capture professional-grade dynamic footage with minimal effort or training. Furthermore, the system has influenced wider drone autonomy trends, serving as a foundational reference for consumer-level object detection, re-identification algorithms, and adaptive flight control behaviors.

While ActiveTrack highlights the benefits of accessible autonomy, it also raises considerations related to privacy, bystander awareness, and responsible use, particularly in public or populated environments. As such, its deployment reinforces the importance of user education, regulatory compliance, and ethical decision-making in everyday drone operations. Overall, DJI ActiveTrack stands as a landmark example of practical, safe, and user-friendly autonomy, bridging the gap between advanced AI technology and everyday civilian drone applications.

Figure 9.3 illustrates the DJI ActiveTrack system, in which a drone autonomously detects, locks onto, and follows a moving subject using computer vision and onboard AI. The drone continuously processes video frames in real time, predicts subject movement, and adjusts flight paths accordingly. Built-in obstacle avoidance sensors ensure safe flight while maintaining camera focus on the target. This system enables smooth autonomous cinematography and real-time tracking without manual piloting.

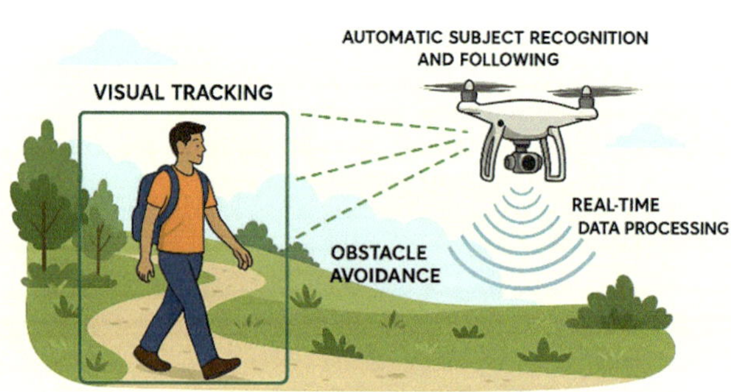

Figure 9.3 DJI ActiveTrack – real-time object tracking and autonomous following.

9.8.4 Indian Army – SWARM-100 Initiative

The SWARM-100 initiative of the Indian Army represents one of the most significant steps in advancing indigenous autonomous swarm warfare capabilities in South Asia. First demonstrated publicly during the Army Day Parade in 2021, the system consists of a coordinated group of 100 autonomous unmanned aerial vehicles, operating together as a distributed and resilient combat formation. Designed through collaboration between the Indian Army and domestic defense innovation startups, SWARM-100 reflects a shift from traditional single-UAV missions to collective, network-based autonomous operations.

The swarm is capable of performing a variety of mission profiles, including surveillance and reconnaissance, precision strikes using small explosive payloads, logistic resupply of critical materials, and decoy or distraction operations intended to overwhelm enemy air defenses. Each drone in the swarm is equipped with onboard processors, navigation sensors, and communication modules that allow it to make local decisions and share status information with other drones in real time. This distributed autonomy ensures that the swarm can reconfigure itself dynamically if some units are lost or disrupted, maintaining mission continuity under contested conditions.

One of the notable characteristics of SWARM-100 is its focus on high-altitude and rugged terrain operations, reflecting India's emphasis on mountain warfare and border security along geographically challenging zones. The system is designed to launch rapidly, navigate variable wind and temperature conditions, and execute coordinated target engagements without requiring continuous manual control. The swarm can be deployed from ground vehicles, forward posts, or portable launch units, offering mobility and tactical agility.

The SWARM-100 program demonstrates India's efforts to strengthen operational capacity through indigenous innovation, reduce dependence on imported UAV systems, and develop autonomous capabilities suited to its unique strategic environment. At the same time, its deployment raises considerations related to rules of engagement, security of communication links, electronic warfare resilience, and responsible operational use in conflict scenarios. As swarm technology evolves, SWARM-100 is expected to serve as a foundation for larger, more capable, multi-purpose autonomous swarms integrated into India's joint warfare doctrine.

Figure 9.4 illustrates the Indian Army's SWARM-100 initiative, which deploys a coordinated autonomous swarm of one hundred drones for reconnaissance, precision engagement, and encirclement missions. A central

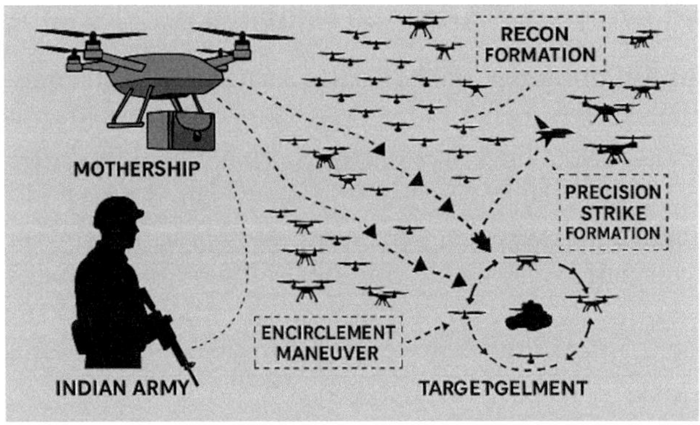

Figure 9.4 Indian Army – SWARM-100 autonomous drone swarm demonstration.

command node or "mothership" UAV issues mission objectives, while each drone uses distributed intelligence and a shared communication mesh to coordinate roles, avoid collisions, and adapt to dynamic battlefield conditions. The swarm can disperse, regroup, and execute multi-axis maneuvers, demonstrating scalable and resilient cooperative autonomy.

9.8.5 United States – Perdix Micro-Drone Swarm

The Perdix micro-drone program developed under US defense research initiatives served as an early and influential demonstration of practical collective autonomy in airborne swarms. Conceived as a low-cost, expendable micro-UAV that could be mass-launched from platforms such as fighter aircraft and tactical vehicles, Perdix highlighted key innovations in distributed decision-making, decentralized sensing, and fault-tolerant coordination. In public demonstrations, dozens to over a hundred Perdix units were released in mid-air and immediately organized into coordinated formations without centralized control, sharing sensor data, self-assigning roles, and dynamically reconfiguring when individual units were removed or failed.

Technically, Perdix exemplified several important design principles for combat-grade swarms: simplicity and affordability of individual units to enable mass employment; robust peer-to-peer communications to preserve cohesion under contested conditions; lightweight onboard processing to execute local autonomy rules; and collective algorithms that produce emergent behaviors (formation keeping, area coverage, target localization) from simple local rules. Operationally, these capabilities pointed to new mission sets distributed reconnaissance across denied areas, cooperative target identification, electronic-attack masking, and saturation tactics that complicate conventional air-defense responses.

The Perdix demonstrations also surfaced practical and doctrinal considerations. Ensuring secure, jam-resistant communications and resilient autonomy under electronic-warfare conditions proved essential; validating predictable swarm behavior under rules of engagement raised verification and legal questions; and logistics for mass production, launch, and replenishment became a central planning issue. Ethically, the program stimulated debate about autonomous lethal effects, human oversight, and command responsibility for swarm actions.

Figure 9.5 illustrates the US Perdix micro-drone swarm deployed from a fighter aircraft in flight. Once released, the micro-drones autonomously stabilize, establish a peer-to-peer communication network, and form coordinated swarm formations. The swarm operates without centralized control, using distributed decision-making to adapt to dynamic mission conditions. This demonstration showcases one of the earliest field-tested examples of cooperative autonomous drone swarming in defense operations.

9.8.6 China – Zhuhai air show swarm demonstrations

China has been a prominent leader in showcasing large-scale drone swarms, with the Zhuhai Air Show serving as the primary venue for public

Figure 9.5 US Perdix micro-drone swarm deployment.

demonstrations of its advanced swarm technology. At these exhibitions, hundreds to over a thousand drones have been launched simultaneously, performing highly synchronized aerial patterns, formation transitions, and cooperative maneuvering routines. While these displays often appear artistic in nature, they are underpinned by sophisticated mesh communication architectures, distributed AI control systems, and real-time swarm coordination algorithms, reflecting dual-use capabilities relevant to both civilian and military domains.

The Zhuhai demonstrations highlight China's emphasis on scalable swarm deployment, where the primary focus is not on the capabilities of any individual drone, but on the collective behavior of the group. Each drone operates with localized autonomy, continuously exchanging positional and state data with neighboring units to maintain formation stability and prevent collisions. This approach allows the swarm to reconfigure itself dynamically, enabling adaptable responses to interference, environmental change, or loss of individual drones. Such mass-coordination control systems could translate into operational advantages in scenarios involving reconnaissance saturation, electronic warfare masking, or multi-directional attack maneuvers.

From a strategic perspective, the scale and precision of the Zhuhai swarm demonstrations signal China's investment in both psychological deterrence and technological messaging. The public nature of these displays communicates technical maturity and national capability to domestic and international audiences. At the same time, the underlying algorithms and communication systems are believed to be evolving toward military applications, such as battlefield surveillance grids, drone-based area denial systems, and coordinated loitering munition swarms.

However, such advancements also bring forward challenges regarding command reliability in electronic warfare environments, secure communication under jamming, swarm predictability under stress, and safety protocols in civilian airspace. As China continues to refine its swarm technologies, the Zhuhai demonstrations serve as both technological milestones and strategic signals, marking ongoing progress toward autonomous, high-density unmanned aerial force structures.

Figure 9.6 illustrates a large-scale coordinated drone swarm as demonstrated during the Zhuhai Air Show in China. Hundreds of small unmanned aerial vehicles form synchronized aerial patterns under distributed control, communicating continuously through a shared mesh network. Each drone maintains awareness of neighboring units, enabling the swarm to reorganize, shift formations, and adapt to environmental changes without centralized command input. The figure emphasizes China's focus on scalable

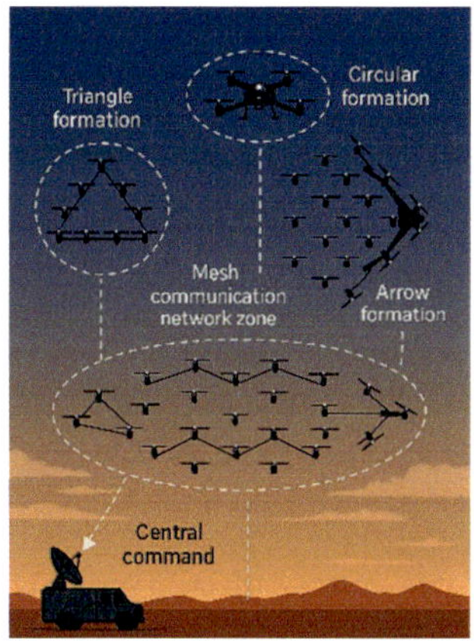

Figure 9.6 China – Zhuhai air show swarm demonstrations.

swarm autonomy, where collective behavior, formation discipline, and real-time coordination are prioritized over individual platform complexity. These demonstrations highlight the strategic potential of high-density drone swarms for surveillance coverage, electronic warfare masking, and psychological signaling in future aerial operations.

9.8.7 Dubai – early adoption of drone taxi demonstrations

Dubai is widely recognized as one of the first cities to publicly test and promote autonomous passenger drone services, positioning itself as a global pioneer in urban air mobility (UAM). Beginning in 2017, the Dubai Roads and Transport Authority (RTA), in collaboration with companies such as EHang and Volocopter, conducted full-scale demonstration flights of drone taxis intended for short-distance intra-city travel. These early trials showcased the feasibility of electric vertical take-off and landing (eVTOL) aircraft operating along designated aerial corridors to transport passengers between strategic urban hubs without conventional runways.

Dubai's approach to drone taxi deployment reflects a policy-first and infrastructure-driven strategy. The city-initiated planning for an integrated network of vertiports – compact aerial mobility stations located at transportation nodes, rooftops, and business districts. These vertiports are designed to provide passenger handling, safety procedures, rapid charging facilities, and digital interfaces linking aerial vehicles to the city's broader public transportation grid. The goal is the development of a multi-layered mobility ecosystem, where ground and air transportation operate in seamless coordination.

The early demonstrations also highlighted Dubai's commitment to regulatory innovation. The city worked proactively with aviation authorities to draft operational safety requirements, passenger protection standards, autonomous traffic management rules, and airspace zoning guidelines for low-altitude flight. Although full commercial rollout is still in development, Dubai's demonstration flights served to build public awareness and global confidence in the practicality and safety of passenger drones.

Figure 9.7 illustrates the early adoption of drone taxi technology in Dubai. A multi-rotor passenger eVTOL aircraft is shown approaching a rooftop vertiport, set against the modern Dubai skyline. These demonstration flights showcased Dubai's pioneering efforts in urban air mobility (UAM), emphasizing infrastructure planning, regulatory readiness, and integration of autonomous aerial transport into future smart city mobility networks.

9.8.8 Paris 2024 Olympic Games – UAM preparedness and air mobility corridors

In preparation for the 2024 Summer Olympic Games, Paris became one of the first major European cities to actively plan for urban air mobility (UAM) demonstration flights, using the global event as a proving ground for drone taxi infrastructure and regulated air mobility corridors. Supported by the French Civil Aviation Authority (DGAC), the airport operator Groupe ADP, and air mobility developers such as Volocopter, Paris established a framework for test vertiports, controlled low-altitude flight lanes, and monitoring systems to evaluate the feasibility of passenger-carrying eVTOL vehicles in a dense urban environment.

Figure 9.7 Dubai – early adoption of drone taxi demonstrations.

A dedicated test site was constructed at the Pontoise – Cormeilles-en-Vexin Aerodrome, serving as a UAM development hub for aircraft testing, safety certification procedures, noise level assessments, passenger handling simulations, and pilot training programs. The site also enabled the development of unmanned traffic management (UTM) protocols to coordinate drone taxis with existing air traffic and ground transport infrastructure. Paris also explored temporary Olympic air corridors connecting major transport nodes, such as Charles de Gaulle Airport, Paris-Le Bourget Airport, and central river-side vertiports along the Seine.

The initiative placed strong emphasis on public acceptance and environmental evaluation. Noise footprint studies were conducted to ensure that eVTOL operations met urban livability standards, and public exhibitions were organized to introduce citizens to aerial transport concepts. The program also prioritized data security, redundancy in autonomous flight controls, and emergency landing protocols, ensuring that early-phase deployments would proceed safely and transparently.

Although widespread commercial deployment will proceed gradually after the Olympic Games, the Paris initiative represents a model for structured UAM rollout, illustrating how large public events can serve as real-world test platforms for next-generation transportation. By aligning regulatory planning, urban design, and public outreach, Paris has positioned itself as a European leader in the integration of air mobility into metropolitan infrastructure.

Figure 9.8 illustrates a conceptual rooftop vertiport designed for urban air mobility operations. The layout includes a dedicated landing and takeoff pad for eVTOL passenger drones, a fast-charging or battery swap zone to support rapid turnaround time, a passenger boarding/safety-check area, and an elevator or transit access point connecting the vertiport to

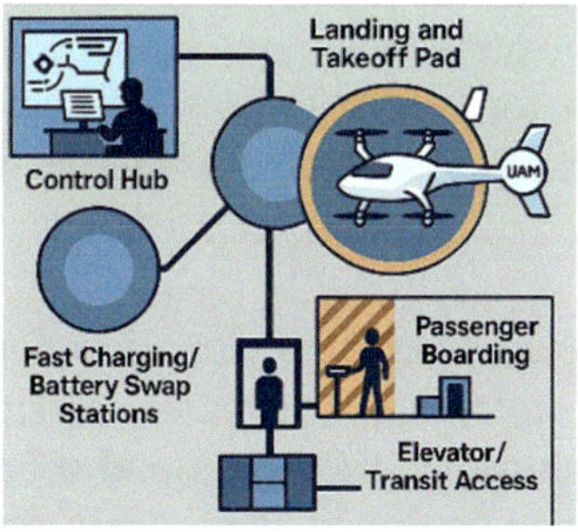

Figure 9.8 Conceptual layout of a rooftop vertiport for urban air mobility.

ground transportation networks. A centralized control hub coordinates arrival sequencing, clearance, and communication with city UAM traffic management systems. This configuration enables seamless, safe, and efficient multimodal transport integration within a smart-city air mobility framework.

9.8.9 Joby Aviation – FAA certification and commercial deployment pathway in the United States

Joby Aviation, a California-based aerospace company, is one of the leading developers of electric vertical take-off and landing (eVTOL) aircraft intended for commercial passenger service in the United States. Joby's eVTOL is designed to operate as a quiet, zero-operational-emissions air taxi, capable of transporting passengers across urban regions, bypassing traffic congestion and reducing travel times. With a targeted range of approximately 150+ km and cruising speeds around 300 km/h, Joby's aircraft is optimized for short to medium intra-city and regional travel.

What distinguishes Joby's progress is its structured and phased approach to FAA (Federal Aviation Administration) certification, which is essential for legal commercial flight operations in the United States. The company is pursuing Type Certification (aircraft meets regulatory airworthiness standards), Production Certification (manufacturing quality compliance), and Part 135 Air Carrier Certification (authorization to operate air taxi services). Joby has already obtained Part 135 certification, allowing it to begin operational training and early pilot service development. Flight testing continues under FAA oversight, including performance validation, safety redundancy demonstration, and noise certification trials.

To support deployment, Joby has partnered with Delta Air Lines to integrate drone air taxi services into airport ground-to-terminal mobility in major US metropolitan regions. This partnership focuses on reducing travel delays during airport transfers – one of the most promising early-phase commercial business models for UAM. Joby is also developing vertiport infrastructure, autonomous traffic management interfaces, and digital passenger booking systems designed to function seamlessly with existing city transport networks.

However, progress is influenced by public safety assurance, regulatory harmonization, airspace integration logistics, and community acceptance, especially regarding flight noise footprints and urban aerial traffic density. Joby's phased certification pathway demonstrates a gradual introduction model, beginning with piloted operations, followed by eventual transition to greater levels of autonomy as regulatory confidence and supporting infrastructure mature.

Figure 9.9 illustrates the internal architecture of the Joby eVTOL air taxi vehicle. The aircraft features six tilt-rotor electric propulsion units, allowing vertical takeoff, hover, and efficient forward flight. A high-voltage battery pack supplies the electric drive system, while distributed electric motors provide thrust redundancy for safety. The cabin accommodates multiple passengers with crashworthy seating structures. Safety provisions include a whole-aircraft parachute system, structural energy-absorbing landing gear, and multiple redundant control circuits. The configuration highlights Joby Aviation's approach to quiet, zero-operational-emission urban air mobility.

9.8.10 United States: Quantum-resilient UAV communication architecture

The United States has been actively developing quantum-resilient communication architectures for unmanned aerial systems to protect them against emerging cyber and electronic warfare threats. As drones become increasingly autonomous and network-dependent, traditional encryption and radio-based control links face rising vulnerabilities – especially with the anticipated future capabilities of quantum computing, which could rapidly break classical cryptographic algorithms. To mitigate these risks, US defense research institutions such as the Air Force Research Laboratory (AFRL), DARPA, and MIT Lincoln Laboratory have introduced secure communication frameworks designed to ensure confidentiality, integrity, and continuity of UAV operations even in contested electromagnetic environments.

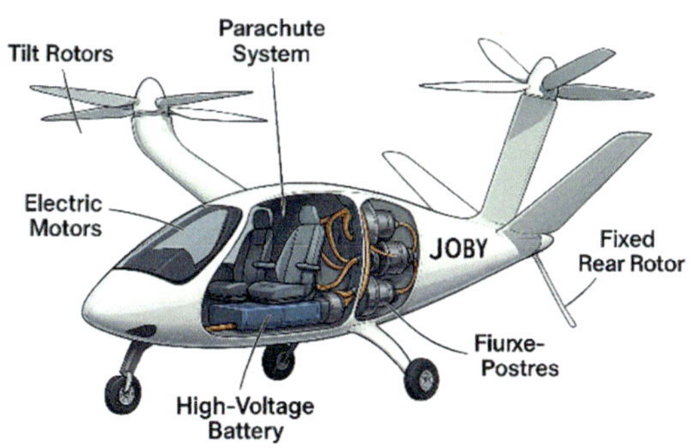

Figure 9.9 Joby eVTOL cutaway diagram (propulsion and safety systems).

A core strategy of this architecture is the integration of post-quantum cryptographic algorithms, which rely on mathematically complex lattice-based encryption that remains resistant to quantum decryption attacks. In addition, experimental systems are incorporating quantum key distribution (QKD), enabling drones and ground control stations to exchange encryption keys using the quantum properties of photons. Any attempt to intercept or tamper with these keys results in detectable disturbances, allowing the system to identify intrusions in real time and switch to alternate secure channels.

The US approach also emphasizes decentralized mesh networking among UAV fleets, enabling drones to maintain operational coordination even if satellite links or command hubs are jammed. In this configuration, individual UAVs function as autonomous communication nodes, relaying data through neighbor-to-neighbor pathways to preserve command-and-control capability. This enhances resilience by preventing the loss of a single communication relay from compromising the entire system.

Operational testing has demonstrated that quantum-resilient communication enables drones to operate reliably in GPS-denied, signal-jammed, or cyber-contested airspace, which is increasingly common in modern military theatres. Such architecture is fundamental for loyal wingman UAVs, swarm drones, reconnaissance assets, and long-range autonomous strike platforms that must maintain mission integrity even under electronic warfare pressure.

While the technology is promising, challenges remain, including the miniaturization of QKD hardware, ensuring energy efficiency for onboard processors, and establishing interoperability standards across allied defense systems. Nonetheless, the US quantum-resilient UAV communication initiative marks a critical step toward securing next-generation autonomous air operations, positioning the architecture as a foundational layer for future air combat networks, unmanned traffic management, and global drone mobility systems.

9.8.11 European Union: Quantum key distribution for drone swarm networks

Within the European Union, research into drone swarm communication security has increasingly focused on the deployment of quantum key distribution (QKD) to enable secure, tamper-evident communication among autonomous UAV networks. Supported by initiatives such as the EU Quantum Flagship Program, defense research agencies, and national laboratories in France, Germany, and the Netherlands, these efforts aim to protect swarm coordination links from interception or signal spoofing – particularly in future operational environments where adversaries may possess advanced electronic warfare or quantum-enabled decryption capabilities.

QKD operates by transmitting encryption keys encoded in the quantum states of photons. Any attempt to intercept or alter these transmissions disturbs the quantum state, immediately revealing a breach. This makes QKD-based communication inherently intrusion-detectable, offering a stronger alternative to classical cryptographic systems, which rely solely on computational complexity. In swarm configurations, QKD can be used to continuously refresh encryption keys between drones, ensuring that communication links remain secure even as the swarm moves, changes formation, or loses individual units.

European researchers have demonstrated airborne QKD links between UAVs using both laser-based free-space communication and optical wireless networking, enabling encrypted swarm coordination without dependence on vulnerable radio-frequency channels. This approach is particularly advantageous for dense urban environments, border surveillance operations, and critical infrastructure protection, where controlled low-altitude airspace must remain secure.

However, the deployment of QKD for UAV swarms also presents challenges, such as ensuring line-of-sight optical alignment, maintaining stability in atmospheric turbulence, and miniaturizing quantum communication modules to fit within SWaP-constrained (size-weight-and-power limited) drone platforms. Ongoing EU research is working toward compact, ruggedized QKD payloads and hybrid optical-RF handover protocols, which allow secure communication even when environmental conditions temporarily disrupt optical links.

The EU's QKD swarm programs represent a broader strategy to create high-trust autonomous air networks that can operate resiliently in digitally contested and surveillance-sensitive environments, forming a critical foundation for future European defense, civil aviation security, and humanitarian drone operations.

9.8.12 Verizon Skyward – 5G drone operations in the United States

Verizon's Skyward platform played a pioneering role in demonstrating how 5G cellular networks can support real-time, long-range, and data-intensive drone operations in the United States. Before its program concluded in 2022, Skyward worked in partnership with commercial operators, industrial clients, and public safety agencies to prove that drones connected through carrier-grade cellular networks could maintain stable command-and-control links, stream high-resolution imagery, and coordinate multi-drone missions beyond visual line of sight (BVLOS). These trials marked a significant step toward integrating drones into national airspace and developing scalable urban and regional drone service networks.

Skyward's core contribution was the operational validation of network-based flight control and telemetry transmission. By leveraging Verizon's 5G infrastructure, drones were able to access low-latency, high-bandwidth uplink and downlink channels, allowing operators to receive continuous situational data, detect hazards, and adjust mission parameters dynamically. This capability proved especially valuable in industrial asset inspection, emergency response, utility monitoring, and real-time event surveillance, where uninterrupted data flow is essential for decision-making and safety.

The program also contributed to the development of unmanned aircraft system traffic management (UTM) integration frameworks in the United States, working alongside the Federal Aviation Administration (FAA) and NASA. Skyward's testing helped shape protocols for flight approvals, dynamic airspace awareness, and automated remote pilot notifications, supporting the broader national goal of establishing safe and standardized UAM operations.

Although Skyward's commercial service was discontinued, the technological and regulatory foundations laid by its initiatives continue to influence current drone communication projects in government, industry, and telecom sectors. The program demonstrated that 5G connectivity can serve as a backbone for large-scale autonomous drone ecosystems, enabling fleets to operate collaboratively, safely, and with high information fidelity.

Skyward's role is best understood as a transitional milestone in the progression toward fully integrated urban air mobility, illustrating how telecommunications infrastructure, aviation standards, and industry partnerships must align to unlock the full potential of connected UAV systems in the United States.

Figure 9.10 illustrates Verizon Skyward's use of 5G cellular networks to support real-time drone operations in the United States. The drone communicates with ground control through a high-bandwidth, low-latency 5G network tower, enabling continuous telemetry streaming, live video transmission, and beyond visual line of sight (BVLOS) command and control. This architecture demonstrates how telecommunications infrastructure forms a foundational layer for scalable, safe, and integrated UAV missions in industrial, emergency response, and urban mobility applications.

Figure 9.10 Verizon Skyward – 5G-enabled drone operations.

9.8.13 Nokia Drone Networks – automated surveillance and emergency response in Europe

Nokia's Drone Networks platform represents one of the most advanced examples of integrated, automated drone operations deployed for public safety, industrial monitoring, and emergency response across Europe. The system combines 5G/LTE private networks, autonomous drone docks, high-definition sensors, and an AI-enabled control platform to allow drones to be launched, flown, monitored, and recovered remotely, without an on-site operator. This model shifts drones from being manually piloted tools to networked aerial assets functioning as part of a continuous surveillance and incident-response infrastructure.

The drones operate from weather-protected docking stations, where they can automatically charge, upload data, and receive updated mission instructions. When triggered either manually by command authorities or automatically through sensor alerts drones can be deployed in seconds to respond to fires, intrusions, industrial accidents, or natural disasters. Using real-time streaming over 5G/LTE, video and telemetry data are transmitted directly to emergency command centers, enabling faster situational awareness, decision-making, and coordinated response.

Nokia Drone Networks also integrates AI-based video analytics, allowing the system to detect anomalies such as unauthorized movement, structural damage, hazardous leaks, or evolving fire boundaries. This enables proactive and predictive emergency monitoring, particularly in ports, industrial factories, critical infrastructure zones, border environments, and urban security monitoring.

In addition to public safety, the system supports search and rescue, environmental inspections, and maritime patrols especially where ground access is limited or hazardous. The reliance on private cellular networks, rather than open public networks, enhances data security, communication resilience, and airspace safety, aligning the platform with European regulatory standards for unmanned aircraft operations.

Nokia Drone Networks demonstrates how network-enabled autonomy can transform drones into 24/7 aerial response units, expanding both the scale and reliability of drone-based surveillance and rescue missions. It also serves as a model for the integration of UAVs within smart-city, industrial safety, and national emergency management frameworks across Europe and beyond.

Figure 9.11 illustrates the Nokia Drone Networks platform, which enables fully automated drone operations using secure 5G/LTE communication links. The drone launches from an autonomous docking station that provides charging, data transfer, and mission updating without the need for an onsite operator. Real-time video and telemetry streams are transmitted to a centralized command center, allowing rapid situational awareness and coordinated emergency response. This network-integrated system supports continuous monitoring of critical infrastructure, urban environments, industrial facilities, and disaster zones across Europe.

9.8.14 India 5G BVLOS drone trials for medical and agricultural delivery

India has undertaken significant national-level initiatives to integrate 5G connectivity with beyond visual line of sight (BVLOS) drone operations, particularly in the fields of medical logistics and precision agriculture. These

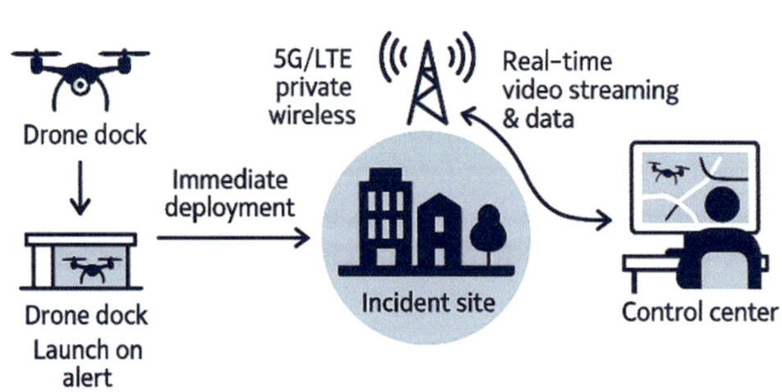

Figure 9.11 Nokia Drone Networks for automated surveillance and emergency response.

trials have been supported through collaborations between the Telecom Regulatory Authority of India (TRAI), Department of Telecommunications (DoT), DGCA, and multiple private drone manufacturers and telecom providers. The primary goal of these trials is to establish reliable long-range drone operations capable of serving remote, rural, and hard-to-reach regions where traditional transportation infrastructure is limited.

In medical applications, BVLOS drones equipped with 5G communication modules have been used to transport vaccines, blood products, diagnostic samples, critical medicines, and life-saving emergency supplies. The high-bandwidth and low-latency characteristics of 5G networks ensure continuous telemetry monitoring, real-time route correction, and stable command-and-control links throughout the mission. These capabilities reduce response time and expand access to emergency healthcare services, especially in mountainous and tribal regions.

In agriculture, BVLOS drones have been deployed for crop monitoring, soil moisture evaluation, fertilizer and pesticide spraying, and crop health analytics. Using real-time aerial imagery streamed over 5G networks, farmers and agri-scientists can make data-driven agronomic decisions, enabling more precise resource usage and improved crop productivity. The extended range of BVLOS operations allows coverage of large farmland regions with fewer takeoffs and landings, making drone-based agriculture more economically viable.

These trials also contribute to the establishment of a national unmanned traffic management (UTM) framework under the Digital Sky platform, supporting automated flight approvals, geofencing, and safe airspace integration. The results highlight India's strategy of using telecom-enabled UAV networks to address challenges in healthcare accessibility and agricultural efficiency.

While the progress is promising, continued development is needed in battery endurance, autonomous obstacle avoidance, rural connectivity reliability, and operator training. Nonetheless, India's 5G BVLOS trials represent a significant step toward scalable drone-based public service networks, positioning the country as an emerging leader in real-world drone utility deployment across social and economic sectors.

9.8.15 NASA Ingenuity and Dragonfly – pioneering drones beyond earth

NASA's Ingenuity and Dragonfly missions represent groundbreaking steps in the use of drones for extraterrestrial exploration, demonstrating how aerial mobility can expand scientific reach far beyond what land-based rovers alone can achieve. These missions illustrate a shift in space exploration strategy toward autonomous, lightweight, and flight-capable robotic systems that can navigate diverse planetary environments and gather data from previously inaccessible terrains.

Ingenuity, launched with the Perseverance rover and first flown in 2021 on Mars, became the first powered aircraft to achieve controlled flight on another planet. Operating in an atmosphere only 1% as dense as Earth's, Ingenuity required ultra-light construction, high-speed counter-rotating rotor blades, and fully autonomous navigation. Originally designed as a short technology demonstration, Ingenuity went on to complete dozens of successful flights, scouting terrain, identifying safe rover travel paths, and proving that aerial reconnaissance can substantially enhance planetary exploration efficiency.

In contrast, Dragonfly is a much larger and scientifically-focused rotorcraft planned to explore Titan, Saturn's largest moon, in the late 2020s. Titan's dense nitrogen atmosphere and low gravity make it well-suited for rotorcraft flight. Dragonfly is designed as an octocopter capable of flying between multiple surface locations to analyze organic compounds, investigate methane-based weather patterns, and study environments that may resemble early Earth conditions. Unlike Ingenuity, which performs short scouting flights, Dragonfly is a mobile science laboratory, capable of performing extended missions and collecting chemical and geological data at diverse exploration sites.

Together, Ingenuity and Dragonfly demonstrate how drones can serve as flexible, intelligent, and mobile scientific explorers, capable of surveying vast landscapes, evaluating hazards, supporting future human missions, and exploring environments that rovers and landers cannot easily reach. Their success signals an emerging era of planetary aerial robotics, in which drones become essential tools for mapping, resource scouting, astrobiology research, and off-world infrastructure planning, ultimately supporting humanity's long-term expansion into the solar system.

9.8.16 Dragonfly – a rotorcraft lander for Titan (planned for 2028 launch)

Dragonfly is a pioneering NASA mission designed to explore Titan, Saturn's largest moon, and is scheduled for launch in 2028. Titan is of profound scientific interest due to its dense nitrogen atmosphere, methane cycle, complex organic chemistry, and surface landscapes that in many ways resemble early Earth. Unlike Mars, Titan's atmospheric pressure and low gravity make it highly suitable for rotorcraft flight, enabling a drone-like aircraft to travel efficiently across diverse terrains. Dragonfly will function not only as a mobile aerial scout but also as a fully equipped scientific lander, capable of conducting in-depth chemical and geological investigations.

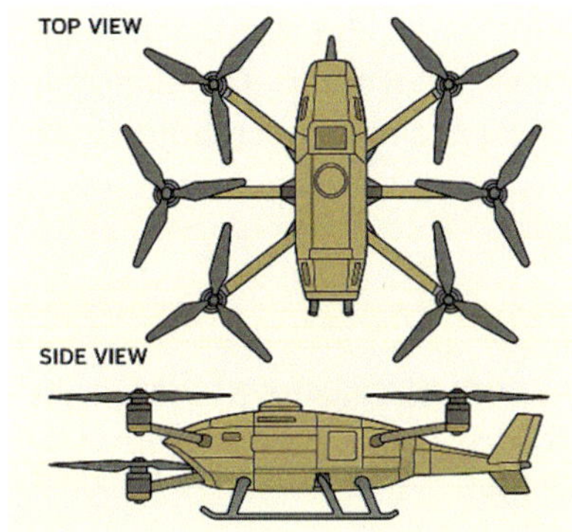

Figure 9.12 Dragonfly rotorcraft design (top and side views).

Dragonfly is an octocopter rotorcraft, designed with eight rotors arranged on four arms to ensure redundancy, stability, and maneuverability in Titan's variable terrain and atmospheric conditions. The spacecraft will perform multiple take-off and landing cycles, allowing it to move between distinct study sites such as dune fields, impact craters, and potential cryovolcanic regions. This mobility marks a significant advancement over stationary landers or wheeled rovers, which are limited to slow, linear movement and restricted exploration areas.

The mission's primary scientific objective is to investigate Titan's prebiotic chemistry – the complex organic molecules that may offer clues about how life can arise in environments beyond Earth. Dragonfly carries instruments for mass spectrometry, seismic sensing, meteorology, gamma-ray neutron spectroscopy, and high-resolution imagery. These tools will allow the spacecraft to analyze surface composition, atmospheric structure, geological history, and potential habitability. By studying Titan's organics and subsurface ice/ocean interface, Dragonfly aims to answer key questions about life's chemical origins in conditions different from Earth's.

Dragonfly is also designed for high autonomy, given the long communication delays between Titan and Earth. It must plan routes, avoid hazards, monitor power reserves, and adapt to weather changes independently. The mission will be powered by a radioisotope thermo-electric generator (RTG), ensuring energy availability during Titan's long nights and low sunlight conditions.

Figure 9.12 shows the NASA Dragonfly rotorcraft designed for Titan surface exploration. The top view highlights the octocopter configuration, with eight rotors arranged symmetrically to provide stability and lift in Titan's dense atmosphere. The side view illustrates the aerodynamic fuselage, landing skids for uneven terrain, and instrument housing for environmental sampling and autonomous navigation. The dual-rotor arm layout ensures redundancy, fault tolerance, and maneuverability during repeated takeoff and landing sequences across diverse surface regions on Titan.

9.9 Summary

This chapter explored the future trajectory of drone technology, emphasizing how advancements in AI-driven autonomy, swarm intelligence, quantum-enhanced computing, and high-speed communication networks are reshaping the role of drones in society. The evolution of drones is moving toward systems capable of independent decision-making, collaborative mission execution, and seamless integration into urban air mobility (UAM) frameworks, where passenger drones and aerial logistics networks may become part of everyday transportation. The introduction of 5G and emerging communication infrastructures is enabling reliable beyond-visual-line-of-sight (BVLOS) operations, while quantum computing promises new breakthroughs in real-time data processing, encryption security, and swarm optimization. Beyond Earth, drones such as NASA's Ingenuity and Dragonfly are opening new frontiers in planetary exploration and prebiotic research. Real-world case studies from military swarm systems (SWARM-100, Perdix, Zhuhai) to civilian and industrial applications like DJI ActiveTrack, Joby Aviation air taxis, Nokia automated drone networks, and India's 5G medical delivery trials demonstrate the versatility and strategic significance of drones across defense, public safety, mobility, agriculture, and space research. Collectively, these developments indicate a future where drones operate as intelligent, networked, and mission-adaptive systems, deeply integrated into both terrestrial and extraterrestrial infrastructure.

Review Questions

1. Define the term autonomous drone and explain how it differs from remotely piloted UAVs.
2. What are the key technological trends shaping the future evolution of drone systems?
3. Explain the significance of AI in enabling drones to make real-time decisions.
4. How do machine learning and neural networks improve drone navigation and task performance?

5. What is meant by swarm intelligence in the context of drone systems?
6. Discuss two major future applications of drone swarm operations.
7. Describe how collaborative AI supports communication and coordination within drone swarms.
8. Compare military and civilian uses of swarm drone systems.
9. What is urban air mobility (UAM) and why is it considered the future of transportation?
10. Identify the main challenges in integrating passenger drones into city environments.
11. Why is public acceptance critical for successful UAM adoption?
12. Describe how quantum computing could enhance drone data processing.
13. Explain the role of quantum algorithms in real-time airborne decision-making.
14. List three potential sectors that may benefit from quantum-enhanced drone operations.
15. Why are 5G networks important for long-range BVLOS drone missions?
16. Explain how low-latency communication improves swarm coordination.
17. Describe how drones can operate as part of a national air traffic management system.
18. How could drones contribute to extraterrestrial research and planetary science?
19. What unique flight challenges must drones overcome on Mars?
20. Why is Titan considered suitable for rotorcraft missions like Dragonfly?
21. Discuss the potential role of drones in space resource mining.
22. Summarize the major tactical advantage of Military Swarm Drone systems.
23. Explain the Loyal Wingman concept and its importance in future air combat.
24. How does DJI ActiveTrack enable autonomous object tracking during flight?

25. What strategic purpose does the Indian Army's SWARM-100 initiative serve?
26. Describe one key lesson learned from the US Perdix micro-drone swarm demonstrations.
27. What do the Zhuhai Air Show swarm displays indicate about China's swarm development strategy?
28. How did Dubai position itself as an early adopter of drone taxi demonstrations?
29. What was the significance of UAM testing for the Paris 2024 Olympic Games?
30. Summarize the certification pathway followed by Joby Aviation in the United States.
31. What problem is quantum-resilient UAV communication architecture designed to prevent?
32. Describe how quantum key distribution (QKD) enhances security in drone swarms.
33. How did Verizon Skyward demonstrate the viability of 5G-enabled drone operations?
34. What makes Nokia Drone Networks suitable for emergency response environments?
35. Explain the importance of 5G BVLOS trials in improving medical logistics in rural India.
36. What scientific objective underlies the NASA Ingenuity mission?
37. Describe the primary exploration goal of the Dragonfly mission to Titan.
38. How might future drones operate as part of smart city infrastructure?
39. Identify two ethical considerations associated with increasingly autonomous drone systems.
40. Predict one realistic advancement we may see in drone capabilities within the next decade.

Bibliography

Agrawal, P. S., Jawarkar, P. S., Dhakate, K. M., Parthani, K. M., & Agnihotri, A. S. (2024). Advancements and challenges in drone technology: A comprehensive review. In *2024 4th International Conference on Pervasive Computing and Social Networking (ICPCSN)* 638-644; IEEE https://doi.org/10.1109/ICPCSN62568.2024.00106

Amburn, P., Campanile, F., Graeff, R., Weyer, R., Wright, D., Vickery, R., ... & Wong, S. (2005). RADAR Analysis for Proposed Predator Design. In *2005 Users Group Conference (DOD-UGC'05)* 65-69; IEEE https://doi.org/10.1109/DODUGC.2005.57.

Arthur, G. (2024). MUM-T: MUM-T makes gradual gains in Asia-Pacific. Asia-Pacific Defence Reporter (2002), 50(3), 26–29. https://search.informit.org/doi/10.3316/informit.T2024072300020990270317459.

Assessment, F. E. (2003). Predator Force Structure Changes at Indian Springs Air Force Auxiliary Field Nevada.

Atiga, O., Tanzubil, B. J., Asaana, J., Ayamba, E. C., Seidu, M. T., & Mohammed, I. (2024). Shortening The Last-Mile: Impact of Zipline Medical Drone Delivery on The Operations of Hard-to-Reach Healthcare Facilities in Northern Ghana. *African Journal of Applied Research*, *10*(1), 178-200 https://doi.org/10.26437/ajar.v10i1.677

Baginova, v. V., & kalmurzaeva, d. K. (2017). Unmanned Aerial Vehicles and Logistics. *World of Transport and Transportation*, *15*(6), 120-124.

Bergmann, K. (2020). RAAF starts to examine trusted autonomy for uninhabited aerial systems. *Asia-Pacific Defence Reporter (2002)*, *46*(10), 26-30. https://search.informit.org/doi/10.3316/informit.633837801916641

Bhattarai, S. (2024). Drone-Enabled Aerial Monitoring for Dynamic Traffic Control Using Multi-Modal Data Synthesis. *Journal of Robotic Process Automation, AI Integration, and Workflow Optimization*, *9*(12), 23-35. https://helexscience.com/index.php/JRPAAIW/article/view/2024-12-10

Blom, J. D. (2010). *Unmanned aerial systems: A historical perspective* (Vol. 45). Fort Leavenworth, KS: Combat Studies Institute Press.

Bouse, D. (1998). Are wildlife films really "nature documentaries"?. *Critical Studies in Media Communication*, *15*(2), 116-140 https://doi.org/10.1080/15295039809367038

Caloianu, A. F. (2023). Türkiye's Energy Strategies and Policies in the Black Sea: Neo-Ottomanism in the Context of the War in Ukraine and the 2023 Elections. *VALENTIN NAUMESCU RALUCA MOLDOVAN ANAMARIA FLORINA CALOIANU*, 175.

Campbell, J. B. (2008). Origins of aerial photographic interpretation, US Army, 1916 to 1918. *Photogrammetric Engineering & Remote Sensing*, *74*(1), 77-93. https://doi.org/10.14358/PERS.74.1.77

Chattopadhyay, S. (2024). AMCA: An Indian Fifth Generation Story; an update. *Vayu Aerospace and Defence Review*, (5), 51-52.

Choi, H. W., Kim, H. J., Kim, S. K., & Na, W. S. (2023). An overview of drone applications in the construction industry. *Drones*, *7*(8), 515. https://doi.org/10.3390/drones7080515

Cooper, J. (2019). Russia's 'Invincible' Weapons: An Update.

Cuerno-Rejado, C., Garcia-Hernandez, L., Sanchez-Carmona, A., Carrio, A., SANCHEZ LOPEZ, J. L., & Campoy, P. (2016). Historical evolution of the unmanned aerial vehicles to the present. *Dyna*, *91*(3).

Cullen, T. M. (2011). *The MQ-9 Reaper remotely piloted aircraft: Humans and machines in action* (Doctoral dissertation, Massachusetts Institute of Technology). http://hdl.handle.net/1721.1/80249

Dancey, P. (2014). *British Aircraft Manufacturers Since 1909*. Fonthill Media.

De Saint-Jean, F. V., & Pouzargue, M. (2013). Comportement dynamique d'un UCAV soumis à un environnementaéroacoustique sévère. *Metallurgical Research & Technology*, *110*(5), 325-330. https://doi.org/10.1051/metal/2013075

Duffield, M. (2015). The digital development-security nexus: Linking cyber-humanitarianism and drone warfare. In *Handbook of international security and development* (pp. 80-94). Edward Elgar Publishing. https://doi.org/10.4337/9781781955536.00012

Duque, L., Seo, J., & Wacker, J. (2018). Synthesis of unmanned aerial vehicle applications for infrastructures. *Journal of Performance of Constructed Facilities*, *32*(4), 04018046. https://doi.org/10.1061/(ASCE)CF.1943-5509.0001185

Fang, W., Liu, F., Yang, F., Shu, L., & Nishio, S. (2010). Energy-efficient cooperative communication for data transmission in wireless sensor networks. *IEEE Transactions on Consumer Electronics*, *56*(4), 2185-2192. http://doi.org/10.1109/TCE.2010.5681089

Franke, U. E. (2013). Welche Drohne passt zu mir? Deutschlands schwierige Entscheidung für ein bewaffnetes „Unmanned Aerial Vehicle “. *Zeitschrift für Außen-und Sicherheitspolitik*, *6*(4), 485-496. https://doi.org/10.1007/s12399-013-0362-z

Gettinger, D. (2024). 3 Defining Drones. *De Gruyter Handbook of Drone Warfare*, *4*, 25.

Grimal, F., & Sundaram, J. (2018). Combat drones: Hives, swarms, and autonomous action?. *Journal of Conflict and Security Law*, *23*(1), 105-135. https://doi.org/10.1093/jcsl/kry008

Hall, R. C. (2014). Reconnaissance drones: Their first use in the Cold War. *Air power history*, *61*(3), 20-27. http://www.jstor.org/stable/26276490

Hansen, J. G., & de Figueiredo, R. P. (2024). Active object detection and tracking using gimbal mechanisms for autonomous drone applications. *Drones*, *8*(2), 55. https://doi.org/10.3390/drones8020055

Hassell Jr, B. E. (2009). Medium and High Altitude Unmanned Aircraft System Acquisition: An Efficiency Study of Magnitude and Capability.

Hwang, W. (2020). Development of Peoples' Republic of China's Unmanned Aerial Vehicles (UAVs) and Its Impact on the East China Sea. *International Journal of China Studies*, *11*(1), 121-144. https://www.proquest.com/scholarly-journals/development-peoples-republic-chinas-unmanned/docview/2434859823/se-2

Jeler, G. E. (2020). Brief historical milestones on the evolution of UAV systems: 1914-1939. *Bulletin of" Carol I" National Defence University (EN)*, (03), 137-143.

Johnston, L. J. (2013). "Predicting the Maximum-Lift Performance of Unmanned Combat Air Vehicle Planforms Using the Euler Equations. In *Proceeding of the EUCASS 2013 Conference* (pp. 1-4).

KA AL-Dosari, K. N. (2025). *Drones for safety and security surveillance systems (D4S): The case of mega sporting events* (Doctoral dissertation, Brunel University London).

Kagawa, T., Ono, F., Shan, L., Takizawa, K., Miura, R., Li, H. B., ... & Kato, S. (2017, December). A study on latency-guaranteed multi-hop wireless communication system for control of robots and drones. In *2017 20th International Symposium on Wireless Personal Multimedia Communications (WPMC)* (pp. 417-421). IEEE. https://doi.org/10.1109/WPMC.2017.8301849

Kanat, K. B., & Cetin, O. A. (2025). Turkey's Leap into Aerospace Autonomy. In *Allies in Arms: Turkish-American Defense Relations* (pp. 163-201). Cham: Springer Nature Switzerland. https://doi.org/10.1007/978-3-031-96062-8_6

Kardasz, P., Doskocz, J., Hejduk, M., Wiejkut, P., & Zarzycki, H. (2016). Drones and possibilities of their using. *J. Civ. Environ. Eng*, *6*(3), 1-7.

Keane, J. F., & Carr, S. S. (2013). A brief history of early unmanned aircraft. *Johns Hopkins APL technical digest*, *32*(3), 558-571.

Kompała, D. (2014). RQ-4 Global Hawk w rozpoznaniu powietrznym. *Zeszyty Naukowe AON*, 164-188.

Lu, H., Shen, H., Tian, B., Zhang, X., Yang, Z., & Zong, Q. (2022). Flight in GPS-denied environment: Autonomous navigation system for micro-aerial vehicle. *Aerospace Science and Technology*, *124*, 107521. https://doi.org/10.1016/j.ast.2022.107521

Madi, R., & Madi, N. (2024, December). Integrating Flying Taxis: Dubai's Pioneering Efforts. In *Global Congress on Emerging Technologies (GCET-2024)* (pp. 191-194). IEEE. https://doi.org/10.1109/GCET64327.2024.10934459

Major, R. (2012). RQ-2 Pioneer: The Flawed System that Redefined US Unmanned Aviation.

Masse, D. (2011). Boeing Phantom Ray Completes First Flight. *Microwave Journal*, *54*(6).

Mayer, M. (2015). The new killer drones: Understanding the strategic implications of next-generation unmanned combat aerial vehicles. *International Affairs*, *91*(4), 765-780. https://doi.org/10.1111/1468-2346.12342

Newcome, L. R. (2004). *Unmanned aviation: a brief history of unmanned aerial vehicles*. AIAA.

Niarchos, M., Stamatiadou, M. E., Dimoulas, C., Veglis, A., & Symeonidis, A. (2022). A semantic preprocessing framework for breaking news detection to support future drone journalism services. *Future Internet*, *14*(1), 26. https://doi.org/10.3390/fi14010026

Okere, E. E., & Balyan, V. (2025). Sixth Generation Enabling Technologies and Machine Learning Intersection: A Performance Optimization Perspective. *Future Internet*, *17*(2), 50. https://doi.org/10.3390/fi17020050

Ouma, J. A., Chappelle, W. L., & Salinas, A. (2011). *Facets of occupational burnout among US Air Force active duty and National Guard/reserve MQ-1 predator and MQ-9 reaper operators* (No. AFRLSAWPTR 20110003).

Oxford Analytica. (2016). Advanced combat drone will boost UK-French industry. *Emerald Expert Briefings*, (oxan-db). https://doi.org/10.1108/OXAN-DB209977

Pliatsios, D., Goudos, S. K., Lagkas, T., Argyriou, V., Boulogeorgos, A. A. A., & Sarigiannidis, P. (2021). Drone-base-station for next-generation internet-of-things: A comparison of swarm intelligence approaches. *IEEE Open Journal of Antennas and Propagation*, *3*, 32-47. https://doi.org/10.1109/OJAP.2021.3133459

Prisacariu, v. (2022). A brief history of UAVs in the 1970s. *Scientific Research & Education in the Air Force-AFASES*, *2022*. https://doi.org//10.19062/2247-3173.2022.23.28

Ram, S. S., Singhal, S., & Ghatak, G. (2022). Optimization of network throughput of joint radar communication system using stochastic geometry. *Frontiers in Signal Processing*, *2*, 835743. https://doi.org/10.3389/frsip.2022.835743

Raman, R., & Datta, U. (2021, April). The role of 'Unmanned Aerial Vehicles' in smart city planning and management. In *International Conference on Unmanned Aerial System in Geomatics* (pp. 99-120). Cham: Springer International Publishing. https://doi.org/10.1007/978-3-031-19309-5_8

Reilly, B. (2021). Skyborg program of record transition may slip past FY-23. *Inside the Air Force*, *32*(32), 1-5. https://www.jstor.org/stable/27053942

Ryan Jr, T. R., & Mittal, V. (2019). Potential for army integration of autonomous systems by warfighting function. *Military Review*, *99*(5), 122. https://www.armyupress.army.mil/Portals/7/military-review/Archives/English/SO-19/Mittal-Autonomous-Systems.pdf

Sapmaz, A. (2023). Rusya federasyonu'nun insansız hava aracı/silahlı insansız hava aracı kabiliyetlerinin değerlendirilmesi. *Türkiye İnsansız Hava Araçları Dergisi*, *5*(1), 11-20. https://doi.org/10.51534/tiha.1199782

Simic Milas, A., Cracknell, A. P., & Warner, T. A. (2018). Drones–the third generation source of remote sensing data. International Journal of Remote Sensing, 39(21), 7125-7137. https://doi.org/10.1080/01431161.2018.1523832

Singh, R., & Kumar, S. (2025). A Comprehensive Insights into Drones: History, Classification, Architecture, Navigation, Applications, Challenges, and Future Trends. *arXiv preprint arXiv:2501.10066*. https://doi.org/10.48550/arXiv.2501.10066

Singhal, A. (2021). Impact of drone technology on human life. BODHI INT'L J. RES. HUMANITIES, ARTS & SCI., 5, 11.

Slocombe, G. (2019). Military UAS/RPAS in Asian airspace. *Defence Review Asia*, *13*(2), 14-16. https://search.informit.org/doi/10.3316/informit.368305014706168

Stensrud, R., Mikkelsen, B., & Valaker, S. (2024). Exploring human-autonomy teaming methods in challenging environments: the case of fighter pilots and loyal wingmen. *Human-Intelligent Systems Integration*, *6*(1), 1-14. https://doi.org/10.1007/s42454-024-00050-y

Stoer, E. H. (2001). *Flying bombs, aerial torpedoes, and Kettering bugs: America's first cruise missiles*. The Florida State University.

Tekinalp, O., & Cavus, N. (2012, September). Multiobjective Conceptual Design of an Unmanned Combat Air Vehicle. In *12th AIAA Aviation Technology, Integration, and Operations (ATIO) Conference and 14th AIAA/ISSMO Multidisciplinary Analysis and Optimization Conference* (p. 5532). https://doi.org/10.2514/6.2012-5532

Uhl, B. (1990, February). The new light weight, high performance reconnaissance camera KRb 8/24 F. In *Airborne Reconnaissance XIII* (Vol. 1156, pp. 49-57). SPIE. https://doi.org/10.1117/12.962470

Valavanis, K. P., & Kontitsis, M. (2007). A historical perspective on unmanned aerial vehicles. In *Advances in Unmanned Aerial Vehicles: State of the Art and the Road to Autonomy* (pp. 15-46). Dordrecht: Springer Netherlands. https://doi.org/10.1007/978-1-4020-6114-1_2

Van Bourgondien, J. (2012). *Analysis of the sustainment organization and process for the Marine Corps RQ-11B Raven Small Unmanned Aircraft System (SUAS)*. Acquisition Research Program. https://dair.nps.edu/handle/123456789/2031

Vogel, D. (2021). Future Combat Air System: too big to fail; differing perceptions and high complexity jeopardise success of Strategic Armament Project. https://nbn-resolving.org/urn:nbn:de:0168-ssoar-71975-6

Waibel, M., Keays, B., & Augugliaro, F. (2017). *Drone shows: Creative potential and best practices*. ETH Zurich.

Wang, H., Cheng, H., Hao, H. (2020). The Use of Unmanned Aerial Vehicle in Military Operations. In: Long, S., Dhillon, B.S. (eds) Man-Machine-Environment System Engineering. MMESE 2020. Lecture Notes in Electrical Engineering, *645*. Springer, Singapore. https://doi.org/10.1007/978-981-15-6978-4_108

Welch, A. (2015). A cost-benefit analysis of Amazon Prime Air. https://scholar.utc.edu/honors-theses/47

Wolf, W. (2021). Off Target: America's Guided Bombs, Missiles and Drones 1917-1950. Fonthill Media.

Wyatt, E. (2013). The DARPA/air force unmanned combat air vehicle (UCAV) program. In *AIAA International Air and Space Symposium and Exposition: The Next 100 Years* (p. 2616). https://doi.org/10.2514/6.2003-2616

Yelwe, R. M. (2025). Light Combat Aircraft (LCA) Tejas vs. JF-17 Thunder: Evaluating performance, capabilities, and strategic importance. *Comparative Strategy*, 1-30. https://doi.org/10.1080/01495933.2025.2501644

Zaloga, S. J. (2011). *Unmanned aerial vehicles: robotic air warfare 1917–2007* (Vol. 144). Bloomsbury publishing.

Zenz, A., & Powles, J. (2024). Resisting technological inevitability: Google Wing's delivery drones and the fight for our skies. *Philosophical Transactions A*, *382*(2285), 20240107. https://doi.org/10.1098/rsta.2024.0107

Index

About the Authors

Dr. Kaushik Kumar holds a Ph.D. in Virtual Manufacturing from Jadavpur University, India, an MBA in Marketing Management from Indira Gandhi National Open University, India, and a Bachelor of Technology from Regional Engineering College (now National Institute of Technology), Warangal, India. He has over 25 years of rich teaching and research experience, complemented by 11 years of professional exposure in a globally reputed manufacturing unit.

Currently, he serves as a Professor in the Department of Mechanical Engineering at Birla Institute of Technology, Mesra, Ranchi, India. His research interests span composites, optimization, non-conventional machining, CAD/CAM, rapid prototyping, and quality management systems. Dr. Kumar has been instrumental in pioneering innovative product development for both industrial and societal applications, earning 45+ patents for his contributions.

He has established a state-of-the-art composite research laboratory and has authored more than 60+ books (including 31 edited volumes). Additionally, he has published 100 book chapters and over 200 research papers in reputed national and international journals. Dr. Kumar has held numerous editorial roles, including editor-in-chief, series editor, guest editor, and editorial board member, and actively serves as a reviewer for multiple international and national journals.

Throughout his career, he has secured sponsored research and consultancy projects worth approximately ₹1 crore from government and international agencies. He has been invited as a keynote speaker, resource person, and expert lecturer at various international and national conferences, workshops, and faculty development programs. He has successfully guided numerous doctoral, master's, and undergraduate students in India and abroad and served as an examiner and reviewer for dissertations and theses internationally.

Prof. Dr. Arzu Eren Şenaras is a Professor in the Department of Econometrics at Bursa Uludağ University, where she also serves as the Head of the Operations Research Division. She is a distinguished academic and researcher specializing in operations research, system dynamics modeling, and simulation-based optimization. Her expertise spans logistics, supply chain management, energy systems, and complex system modeling, with a strong focus on practical applications of quantitative methods.

Prof. Dr. Eren Şenaras earned her bachelor's degree in Econometrics from Uludağ University in 2007, followed by a master's degree, during which she developed a novel simulation application for revenue determination in intercity bus routes. She completed her Ph.D. in Econometrics in 2016, focusing on the dynamic behaviors of exchange rates through system dynamics modeling. Since beginning her academic career as a research assistant in 2008, she has built a robust research and teaching portfolio, advancing to associate professor in 2018 before attaining full professorship.

Her research contributions include over 20 peer-reviewed journal articles, numerous conference presentations, and interdisciplinary projects such as agent-based simulations for disaster-prepared food supply chains and modeling Turkey's electricity market. She has applied techniques like dynamic programming, causal loop diagrams, and ant colony optimization to real-world challenges in manufacturing and logistics. Beyond research, Prof. Dr. Eren Şenaras is a committed mentor, guiding graduate students in innovative projects that address sustainability, digital transformation, and practical problem-solving.

Dr. Chikesh Ranjan is currently serving as a Project Engineer in the Department of Mechanical Engineering at the National Institute of Technology, Rourkela, Odisha, India. He earned his Ph.D. in Mechanical Engineering from Birla Institute of Technology, Mesra, Ranchi in 2024, with a research focus on the tribological properties and thermal degradation kinetics of natural fiber-reinforced polymeric composites. Prior to this, he completed his Master of Engineering in Mechanical Engineering (Design of Mechanical Equipment) with distinction from the same institute in 2013 and a Bachelor of Engineering in Mechanical Engineering from Marathwada Institute of Technology, Aurangabad in 2010.

Dr. Ranjan has extensive academic and research experience, having worked as an Assistant Professor at R.T.C Institute of Technology, Ranchi from 2013 to 2020, where he also served as Program Officer for the National Service Scheme (NSS). He has a prolific scholarly output, with over 24 Scopus-indexed book chapters, 13 SCOPUS/WoS publications, four SCI/SCIE journal articles, and more than 14 authored books, both national and international, focusing on areas such as drones, robotics, polymeric composites, additive manufacturing, and smart materials.

His inventive contributions include multiple national and international patents, encompassing designs and applications in drones, renewable energy devices, automated systems, and functionally graded composites. He is an active educator, having organized and coordinated numerous workshops, boot camps, and training programs on drone technology, CAD/CAM, robotics, and automation. He is proficient in software tools such as AUTOCAD, CREO, CATIA, and ANSYS and has a keen interest in integrating artificial intelligence and optimization techniques into mechanical systems.

Dr. Ranjan's research and professional profiles are available on SCOPUS, Web of Science, ORCID, Google Scholar, and Vidwan. His work bridges theoretical research with practical engineering solutions, fostering innovation in mechanical engineering and allied technologies.

Dr. Onur Mesut Şenaras is a distinguished engineer and academic specializing in automotive engineering, industrial optimization, and lean manufacturing. He earned his B.Sc. in Industrial Engineering from Dokuz Eylül University (2004), his M.Sc. in Automotive Engineering from Uludağ University (2013), and his Ph.D. in Automotive Engineering from Uludağ University (2024).

With nearly two decades of professional experience, Dr. Şenaras began his career as Controlling and Budget Responsible at Vestel Beyaz Eşya (2005–2007) and subsequently joined Oyak Renault, holding key roles such as Body Performance Responsible, Kaizen Pilot, and Design Automation Leader. His contributions have driven digitalization, automation, and system optimization within the automotive sector. He has received multiple prestigious awards, including the Renault Ergonomics Award (2010), Renault Investment Optimization Award (2011), and several factory ergonomics awards, and holds a national patent in ergonomics.

As a researcher and author, Dr. Şenaras focuses on automated guided vehicles (AGVs), artificial intelligence, system dynamics, and simulation. He co-authored *Expulsion in Resistance Spot Welding* (2017) and has contributed chapters on flexible manufacturing systems and subway passenger flow modeling. At Uludağ University, he mentors students and teaches courses on lean manufacturing, logistics management, and genetic algorithms, bridging cutting-edge research with practical industry applications.

Dr. Şenaras's professional and academic work significantly advances the fields of automotive engineering and industrial systems, establishing him as a leading figure in integrating innovation, research, and practical solutions.

For Product Safety Concerns and Information please contact our EU representative GPSR@taylorandfrancis.com Taylor & Francis Verlag GmbH, Kaufingerstraße 24, 80331 München, Germany